Pvt. Ltd.

VIRUS DISEASES OF CITRUS & MANAGEMENT

Prof. Y.S. Ahlawat
Advanced Center for Plant Virology
Indian Agricultural Research Institute
New Delhi – 110 012

2012

Studium Press (India) Pvt. Ltd.

VIRUS DISEASES OF CITRUS
&
MANAGEMENT

ISBN: 978-93-80012-24-7

Published by:

Studium Press (India) Pvt. Ltd.
4735/22, 2nd Floor, Prakash Deep Building (Near PNB),
Ansari Road, Darya Ganj, New Delhi-110 002
Tel.: +91-11-4324200-15 (15 lines); Fax: +91-11-43240215;
jngovil@gmail.com; jngovil@hotmail.com

Printed at:
Salasar Imaging Systems
Delhi-110035 (India)

Foreword

Citrus *is among the most important fruit crops* and it is produced all over the world in about 140 countries. But citrus is commercially grown in about 50 countries mostly situated in Northern hemisphere. World production of citrus fruits is about 90 million tons. However, lemons and limes production is projected as 10.6 million MT in 2010, an increase of 15% from 1996-98 level. The estimated production of citrus fruits in India is 3.79 million tons. Production of citrus fruits has experienced continuous increase from 1961 onward.

Diseases particularly of virus nature posed serious threat to citrus industry as these pathogens are difficult to manage. The impact of growing disease-free citrus through certification and other techniques like shoot tip grafting has been experimentally demonstrated throughout world. Reliable techniques of virus diagnosis have been developed all over the world and now it is a major component in certification programmes.

The book entitled "VIRUS DISEASES OF CITRUS & MANAGEMENT" written by Professor Y.S. Ahlawat is aimed to describe citrus diseases caused by viruses and other *budwood transmitted* pathogens will certainly be very helpful to advance citrus production *mainly by increasing the awareness' of the importance* of growing virus-free plantations of citrus in developing countries. The life long experience of the author in this particular field and his original discoveries encouraged him to bring out this book.

The book has eight major chapters on introduction, virus diseases with known aetiology (17 diseases), virus-*like* diseases of unknown aetiology (15 diseases), diseases where virus etiology is suspected (8 diseases), two diseases caused by prokaryotes, three diseases caused by Mollicutes, seven diseases caused by viroids and the last chapter is on management including success stories. The description of each disease and the pathogen associated with it have been described in relation to their geographical distribution, economic importance, disease symptoms, transmission, vector(s), host range, indicator host(s), rootstock-scion effect, nature and structure of associated pathogen, their genome organization, methods of detection and management. The book contains 37 figures, 53 colour plates and over 1000 relevant references.

The book is designed to upgrade the knowledge of graduate and post-graduate students, teachers and farm mangers. It will also be a useful document in developing Pest Risk Analysis in National Plant Quarantine system and management of citrus diseases in a holistic manner.

M. Bar Joseph

Dr. Moshe Bar-Joseph, GimaloTec,
Hazanchanim 8, Rehovot 76211
Formerly, Head, The S. Tolkowsky Laboratory of Citrus Disease Research,
The Volcani Center, A.R.O. Bet Dagan, Israel

Acknowledgement

I wish to thank to the Department of Science and Technology, Govt. of India who sanctioned this project for funding under USERS scheme. I am indeed thankful to Dr. S.S. Kohli, Director and Scientist F, Department of Science and Technology for all of his help regarding finances etc. always in time. I am indebted to my students, Drs. D.K. Ghosh, R.P. Pant, and D.R. Das for their help in various ways during this project. I also thank to our student Dr. Avijit Roy, University of Florida, CREC, Lake Alfred, Florida for suggesting addition/alteration in some part of the book. I wish to thank Dr. V.K. Baranwal who provided some interesting material for the book. I am indeed grateful to my professional friends, Drs/Professors C.N. Roistacher, Department of Plant Pathology, Universty of California, Riverside, USA, J.M. Bove, Laboratoire cellulaire et Moleculaire, INRA, Bordeaux, France, Nuria Duran-Vila, IVIA, Moncada, Spain, M. Bar-Joseph, department of Virology, ARO, Volcani Centre, Bet-Dagan, Israel, R.H. Brlansky, S. Gowda and Megh Singh from University of Florida, CREC, Lake Alfred, Florida and G.W. Muller, Centro de citricultura Sylvio Moreira-IAC, Cordeiropolis, SP, Brazil for their valuable suggestions and providing required help. I am also thankful to Head of the Division of Plant pathology and Director IARI, New Delhi and administrative staff involved for various activities of project. At last but not the least I wish to put on record my sincere thanks to my wife Smt. Urmila Ahlawat, My son Vivek and my daughter Dr. Anshu for helping me in various ways.

About the Author

Yashvir S. Ahlawat was born in a village Kandera in Baghpat district of Uttar Pradesh in 1943. He obtained his M.Sc. and Ph.D degrees from Agra University. He carried out his research assignments at Indian Agricultural Research Institute, (IARI) New Delhi and its Regional Stations in Shimla and Kalimpong. During his research career he discovered several virus and phytoplasma diseases infecting agricultural and horticultural crops. However, his major interest had been in virus and virus like diseases of citrus. At IARI, Prof. Ahlawat worked at the advanced centre for Plant Virology developed with the assistance of UNDP of FAO and headed Virology Unit of IARI. During his 40 years of research career, Dr. Ahlawat guided 10 Ph.D. and 4 M.Sc. students. He published over 150 research papers in reputed scientific journals. He handled both National and International projects on virus diagnostics, HRD, and National Facility for Virus diagnosis and quality control of tissue culture-raised plants. Under HRD project, Dr. Ahlawat conducted several group and hands- on trainings on various aspects of plant virology including one refresher course for senior professors from Indian Universities on molecular diagnosis of plant viruses. The research contributions made by Prof. Ahlawat have been recognized nationally and internationally. Nationally, he received several honors and awards from Indian Council of Agricultural Research (ICAR), Indian Phytopathological Society (IPS), and Indian Society of Citriculture (ISC) for outstanding research contributions in plant virology. Internationally, he was awarded **James M. and Adeline Wallace** Research award 1998 by International Organization of Citrus Virologists, Universty of California for his outstanding contribution in citriculture. Dr. Ahlawat is a member and fellow of several national and international professional societies and served them in various capacities. He was elected as President, IPS in 2005 and served as Secretary General and Vice-President in Indian Society of Citriculture. As FAO Fellow, Dr Ahlawat worked at the University of California, Riverside and at Instituto Biologico at Sao Paulo, Brazil. He visited University of California again as IBPGR Expert to develop guidelines for safe movement of citrus germplasm. He participated and chaired scientific sessions in several national and international symposia in India and abroad. Professor Ahlawat visited INRA at Bordeaux in France twice as visiting Professor. He was consultant to the Royal Govt. of Bhutan and visited University of Florida, CREC, Lake Alfred recently. He edited 8 books, contributed several book chapters and popular articles. After his superannuation in 2005, he had been working as Emeritus Scientist ICAR and currently working as ES, Department of Science and Technology, Govt. of India.

Preface

I have been actively working on virus and virus-like diseases of citrus for the last 35 years. I am a regular member of International Organization of Citrus Virologist (IOCV) and am actively associated with Indian Society of Citriculture since its inception. It had been long felt need by citrus virologists in India and abroad to bring out a comprehensive publication on virus and virus-like pathogens infecting citrus with special emphasis to the research work done in the Indian subcontinent which is considered as home of citrus. Therefore, I decided to write my experiences on citrus viruses in India and abroad. While working at the University of California, Riverside, I experienced on citrus diseases like psorosis, vein enation, yellow vein, blind pocket, concave gum, cristacortis, stubborn, exocortis, cachexia and the certification programmes maintained by the University. During my recent visit to University of Florida, CREC, Lake Alfred, Professor R.H. Brlansky showed me citrus blight and its field detection by pressure injection and huanglonbing disease in field. While working at INRA, Bordeaux with Professor J.M. Bove, I was exposed to several diseases like huanglongbing, tristeza strains, bud union crease, witches' broom and several others. I also visited Corsica with Professor Bove to see certification programmes and trees cleaned by STG. I visited Brazil for a period of one month and was shown declinio disease in field along with demonstration for its detection by pressure injection by late Dr. Victoria Rossetti. In Brazil, I was also shown diseases like Bahia psorosis, leprosis, leaf curl, variegated chlorosis and cross protected trees against tristeza disease by Dr. G.W. Muller and other scientists. During my research career, it has been my privilege to work on several National and International projects on human resource development and diagnostics in the field of citrus virology. I gained long exposure on virus and virus like diseases working on Advanced Centre for Plant Virology at Indian Agricultural Research Institute, New Delhi and its Regional stations at Shimla and Kalimpong. To avoid introduction of viruses through the movement of citrus germplasm, IOCV, FAO, and IBPGR jointly published *FAO/IBPGR Technical Guidelines for the Safe Movement of Citrus Germplasm* (Frison and Taher, 1991) and I was also a member of this team. This document is serving citrus industry to check virus movement with planting material or through citrus germplasm from one country to another or from one region to another region.

The first chapter of the book provides brief introduction about classification of citrus, its species and relatives. Information has been provided on economic importance of citrus, citrus producing countries and world citrus production. A brief introduction of viruses has been given along with the major pathogens infecting citrus. Important viruses associated with diseases in citrus trees are (1) Ilarviruses associated with citrus crinkly leaf, citrus variegation and citrus leaf rugose (II) the Rhabdoviruses associated with citus leprosies and zonate chlorosis and Yellow vein mosaic of Pummelo (III) Ophiovirus associated with psorosis diseases (IV) Satsuma dwarf virus group (V) tatter leaf-citrange stunt (VI) Closterovirus associated with tristeza disease (VII) badnavirus associated with citrus yellow mosaic disease in India and (VIII) mandrivius associated with Indian citrus ring spot disease. In addition to viruses, other serious diseases in citrus are caused by bacteria (huanglongbing and variegated chlorosis), by Mollicutes (citrus stubborn, lime witches broom and citrus rubberywood) and by viroid pathogens (exocortis, cachexia, and others). Viruses spread through insect vectors which mostly belong to Order Homoptera, families Aphididae (200 spp. transmit 160 viruses), Psyllidae (6 spp. transmit phytoplasmas and bacteria-like organisms), Coccidae (15 spp. transmit 2 viruses), Aleyrodidae (4 spp. transmit 70 viruses), Auchenorryncha (151 spp. transmit 38 viruses, 31 Phytoplasmas, 4 spiroplasmas and 4 Rickettsia-like organisms), Order Coleoptera (74 spp. transmit 45 viruses), Thysanoptera (9 spp. transmit 4 viruses). Information has also been provided on citrus germplasm and repository, National and International organizations dealing with citrus virology.

In Chapter-2, virus diseases with known etiology have been described in detail. 17 virus diseases have been described in this chapter. These are: Citrus psorosis, tristeza, crinkly leaf, infectious variegation, leaf rugose, leprosis and zonate chlorosis, Satsuma dwarf, mosaic, Navel orange infectious mottling, vein enation, woody gall, tatter leaf = citrange stunt, Indian citrus ring spot, ring spot, yellow mosaic, Pummelo yellow vein mosaic, Yellow vein clearing, dweet mottle = citrus leaf blotch. These diseases have been described in relation to their distribution and economic importance. Descriptions have also been provided on disease symptoms, modes of transmission of viruses by various methods including their vector(s), host range, indicator host(s) and rootstock-scion effect. The nature and structure of the associated viruses including their genome organization have also been discussed. Methods of detection and management have been described in detail.

Chapter-3 describes 15 diseases of citrus considered to be of virus nature but their etiology is not established as yet. These diseases are: Citrus blight, concave gum-blind pocket, impietratura, chlorotic dwarf, yellow vein, leaf curl, cristacortis, leathery leaf, multiple sprouting, citrus dieback, bud-union crease, measles, Kassala, Rio grande gummosis, and citrus sudden death. These diseases have been described about their distribution, economic importance of some diseases like citrus blight in Florida and Brazil. Descriptions have been given on disease symptoms, transmission, methods of diagnosis and management if known. In this group, either the disease has been described recently or no further information could be generated on already known diseases due to their negligible impact on production or being limited in their distribution.

In Chapter-4, eight diseases have been tabulated. Viral nature of these diseases has been established mostly due to their transmission ability like virus diseases. Virus-like particles have been observed in electron microscope in two of such diseases, Algerian navel orange and Yellow mottle. Other diseases in this group are: Brittle twig yellows, fatal yellows, fovea, grapefruite bark scaling, Milam lemon stem pitting and Nagami kumquat.

Chapter-5 carries information on citrus diseases caused by Prokaryotes. Two diseases, of serious nature, Huanglongbing and variegated chlorosis have been described in this chapter. These diseases have been described for their distribution, economic importance, symptoms, transmission, vector(s), their distribution and pathogen-vector relationship. Pathogen-host interaction and reliable indicator plants have been discussed in detail. Information has been provided on description and classification of associated bacteria including their genome organization. Details of biological, serological and molecular methods of diagnosis and approaches for disease management have been listed.

In chapter-6, three diseases caused by Mollicutes have been described. These diseases are: Citrus stubborn, Witches' broom and rubberywood. Detailed information has been provided on distribution and economic impotance of these diseases. This chapter carries information on disease symptoms, transmission, vector(s) host range including wild and natural hosts, indicators, pathogens associated with these diseases, their taxonomic position, diagnosis and management.

Chapter-7 provides information on diseases caused by low molecular weight RNAs, the viroids. Latest classification of viroids infecting plants has been given. Following diseases caused by viroids have been discussed: Citrus exocortis, cachexia, yellow corky vein, gummy bark, gum pocket-gummy pitting, shell bark and citrus bent leaf. Detailed information has been provided on distribution and economic impotance, symptoms, transmission, host range, indicators, pathogens associated with these diseases, their taxonomic position, diagnosis and management.

Information has also been provided on viroids responsible for dwarfing of citrus trees.

Chapter-8 will provide information on registration, certification, production of virus-free planting material through shoot-tip grafting, production of nucellar clones, methods of diagnosis, disease management by cultural practices, cross protection and through transgenic protection, and finally the biological control of insect vectors including success stories.

At the end of each chapter related photographs have been given. In total 37 Figures and 53 colour plates showing symptoms of diseases on various hosts have been put in this publication and most of them were collected by me personally while visiting to various countries or downloaded from www.ivia.es.iocv.

I am fully conscious that in this publication, there may be shortcomings and defects. I therefore, request my colleagues working in plant virology to communicate/suggest the defects, if any, to me so that I could attempt to rectify them in future editions. However, this publication should be of value to the students and teachers of plant virology, plant Pathology, and horticulture for diagnosis and management of citrus diseases and to study virus and virus-like pathogens associated with citrus diseases.

I wish to express my deep appreciation to my colleagues in India and professional friends all over the world who helped in various ways to bring out this publication. They will be duly acknowledged.

Y.S. Ahlawat

Contents

List of Coloured Plates

1

INTRODUCTION

Citrus belongs to family Rutaceae which is represented by approximately 150 genera and 1600 species of plants found in tropical and warm regions of the world. Citrus is believed to have originated into parts of Southeast Asia bordering Northeastern India, Myanmar (Burma) and the Yunnan province of China (De Araujo *et al.*, 2003; Nocolosi *et al.*, 2000). Rutaceae family is divided into seven subfamilies comprising of 11 tribes and 93 genera. Three of these subfamilies are represented in Southeast Asia namely Aurantioideae (33 genera), Rutoideae (30 genera) and Toddalioideae (25 genera). Remaining four subfamilies are found in the new World, Australia and other parts of Asia. Aurantioideae or the orange subfamily has 33 genera classified in two tribes and six subtribes which contain over 215 species and 65 varieties of plants. The genus *Citrus* is the most popular genus in subfamily Aurantioideae. Swingle and Reece (1967) and Swingle (1948) arranged the genera of this subfamily as "minor citroids" and "near citrus". Classification of citrus is given below:

Kingdom	:	Plantae
Division	:	Magnoliophyta
Class	:	Magloliopsida
Subclass	:	Rosidae
Order	:	Sapindales
Family	:	Rutaceae
Subfamily	:	Aurantioideae
Tribe	:	Citrae
Genus	:	Citrus

Among group 3 *i.e.* True citrus, following genera and species are included: *Fortunella* (4 spp.), *Clymenia* (2 spp.), *Microcitrus* (7 spp.), *Citrus* subgenus, *Citrus* (10 spp.), and *Citrus* subgenus *Papeda* (6 spp.). *Eremocitrus* and *Microcitrus* are other genera in this family. Important species and hybrids are: *Citrus aurantifolia* (Key lime) *Citrus maxima* (Pummelo), *Citrus medica* (Citron), *Citrus reticulata* (Mandarin orange), *Citrus aurantium* (Bitter or sour orange), *Citrus latifolia* (Persian lime), *Citrus limon* (Lemon), *Citrus limonia* (Rangpur lime), *Citrus paradisi* (Grapefruit), *Citrus sinensis* (Sweet orange) and *Citrus tangerine* (Tangerine).

Citrus fruits are produced all around the world. Some of the citrus cultivars are considered to be of religious significance in several countries in Southeast Asia and in some other countries. Citron cultivar Buddha's hand is considered to be highly religious in several Asian countries. This is a bright yellow fruit and is an ancient mutation of the citron. It is extremely aromatic and used in fragrant ornamants (Plate 1a). Etrog citron- a religious citrus fruit is used by Jewish during Sukkot festival, citron (*Citrus medica*) is sold in market for religious purposes in Germany (Plate 1c), and Red finger lime (*Citrus australiasica*) is a rare delicacy in Australia (Plate 1b). About 140 countries produce various kinds of citrus fruits. But Citrus is commercially grown in about 50 countries. Major *Citrus* producing countries are situated in both Northern and Southern Hemispheres. However, about 70% of total citrus fruits are produced in the Northern Hemisphere. Main citrus fruit producing countries are Brazil, the Mediterranean countries, the United States and countries in Southeast Asia including China and India. These countries represent more than two thirds of global citrus fruit production.

Citrus is one of the choicest fruit having high consumer's preference both as fresh fruit as well as for its refreshing fresh and processed juice. World citrus production for the year 2005 was estimated as 94.8 million tons (FAOSTAT-FAO Statistical Database, 2005). The country-wise production of citrus is given in Table 1. World production of citrus fruits has experienced continuous growth from 1960 onward as shown in Fig. 1. In 2009, citrus

Table 1: World citrus production as per FAOSTAT-FAO statistical Database 2005 (000MT)

Country	*Total citrus production*
Algeria	542.7
Argentina	2670.0
Australia	716.1
Brazil	18902.5
Chile	312.0
China	15227.9
Colombia	330.0
Cuba	216.0
Cyprus	178.7
Egypt	2706.3
Greece	861.0
India	4662.0

Table 1: (*Contd...*)

Table 1: *(Contd...)*

Country	*Total citrus production*
Indonesia	1311.7
Iran	3037.0
Israel	639.9
Italy	3320.9
Japan	1341.0
Korea Republic	594.0
Lebanon	339.0
Mexico	6910.0
Morocco	1320.9
Pakistan	504.5
Paraguay	289.6
Peru	754.0
Spain	6181.3
South Africa	1543.0
Tunisia	307.9
Turkey	2316.8
U.S.A.	10498.5

production increased upto 115 million tons (FAO, 2009). This rise in citrus production is mainly due to increase in cultivation areas and change in consumer's preference towards health and the rising income. Different types of citrus fruits grown in different countries are shown in Table 2.

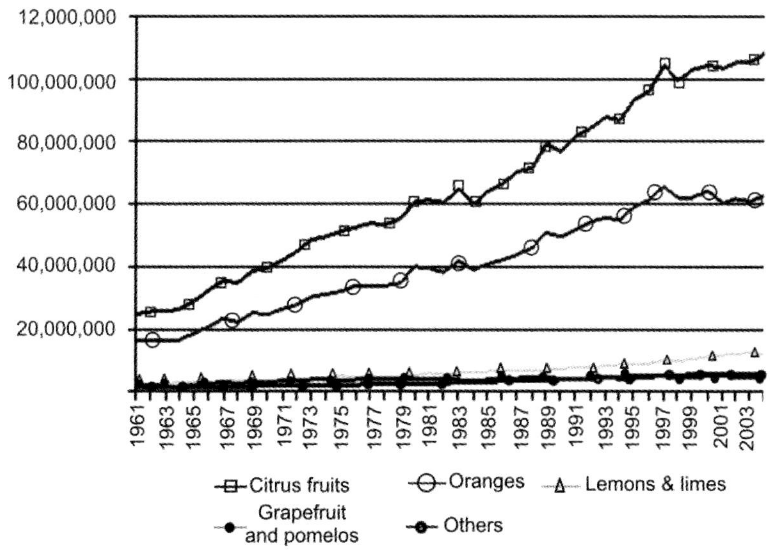

Fig. 1: World citrus fruit production 1961-2004 (in tons)
Source: UNCTAD from FAO data

In Asian countries, citrus fruit production is mainly consumed domestically but as per FAO records, fresh orange consumption is declining in some countries mainly due to replacement by orange juice consumption

Table 2: Different types of citrus fruits and major producing countries as per 2004 FAO data

Oranges	Spain, South Africa, United States, Greece, Morocco, Netherlands, Turkey, Egypt, Australia, Italy
Small citrus	China, Israel, South Africa, Cyprus, India, Netherlands, Pakistan, United States, Spain, Mexico
Lemons and limes	Spain, Argentina, Mexico, Turkey, United States, South Africa, Netherlands, Brazil, Italy, India, Greece
Grapefruit	United States, South Africa, Israel, Turkey, Netherlands, Belgium, Spain, Argentina, Cyprus, Bahamas

and improvements in transportation and storage. However, fresh orange consumption has increased in many developing countries, especially in emerging economies such as Mexico, India, Argentina, Brazil and China. United States of America is the major processed orange consuming country in the world followed by European Union.

India is the second largest producer of different fruits after China, with a production of 44.04 million tons of fruits from an area of 3.72 million hectares. In India, citrus is the largest fruit industry after banana and mango. The estimated production of citrus fruits in India is 9.63 million tons from an area of 0.98 million ha (National Horticulture Board (NHB), 2010). Although citrus fruits are grown throughout the country, but major citrus fruit growing states in India are: Andhra Pradesh, Bihar, Gujarat, Karnataka, Maharashtra, Punjab, Tamil Nadu and Uttar Pradesh. Citrus fruits rank 3rd in area and production accounting for about 12 and 10.4 per cent of the total area and production respectively. Lime, lemons, sweet oranges and mandarins cover bulk of the area under these fruits and are grown mainly in Maharashtra, Andhra Pradesh, Karnataka, Northeastern States, Punjab, Orissa and Madhya Pradesh. Commercially mandarins, sweet oranges and acid limes are most important citrus fruits produced in India. Northeastern region of India is also growing good quality mandarins but commercial cultivation is lacking.

India is the major producer of limes and lemons (Table 3). It tops the production list with ~16% of the world's overall lemon and lime output followed by Mexico (~14.5%), Argentina (~10%), Brazil (~8%) and Spain (~7%). Exact origin of lemons remained a mystery, though it is widely presumed that lemons first grew in India, Northern Burma, and China. In South and Southeast Asia, lemons are known for its antiseptic properties and are also used as an antidote for various poisons. Lemons were later introduced to Persia and then to Iraq and Egypt around AD 700. Lemons were first recorded in literature in a tenth century Arabic treatise on farming, and were also used as an ornamental plant in early Islamic gardens. Lemons were distributed widely throughout the Arab world and the Mediterranean region between AD 1000 and AD 1150.

Table 3: Top lemons and limes producers — 2007 FAO Estimate

Country	*Production (tonnes)*
India	2,060,000
Mexico	1,880,000
Argentina	1,260,000
Brazil	1,060,000
Spain	880,000
People's Republic of China	745,100
United States	722,000
Turkey	706,652
Iran	615,000
Italy	546,584
World	**13,032,388**

Source: Food and Agricultural Organization of United Nations; Economic and Social Department; The Statistical Division

Lemons, however, entered Europe (near southern Italy) not later than the first century AD, during the time of ancient Rome. The first real lemon cultivation in Europe began in Geneva in the middle of the fifteenth century. It was later introduced to the Americas in 1493 when Christopher Columbus brought lemon seeds to Hispaniola along his voyages. Spanish conquest throughout the New World helped spread lemon seeds. It was mainly used as ornamental and in medicine. In 1700s and late 1800s, lemons were increasingly planted in America and were used in cooking and flavouring purposes. Several varieties of lemon are cultivated such as hill lemon, Meyer lemon, Lisbon lemon, Eureka lemon, Tahiti or Persian lemon etc. Ten top countries producing lemons and limes are given in Table 3 as per FAO estimates for the year 2007. However, world lemon and lime production is projected at 10.6 million MT in 2010, an increase of 15 percent from the 1996–98 level.

The Latin word *citrus* was borrowed from ancient Greek *kedros* "cedar, juniper" probably through Etruscan. Taxonomy and systematics of the genus *Citrus* are complex and precise number of natural species is unclear, as many of the named species are clonally propagated hybrids, and there is genetic evidence that even some wild, true-breeding species are of hybrid origin (De Araujo *et al.*, 2003). Natural and cultivated hybrids include commercially important fruits such as the oranges, grapefruit, lemons, some limes, and some tangerines. Research suggests that closely related genus *Fortunella* (kumquats), and perhaps *Poncirus* and the Australian *Microcitrus* and *Eremocitrus*, should be included in *Citrus*; most botanists now classify *Microcitrus* and *Eremocitrus* as part of the genus *Citrus* (Nicolosi *et al.*, 2000). Various species, cultivars and hybrids of citrus are shown in Annexure-1. Southeast Asia especially India and neighboring China are known as the original home of several *Citrus* species and hence this region is rich in citrus germplasm. Over the years, citrus germplasm has been collected worldwide and is being maintained in several countries for citrus improvement. Germplasm available in the region of origin (India and China) is given in Annexure - 2. Indigenous citrus in China includes three genera: *Citrus, Poncirus* and *Fortunella*. To avoid introduction of viruses through the movement of citrus germplasm, the International Organization of *Citrus* Virologist, (IOCV), Food and Agriculture Organization of the United Nations (FAO), and International Bureau for plant Genetic Resources (IBPGR) jointly published *FAO/IBPGR Technical Guidelines for the Safe Movement of Citrus Germplasm* (Frison and Taher, 1991) and the author was also a member of this team. This document is serving citrus industry to check virus movement with planting material or through citrus germplasm from one country to another or from one region to another region.

Citrus fruits are the first fruit crop in international trade in terms of value. There are two clearly differentiated markets in citrus sector: fresh citrus fruits market, with a predominance of oranges, and processed citrus products market, mainly orange juice. A major development over the last two decades of the 20th century was the growth in trade in small citrus fruits, which include tangerines, clementines, mandarins and satsumas at the expense of fresh oranges. Citrus juice also has medical uses; lemon juice is used to relieve the pain of bee stings. Oranges were historically used for their high content of vitamin C, which prevents scurvy. Scurvy is caused by vitamin C deficiency, and can be prevented by having 10 milligrams of vitamin C a day. Before the development of fermentation-based processes, lemons were the primary commercial source of citric acid.

Viruses infect humans, animals, insects, plants, bacteria and fungi. First published report of a plant virus disease was breaking in flower colour of tulips in 1576 (Mc Kay and Warner, 1933) - a disease now known to be caused by an aphid transmitted virus. Classification and nomenclature are standardized by the International committee on taxonomy of viruses (ICTV) through reports published periodically. ICTV recognized a global total of about 2000 species of viruses as of 2005 (Fauquet *et al.*, 2005). So far about 800 plant viruses have been described. Viruses are also divided into genera and species similar to bacteria. Viruses are composed of nucleic acid genome surrounded by a protein coat called a capsid. Together, the genome and capsid are referred to as the nucleocapsid. Genomes of viruses are either RNA or DNA. Viral capsids are composed of many individual subunits called capsomeres. Capsomeres assemble into an icosahedral or irregular shaped capsid. Some viruses have lipids containing envelop that surrounds capsid. The entire virus, including nucleic acid, capsid envelop is called the virion, or virus particle. Viruses may be single stranded DNA viruses (Families: Geminiviridae, Nanoviridae), double stranded DNA viruses (Families: Caulimoviridae), Single stranded RNA reverse transcribing viruses (Families: Pseudoviridae, Metaviridae), double stranded RNA viruses (Families: Reoviridae, Partiviridae), the negative single stranded RNA viruses (Families: Rhabdoviridae, Bunyaviridae), the positive single–stranded RNA viruses (Families: Sequiviridae, Comoviridae, Potyviridae, Luteoviridae, Tombusviridae, Bromoviridae, Tymoviridae, Closteroviridae and Flexiviridae).

Viruses spread through insect vectors which mostly belong to Order Homoptera, families Aphididae (200 spp. transmit 160 viruses), Psyllidae (6 spp. transmit phytoplasmas and bacteria-like organisms), Coccidae (15 spp. transmit 2 viruses), Aleyrodidae (4 spp. transmit 70 viruses), Auchenorryncha (151 spp. transmit 38 viruses, 31 Phytoplasmas, 4 Spiroplasmas and 4 Rickettsia-like organisms), Order Coleoptera (74 spp. transmit 45 viruses), Thysanoptera (9 spp. transmit 4 viruses) (Srivastava and Ahlawat, 2010).

In citrus, more than 40 virus and virus-like diseases have been described from most citrus growing countries. In majority of cases, etiology of these diseases has been established and the associated viruses have been characterized and their taxonomy established. During the last three decades, advanced methods of virus diagnosis have been developed worldwide including India (Ahlawat *et al.*, 1992, 2008). In India, virus and virus-like diseases of citrus have been reviewed from time to time and geographical distribution of virus and virus-like diseases of citrus in India has been shown in Table 4 (Ahlawat and Pant, 2003a, b; Ahlawat and Pant, 1999, 2010). However, recently discovered virus diseases like citrus yellow mosaic, Indian citrus ring spot, yellow vein clearing etc. have not been shown in Table 4. These have been shown in Table 6.

Table 4: Distribution of virus and virus-like diseases of citrus in India

Area	*Disease*																
	CL	*EX*	*GR1*	*IM*	*IV*	*LC*	*LL*	*MO*	*MS*	*OS*	*PS*	*RW*	*SY*	*TR*	*TC*	*WT*[2]	*YM*[3]
Andhra Pradesh	–	–	–	–	–	–	–	–	–	–	–	+	S[4]	+	–	–	–
Hyderabad	–	–	+	+	–	–	–	+	–	–	–	–	–	–	–	–	–
Kodur	–	+	+	–	–	–	–	–	–	–	–	–	–	+	–	–	–
Tirupati	–	–	–	–	–	–	–	+	–	+	–	–	–	–	–	–	+
Arunachal Pradesh	–	–	–	–	–	–	–	–	–	–	–	–	–	–	+	–	–
Along	–	–	+	–	–	–	–	–	–	–	–	–	–	+	–	–	–
Baser	–	–	+	–	–	–	–	–	–	–	–	–	–	–	–	–	–
Assam	–	–	–	–	–	–	+	+	–	–	–	+	–	S	+	–	–
Burnihat	–	–	+	–	+	–	–	–	–	–	–	–	–	+	–	–	–
Ranchi	–	–	–	–	–	–	–	–	–	–	–	–	–	–	–	+	–
Bihar	–	–	–	–	–	–	–	–	–	–	–	–	–	–	–	–	–
Chianki	–	–	+	–	–	–	–	–	–	–	–	–	–	–	–	–	–
Ranchi	–	–	+	–	–	–	–	–	–	–	–	–	–	+	–	–	–
Delhi	–	+	+	+	+	–	–	+	–	–	+	–	–	+	–	–	–

Table 4: (*Contd...*)

Table 4: *(Contd...)*

Area	*Disease*																
	CL	*EX*	*GR1*	*IM*	*IV*	*LC*	*LL*	*MO*	*MS*	*OS*	*PS*	*RW*	*SY*	*TR*	*TC*	*WT*[2]	*YM*[3]
Gujarat	–	–	–	–	–	–	–	–	–	–	–	–	–	–	–	–	–
Anand	–	–	+	–	–	–	–	–	–	–	–	–	–	–	–	–	–
Haryana	–	–	–	–	–	–	–	–	–	–	–	–	–	+	–	–	–
Chandigarh	–	–	+	–	–	–	–	–	–	–	–	–	–	+	–	–	–
Himachal Pradesh	–	–	–	–	–	–	–	–	–	–	–	–	–	–	–	–	–
Dhaulakuan	–	–	+	–	–	–	–	–	–	–	–	–	–	+	–	–	–
Jammu and Kashmir	–	–	–	–	–	–	–	–	–	–	–	–	–	–	–	–	–
Jammu	–	–	+	–	–	–	–	–	–	–	–	–	–	+	–	–	–
Karnataka	–	–	–	–	–	–	–	–	–	–	–	–	–	–	(+)[5]	–	–
Chetali	–	–	+	–	–	–	–	–	–	–	–	–	–	+	–	–	–
Gonicoppal	–	–	+	–	–	–	–	–	–	–	–	–	–	+	–	–	–
Kerala	–	–	–	–	–	–	–	–	–	–	–	–	–	–	(+)	–	–
Ambalvoyal	–	–	+	–	–	–	–	–	–	–	–	–	–	+	–	–	–
Nellyampathy	–	–	+	–	–	–	–	–	–	–	–	–	–	+	–	–	–
Madhya Pradesh	–	–	–	–	–	–	–	–	–	–	–	–	–	–	–	–	–
Jabalpur	–	–	+	–	–	–	–	–	–	–	–	–	–	+	–	–	–
Prabat Pattan	–	–	–	–	–	–	–	–	–	–	–	–	–	+	–	–	–
Sausar	–	–	+	–	–	–	–	–	–	–	–	–	–	–	–	–	–
Maharashtra	–	–	–	–	–	–	–	–	–	–	–	–	–	S	+	–	–
Aurangabad	–	–	+	–	–	–	–	–	–	–	–	–	–	–	–	–	–
Nagpur	–	–	+	–	–	–	–	–	–	–	–	–	–	+	–	–	–
Poona	–	+	+	–	+	–	–	+	–	–	+	–	+	+	–	+	+
Shrirampur	+	+	+	–	–	+	–	–	–	–	–	+	–	+	–	+	–
Meghalaya	–	–	–	–	–	–	–	–	–	–	–	–	–	–	–	–	–
Shilong	–	–	+	–	–	–	–	–	–	–	–	–	–	–	–	–	–
Orissa	–	–	–	–	–	–	–	–	–	–	–	–	+	–	–	–	–
Guma	–	–	+	–	–	–	–	–	–	–	–	–	–	–	+	–	–
Udaigiri	–	–	+	–	–	–	–	–	–	–	–	–	–	–	+	–	–
Punjab	–	–	–	–	–	–	–	–	–	–	–	–	–	–	–	–	–
Aborar	–	+	+	–	–	–	–	–	–	–	–	–	–	–	+	–	–
Rajasthan	–	–	–	–	–	–	–	–	–	–	–	–	–	S	(+)	–	–
Bhawani Mandi	–	–	+	–	–	–	–	–	–	–	–	–	–	+	–	–	–
Jhalawar	–	–	+	–	–	–	–	–	–	–	–	–	–	–	–	–	–
Jodhpur	–	–	+	–	–	–	–	–	–	–	–	–	–	+	–	–	–
Kota	–	+	+	–	–	–	–	–	+	–	–	+	–	+	–	+	–
Pilani	–	–	+	–	–	–	–	–	–	–	–	–	–	–	–	–	–
Sriganganagar	–	+	+	–	–	–	–	–	–	–	–	+	–	+	–	+	–
Udaipur	–	–	–	–	–	–	–	–	–	–	–	+	–	–	–	–	–
Sikkim	–	–	–	–	–	–	–	–	–	–	–	–	–	S	+	–	–
Gayalsing	+	–	+	+	+	–	+	–	–	–	–	+	–	+	–	–	–
Namchi	+	–	+	+	+	–	+	–	–	–	–	+	–	+	–	–	–
Rangpo	+	–	+	+	+	–	+	–	–	–	–	+	–	+	–	–	–

Table 4: *(Contd...)*

Table 4: (*Contd...*)

Area	*Disease*																
	CL	*EX*	*GR*[1]	*IM*	*IV*	*LC*	*LL*	*MO*	*MS*	*OS*	*PS*	*RW*	*SY*	*TR*	*TC*	*WT*[2]	*YM*[3]
Tamil Nadu	–	–	–	–	–	–	–	–	–	–	–	–	–	–	(+)	–	–
Coimbatore	–	+	+	–	–	–	–	–	–	–	–	–	–	+	–	–	–
Uttar Pradesh	–	–	–	–	–	–	–	–	–	–	–	–	–	–	–	–	–
Gopeshwar	–	–	–	–	–	–	–	+	–	–	+	–	–	–	–	–	–
Khanpur	–	–	+	–	–	–	–	–	–	–	–	–	–	–	–	–	–
Pantnagar	–	–	+	–	–	–	–	–	–	–	–	–	–	+	–	–	–
Saharanpur	+	–	+	–	–	–	–	–	–	–	–	–	–	+	–	–	–
West Bengal	–	–	–	–	–	–	–	–	–	–	–	–	–	S	+	–	–
Darjeeling	+	–	+	+	+	–	+	–	–	–	–	+	+	+	–	+	–
Kalimpong	+	–	+	+	+	–	+	–	–	–	–	+	+	+	–	+	–

Key: CL = Crinkly leaf, EX = Exocortis; GR = Greening; IM = Impietratura; IV = Infectious variegation; LC = Leaf curl; LL = Leathery leaf; MO = Mosaic; MS = Multiple sprouting; OS = Oily specks; PS = Psorosis A; RW = Rubbery wood; SY = Seedling yellows; TR = Tristeza; TC = Presence of *T. citricida*; WT = Wound tumour; YM = Yellow midvein

Notes: [1]Presence of greening is based on symptoms in most cases and must be confirmed by electron microscopy and/or DNA hybridization; [2] Vein enation-woody gall; [3] Yellow vein; [4] S means that severe strains are present; [5] (+) indicates that presence is inferred from reports that *T. citricida* occurs in southern parts of India (Raychaudhuri, Nariani and Ahlawat, 1977; Ahlawat and Raychaudhuri, 1988).

Source: FAO Corporate document Repository, Regional office for the near East

Other important pathogens infecting citrus are phytoplasmas earlier known as Mycoplasma-like organisms (MLOs). Phytoplasmas have pleomorphic morphology and have been classified under the class Mollicute. These pathogens are mainly transmitted by leafhopper or psyllid vectors but not mechanically. So far only one member of class Mollicute, the citrus stubborn pathogen has been cultured in synthetic medium. Based on its helical morphology, this organism was named as *Spiroplasma citri*. Typical symptoms of stubborn disease appear when innocuous insect vector, *Euscelis plebejus* is allowed to feed on healthy plants. The insects were made innocuous by injecting the culture of *S. citri* and were then fed to healthy plants (Calavan, 1980).

Although viruses and phytoplasmas had been well established as causal pathogens of plant diseases but still there were some virus-like diseases such as potato spindle tuber and citrus exocortis etc. which still remained with uncertain etiology till Diener reported for the first time that the causal incitant of potato spindle tuber consists only pathogenic RNA of low molecular weight and termed it as *viroid* (Diener, 1971). These pathogens were much smaller than viruses (250–400 nucleotides). After the discovery of viroids, diseases like citrus exocortis, cachexia and a few others were shown to be caused by viroids.

There are another type of genetic RNAs, incapable of replicating by themselves and reside in other viruses and depend upon them for their replication, can cause diseases; they are called Virusoids or Satellite RNAs. Virusoids are genetic RNAs but have no independent existence. The size of these RNAs is 220–338 nucleotides long, circular, live within capsid core of another virus, called helper virus. They totally depend on the helper virus for their replication.They can replicate in the cytoplasm using an RNA-dependent RNA polymerase. This enzymatic activity is common in plants but not found in animal cells. They depend on a helper virus for replication. This helper virus also encapsidates them, *e.g.*: Subterranean clover mottle virus satellite RNA: Helper–*Sobemovirus*. Virusoids are found in bacteria, plants, fungi, invertebrates and vertebrates. Satellite genomes encode proteins, satellite viruses encode capsid proteins but are still dependent upon a helper virus for replication. Examples include Barley yellow dwarf virus satellite RNA: Helper – *Luteovirus,* Tobacco ringspot virus satellite RNA: Helper – *Nepovirus*. These agents may modify the symptoms of infection by their helper virus. They do no interfere with the replication of their helper virus and are therefore differentiated from defective interfering particles that are associated with many viral infections. They can be spread by vegetative propagation, within seeds or by direct inoculation either by insects or man. There are similar infectious agents which infect animals, *e.g.* new satellite 2 transcript. One such agent infecting humans is the hepatitis delta virus (HDV). There is another class of infectious agents, causing serious diseases, which are neither DNA nor RNA, but conformationally changed proteins, which can propagate by inducing conformational changes in other proteins of the same kind, thus their number increases with time; this is a class by itself but unique and they are called Prions (www.microbiologybytes.com/virology/Viroids.html).

Citrus psorosis that caused specific leaf symptoms and bark lesions on trunk and branches of sweet orange (*Citrus sinensis*) was first demonstrated to be a virus disease (Fawcett, 1933). Since then, citrus virus diseases have accumulated greatly. Psorosis virus was also the first virus disease in citrus against which budwood certification programme for its management was developed in 1938 in California and was followed by several countries where citrus is an important industry.

Failure of sweet orange trees on sour orange rootstock (quick decline) was initially considered due to incompatibility factor. However, demonstration by Fawcett and Wallace (1946) and Meneghini (1946) revealed that quick decline in California and tristeza in Brazil were virus diseases and the two were identical. Citrus tristeza virus has wiped out citrus industry of many countries and still causing a loss of $500 million to 1.5 billion per year to the world citrus industry. Many of the important virus diseases now recognized in citrus appear to have originated in India and the orient, and to have moved with citrus as it was carried to other parts of the world through traders (Reuther *et al.*, 1978).

In India, citrus dieback had been a baffling problem for more than seven decades and is largely responsible for decline of citrus industry in India. This disease causes defoliation of young branches and dieback of twigs from the tip. Thousands of trees have died in India due to this malady (Raychaudhuri *et al.*, 1969). This problem has been demonstrated to be caused by both biological and abiological factors (Ahlawat and Pant, 2003; Ahlawat, 2007, 1997a; Raychaudhuri and Ahlawat, 1982; Shyam Singh and Ghosh, 1999). Recent investigations revealed that major cause of dieback syndrome is the infection of some viruses, huanglongbing (greening) bacterium (*Candidus* Liberibacter asiaticus) and some viroids either in a single infection or in mixed infection (Ahlawat, 1997a, 1997b). Indian Council of agricultural Research, New Delhi established a National Research Centre for Citrus (NRCC) at Nagpur in 1986 for the improvement of citrus industry in India. This centre has collected large number of wild and cultivated indigenous germplasm. Different citrus species are being cultivated on 57.9 thousand ha with a production of 300 lakh tones in Northeastern India- the home of citrus. The average productivity in this region is 5.3 t/ha which is much lower than national level of 9.4 t/ha. This is mostly because of non-availability of healthy quality planting material to the growers in that region. In India, rough lemon (*Citrus jambhiri* Lush) is the leading rootstock but fruit quality on this rootstock is poor and it is susceptible to *Phytophthora* and nematodes. This is also one of the important factors for poor productivity of citrus fruits. Other rootstocks like Rangpur lime, Cleopatra mandarin, Sour orange and Troyer citrange must be evaluated for citrus species grown in India. During recent years, NRCC, Nagpur developed a repository of indigenous and exotic citrus germplasm to improve citrus industry in India. Following wild germplasm from Northeastern region of the country has been collected for evaluation.

1. Satkara (*C. macroptera*)
2. Soh Shyrkhoit (*C. latipes*)
3. Citron (*C. medica*)
4. Tasi (*C. sinensis*)
5. Soh Nairiang (*C. sinensis*)
6. Soh Bitara (*C. sinensis*)
7. Calamondin (*C. medurensis*)

8. Soh Khyllah (Hybrid)
9. Galgal (*C. pseudolimon*)
10. Pummelo (*C. grandis*)
11. Sohmyndong (*C. jambhiri*)
12. Ada Jamir (*C. assamensis*)
13. Memon Narang (*C. indica*)
14. Karun Jamir (*C. aurantium*)

A repository of citrus was established in 1987 on the campus of University of California, Riverside.The mission of the Repository is to acquire, preserve, distribute, and evaluate germplasm of *Citrus*. There are over 1,000 accessions in the CVC and Citrus Relatives Collection and about 400 accessions in the Protected Collection which are available for distribution as pathogen-tested budwood. Currently, 28 of the 33 genera in the subfamily *Aurantioideae* are represented in the various collections. Phylogenetic relationship between different cultivars of citrus has not been clearly understood because of hybridization, nucellar embryony and somatic mutations. The most critical issue facing the Repository is the threat of the Asian citrus psyllid and huanglongbing (HLB) for which a constant vigilance is required.

Economic importance of citrus viruses has been realized world over. Therefore, for the first time delegates from 11 countries participated in a conference on citrus virus diseases in Riverside, California in 1957. 35 papers were published in the proceedings of that conference. At this meeting in California, an organization now known as 'International Oragnization of *Citrus* Virologists' (IOCV) was established. Soon after IOCV was founded in 1957, a committee was assigned the work of listing and abstracting all publications related to virus and virus-like diseases under the leadership of Professor J.M. Bove and R. Vogel-two scientists from France. With the dedicated efforts of these leaders and IOCV members, three volumes "Description and illustration of virus and virus-like diseases of citrus" have been compiled and printed. This bibliography is being kept current and has loose leaf volumes with more than 400 removable identifiable colour slides and grouped by subject to accompany the printed information for each disease. It is indeed a valuable aid for persons engaged in research on virus diseases of citrus world over. Nowhere else in the field of plant pathology such a comprehensive description has been prepared for diseases that affect a single crop. IOCV has organized 18 conferences since its inception in 1957 (Table 5). These conferences are held once in three years in different citrus growing countries. The 12th IOCV conference was held in India at Indian

Table 5: IOCV Conferences and their locations

Date	*Conference location*	*Pre or Post-conference tours or visits*
1957	California	California
1960	Florida	Florida
1963	Brazil	Brazil, Argentina
1966	Italy	Italy, Corsica, France, Spain, Morocco
1969	Japan	Japan, Taiwan, Hong Kong, Philippines, New Territories
1972	Swaziland	Swaziland, South Africa, Reunion, Mauritius, Malagasy
1975	Greece	Israel, Greece, Crete, Egypt
1979	Australia	Australia
1983	Argentina	Brazil, Argentina
1986	Spain	Spain
1989	Florida	Florida, California
1992	India	Oman, India
1995	China (Fouzhou)	China (Chengdu, Neijang, Chongqing, Bei Bei)
1998	Brazil	Uruguay, Belize
2001	Cyprus	Egypt
2004	Mexico (Monterrey)	Texas, Veracruz
2007	Turkey	Turkey
2010	Campinas, Brazil	

Source: <www.ivia.es/iocv>

Agricultural Research Institute, New Delhi in 1992. A continuing evolution of progress and problems of citrus viruses have been reviewed in 13th IOCV conference held in China (Lee and Garnsey, 1996). The 18th IOCV Conference was held in campinas, Brazil in 2010. The first International Citrus Symposium was also held at the University of California, Riverside, USA on March 16–26, 1968. In this symposium 16 papers were presented on virus diseases of citrus. Out of 16 papers presented, 8 were presented by USA scientists and one each from South Africa, Philippines, India, Israel, Japan, Peru, Brazil and Cyprus (Chapman, 1969). The EcoPort website was launched in 2000 to provide a knowledge database of biodiversity in citrus. As of September, 2005, the EcoPort website contained over one half million references,over 97,000 pictures and over 200 slide shows, of which 59 pertain to citrus and citrus virus and virus-like diseases. All slide shows and pictures (in JPEG and GIF format) are accessible online and can be downloaded. To access individual pictures and slide shows in EcoPort, use the internet browser and type: http://www.EcoPort.org. All slide presentations can be accessed directly at the following URL: http://ecoport.org/ep?Search (Roistacher *et al.*, 2005)

Among citrus virus disease, about 17 are well established and genomes of their associated viruses have been fully characterized. About 13 diseases considered to be virus-like based on their transmission and absence of other disease inciting agents. Important viruses causing diseases in citrus trees are (I) Ilarviruses causing citrus crinkly leaf, citrus variegation and citrus leaf rugose (II) the Rhabdoviruses causing citus leprosis and zonate chlorosis and Yellow vein mosaic of Pummelo (III) Ophiovirus causing psorosis diseases (IV) Satsuma dwarf virus group (V) tatter leaf-citrange stunt (VI) closterovirus causing tristeza disease (VII) badnavirus causing citrus yellow mosaic disease and (VIII) mandrivius causing Indian citrus ring spot disease. In addition to viruses, other serious diseases in citrus are caused by bacteria (huanglongbing and variegated chlorosis), by Mollicutes (citrus stubborn, lime witches broom) and by viroid pathogens (exocortis, cachexia, and others). Effect of viroids on dwarfing of citrus trees is interesting and will also be described in this book.

Virus and virus-like diseases recently discovered in India by the author and perhaps new to the world have been shown in Table 6 below:

Table 6: Distribution of newly discovered citrus diseases in India

Places in states of India	*ICRSMV*	*CYMBV*	*CYVCV*	*PYVMRV*	*CYCVVd*
Andhra Pradesh	+	+	–	+	+
Assam	–	+	–	–	+
Delhi	+	–	–	–	–
Gujarat	+	–	+	–	–
Hariyana	+	–	–	–	–
Himachal Pradesh	–	–	–	–	–
Jammu & Kashmir	+	–	–	–	–
Karnataka	–	+	–	+	+
Madhya Pradesh	–	–	–	–	+
Maharashtra	+	+	–	–	+
Northeastern States	–	+	–	–	+
Orrisa	–	–	–	–	–
Punjab	+	–	+	–	+
Rajasthan	+	+	–	–	+
Sikkim	–	–	–	–	–
Tamil Nadu	–	+	–	–	–
Uttar Pradesh	+	–	–	–	+
West Bengal	–	–	–	–	+

ICRSMV = Indian citrus ringspot mandrivirus
CYMBV = Citrus yellow mosaic badnavirus
CYVCV = Citrus yellow vein clearing virus
PYVMRV = Pummelo yellow vein mosaic rhabdovirus
CYVCVd = Citrus yellow corky vein viroids

REFERENCES

Ahlawat, Y.S. 1997a. Status of virus and virus-like pathogens infecting citrus in India and future strategies for citrus improvement. *Indian Phytopath* 50: 192-193.

Ahlawat, Y.S. 1997b. Viruses, greening bacterium and viroids associated with citrus (*Citrus* spp.). *Indian J. Agric. Sci.*, 67: 51-57.

Ahlawat, Y.S. 2007. Citrus decline: Problems and Preventions. *Indian Phytopath.* 60: 1-12.

Ahlawat, Y.S. and Pant, R.P. 1999. Viruses, viroids, phytoplasma and greening bacterium associated diseases of citrus. pp. 388-425. *In*: Diseases of horticultural crops-fruits (*Eds.*) L.R. Verma and R.C. Sharma, Indus Publishing Co., New Delhi

Ahlawat, Y.S. and Pant, R.P. 2003a. Major virus and virus-like diseases of citrus in India, their diagnosis and management. *Plant Pathol* 2: 447-474.

Ahlawat, Y.S. and Pant, R.P. 2003b. Virus and virus-like diseases of citrus – the Indian scenario. pp. 569-588. *In*: Advances in citriculture (*Eds.*) Singh, S., Shivankar, V.J., Srivastava, A.K. and Singh, I.P. Jagmander Book Agency, New Delhi.

Ahlawat, Y.S. and Pant, R.P. 2010. Virus and virus-like diseases of citrus in India and their management. pp. 71-94. *In*: IPM system in Agriculture, Virus and virus-like diseases of plants Vol. 10 (*Eds.*) Upadhyay, R.K., Ahlawat, Y.S, Dubey, O.P., Baranwal, V.K. and Srivastava, K.P. Aditya Books Pvt. Ltd. New Delhi, Bangalore, Kolkata.

Ahlawat, Y.S. and Raychaudhuri, S.P. 1988. Status of citrus tristiza and dieback diseases in India and their diagnosis. pp. 871–879. *In*: Proc. 6th Inteml. *Citrus* Cong.

Ahlawat, Y.S., Chakraborty, N.K.,Varma, A. and Raychaudhuri, S.P. 1992. Diagnosis of diseases caused by viruses and fastidious prokaryotes in citrus in India. *In*: *Proc. Intl. Soc. Citriculture* 2: 749-751.

Ahlawat, Y.S., Das, A.K., Ghosh, D.K. and Baranwal, V.K. 2008. Recent trends in diagnosis and management of citrus diseases. pp. 119-130. *In*: Proc. National symp. Citriculture; Emerging trends, held 24-26 July, 2008 at NRCC, Nagpur.

Calavan, E.C. 1980. Stubborn. *In*: Description and illustration of virus and virus-like diseases of citrus (*Eds*. Bové, J.M. and Vogel, R.). SETCO-IRFA, Paris, France.

Chapman, H.D. 1969. Vol. III. *In*: Proc. 1st Intl. citrus symposium, p. 1839.

De Araujo, de Queiroz, E.F., Paganucci, L. and Machado, M.A. 2003. What is Citrus? Taxonomic implications from a study of cp-DNA evolution in the tribe Citreae (Rutaceae subfamily Aurantioideae). *Organisms Diversity & Evolution* 3: 55–62.

Diener, T.O. 1971. Potato spindle tuber 'virus'. IV. A replicating, low molecular weight RNA. *Virology* 45: 411–428.

FAOSTAT-FAO Statistical Database 2005.

FAO estimates 2009. *In*: FAOSTAT. http://faostat.fao.org/site/567/default.aspx#ancor

Fauquet, C.M., Mayo, M.A., Maniloff, J., Desselberger, U. and Ball, L.A., (*Eds.*) 2005. Virus taxonomy: VIIIth Report of the International Committee on Taxonomy of Viruses. San Diego: Elsevier Academic Press. p. 1162.

Fawcett, H.S. 1933. New symptoms of psorosis indicating a virus disease of citrus. (Abs). *Phytopathology* 23: 930.

Fawcett, H.S. and Wallace, J.M. 1946. Evidence of virus nature of citrus quick decline. *Calif. Citrog*. 32: 88–89.

Frison, E.A. and Taher, M.M. 1991. FAO/IBPGR Technical Guidelines for the Safe Movement of *Citrus* Germplasm. FAO, IOCV, IPGRI.

Lee, R.F. and Garnsey, S.M. 1996. Citrus virus and virus-like pathogens: a continueing evolution of progress and Problems. p. 1–7 *In*: Proc. 13th Conf. IOCV, IOCV, Riverside, CA.

Mc Kay, M.B. and Warner, M.F. 1933. Historical sketch of tulip mosaic or breaking. The oldest known plant virus disease. *Natl. Hort. Mag*. 3: 179–216.

Meneghini, M. 1946. Sobre a natureza e transmissibilidade do doencia"tristeza" dos citrus. *O. Biologico* 12: 285–287.

NHB estimates 2010. *In*: http://nhb.gov.in/statistics/area-production-statistics.html

Nicolosi, E., Deng, Z.N., Zentile, A., Lamalfa, S. Continella, A.G. and Tribulato, E. 2000. *Citrus* phylogeny and genetic origin of important species as investigated by molecular markers. *Theoretical and Applied Genetics* 100: 1155–1166.

Raychaudhuri, S.P. and Ahlawat, Y.S. 1982. Problem of citrus diseases in India. Surabhi Printers & Publishers, New Delhi.

Raychaudhuri, S.P., Nariani, T.K. and Ahlawat, Y.S. 1977. Dieback of citrus in India. *In*: Proc. 3rd *释Internl. Soc. Citriculture* 3: 914-918.

Raychaudhuri, S.P., Nariani, T.K. and Lele, V.C. 1969. Citrus dieback problem in India. *In*: Proc. First Intl. Citrus Symposium, Vol. III, pp. 1433–1437, March 16–26, 1968 at UCR, CA.

Reuther, W., Calavan, E.C. and Carman, G.E. 1978. The citrus Industry (Revised *Ed.*). *Univ. Calif. Div. Agric. Sci.*

Roistacher, C.N., Lee, R.F., Putter, C.A.J. and Putter, R.C.J. 2005. EcoPort Slide Shows on the Internet Related to *Citrus* and *Citrus* Diseases. Short Communication. pp. 473-476. *In*: Proc. 16th Conf. IOCV Conf., IOCV, Riverside, CA.

Shyam Singh and Ghosh, S.P. 1999. Hi-Tech Citrus Management. *In*: *Proc. Intl. Symp. on Citriculture*, Nov. 23–27, 1999. NRC for citrus, Nagpur.

Srivastava, K.P. and Ahlawat, Y.S. 2010. Ecofriendly approaches for management of insect vectors of plant viruses. pp. 327–351. *In*: IPM system in Agriculture, Virus and virus-like diseases of plants Vol. 10 (*Eds.*) Upadhyay, R.K., Ahlawat, Y.S., Dubey, O.P., Baranwal, V.K. and Srivastava, K.P. Aditya Books Pvt. Ltd. New Delhi, Bangalore, Kolkata.

Swingle, W.T. 1948. The botany of citrus and its wild relatives of the orange subfamily. *In*: Webber, H.J. and Batchelor, L.D. (*Eds.*). The citrus Industry. Univ. California,Press California pp. 129–174.

Swingle, W.T. and Reece, P.C. 1967. The botany of citrus and its wild Relatives. *In*: Reuther, W., Webber, H. and Batchelor, L. (*Eds.*). The citrus Industry Vol. 1. Revised ed. Univ. California press, California, pp. 190–430.

2

VIRUS DISEASES WITH KNOWN AETIOLOGY

Among the virus diseases of citrus, following 17 diseases have been well studied and their associated viruses have been identified and characterized and methods of virus diagnosis and management practices have been evolved. These diseases will be discussed individually in the following pages.

1. **Citrus psorosis**
2. **Citrus tristeza**
3. **Crinkly leaf**
4. **Infectious variegation**
5. **Citrus leaf rugose**
6. **Citrus leprosis and zonate chlorosis**
7. **Satsuma dwarf**
8. **Citrus mosaic**
9. **Navel orange infectious mottling**
10. **Vein enation and woody galls**
11. **Tatter leaf = citrange stunt**
12. **Indian citrus ringspot**
13. **Citrus ringspot**
14. **Citrus yellow mosaic**
15. **Pummelo yellow vein mosaic**
16. **Citrus yellow vein clearing**
17. **Citrus leaf blotch**

Citrus Psorosis

Citrus is one of the oldest crops grown widely but no virus was known to affect it until Fawcett in 1933 demonstrated the virus nature of psorosis (Fawcett, 1933; Wallace, 1945). The name "psorosis" was coined for a local lesion forming disease (Swingley and Webber, 1896). The name "psorosis-A" was first used by Fawcett in 1932. Two types of psorosis have been proposed, psorosis-A and Psorosis-B (Fawcett, 1933; Wallace, 1945). Although both had similar bark lesion symptoms, they produced different symptoms on leaves and fruits. In cross protection studies, psorosis-A protects psorosis-B infection (Wallace, 1957). He suggested that psorosis-A contained psorosis-B component and both were systemic in infected trees.

There are several other diseases which produce young leaf symptoms similar to psorosis after graft inoculation and were put in psorosis complex. These diseases are concave gum, blind pocket, crinkly leaf and infectious variegation. Comparative symptomatology of these diseases has been reported by Fawcett and Bitancourt (1943). Later investigations confirmed that crinkly leaf and infectious variegation are ilarviruses and distinct from psorosis (Timmer and Benatena, 1977). Diseases which are now put under psorosis group are psorosis-A, psorosis-B, ringspot, and necrotic strain of ringspot (Timmer, 1974). Psorosis-A is considered to be mild strain, whereas, psorosis-B is a severe strain of psorosis disease complex. Psorosis-B is synonymous to ringspot. To consider a disease to be psorosis following points are important (Roistacher, 1993).

1. Infected trees should show classical scaly bark lesions.
2. Wood staining in cut limbs/trunk of sweet orange should show bark lesions.
3. Inoculated indicator plants (sweetorange, mandarin, citron) should show a shock reaction.
4. It should protect against psorosis-B lesion inoculum.
5. Infected field trees must show mature leaf symptoms (ring pattern).
6. Inoculated sweet orange plants should show raised blisters on leaves, thorns and stems.
7. The disease should infect *Chenopodium amaranticolor* upon inoculation.
8. Psorosis-A should show protection against psorosis-B challenge inoculum.

In India, psorosis-A disease has been reported on Mosambi sweet orange (Ahlawat, 1989). This disease showed all the three types of reactions known for psorosis-A *i.e.* shock reaction, young leaf vein flecking and rings or blotches on mature leaves when inoculated to Mosambi sweet orange. However, no bark lesion symptoms were observed on infected trees even in old groves and young leaf symptoms unusually persisted for several weeks (Ahlawat, 1989). In ultrathin sections, inclusions resembling P_2 like protein were observed in phloem cells of infected plants. Later studies revealed association of filamentous virus particles in mature leaves of Indian isolate (Ahlawat and Chakraborty, 1990). Later investigations confirmed that this disease is distinct from psorosis and was designated as Indian citrus ringspot although leaf symptoms resemble to psorosis except that young leaf symptoms persisted for longer duration and could be sometimes seen in mature leaves. Citrus psorosis complex now contains Psorosis - A and B (Whiteside *et al.*, 1988; da Graca *et al.*, 1991).

SYMPTOMS

Characteristic symptoms of psorosis-A disease in field trees are bark scaling and wood staining of stem (Plate 2). Bark scaling are typical diagnostic symptoms provided that the tree is not infected by other pathogens like phytophthora and other diseases showing similar symptoms like shell bark of lemons or leprosis. However, these diseases can be differentiated by indexing on sweet orange indicator seedlings. Leaf symptoms may develop on infected field trees as clearing and flecking of veins in the whole or part of the young leaves in new growth flush. Normally all the leaves of new flush will show vein clearing symptoms preferably at low temperature. These symptoms can be clearly seen when the leaf is shaded from direct sun and viewed against the light. Symptoms on young leaves usually disappear as soon as such leaves mature or even earlier, but in psorosis-B infection, leaf symptoms persist in mature leaves and can also occur on fruits. In addition of vein clearing symptoms a few leaves also develop oak leaf pattern in field trees if these are infected with concave gum. These effects will disappear as the leaves become mature.

Axillary shoots of experimentally inoculated sweet orange seedlings curve downward with smalling of leaves. These leaves show necrosis and fell down. This reaction is known as "shock reaction". Infected plants are not killed by shock reaction and in due course normal new leaves will appear but with typical vein flecking symptoms. However, young leaf symptoms of psorosis were not observed in Bahia, Brazil (Passos, 1963). Symptoms on mature leaves appear as irregular blotching or ring pattern. These symptoms persist on mature leaves. Underside of affected mature leaves may show brownish eruption (Fawcett and Klotz, 1938). Bark lesions develop after 6yr of inoculation. These lesions start as pimply eruptions on main trunk or major branches but subsequently cover the main stem and branches and also grow in size. As the lesions grow, gum may exude from the margin of the lesions. The bark is break away as these lesions get bigger in size. In microscopic examination, gum layers are seen in the xylem.

Following are the major symptoms of psorosis disease which can be useful for field and laboratory identification.

1. Shock reaction of young flushes in Sweet orange and some other *Citrus* spp.
2. Vein clearing or flecking or oak leaf pattern on young leaves which persists for a short period.
3. Blotches or spots on mature leaves.
4. Bark scaling on stem and branches in field affected trees.
5. Wood impregnated with gum, which forms an irregular circular pattern when viewed in the cross section of the trunk.

TRANSMISSION

Citrus psorosis virus (CPV) is readily transmitted by grafting. Some isolates of CPV are transmitted by mechanical inoculation on herbaceous hosts like *Chenopodium quinoa* and *C. amaranticolor* where it produces local lesions. Mechanical transmission from citrus to citrus is done by knife or razor slashing into stem. Psorosis virus is suspected to be seed transmitted in Carrizo citrange (Bridges *et al.*, 1965) but authentic evidence is lacking (Roistacher, 1993). Natural spread of psorosis has been observed in Argentina (Benatena and Portello, 1984), and Uruguay. A sucking insect probably an aphid has been suspected as vector. However, experimental results are required to confirm insect transmission. Transmission of psorosis virus by dodder (*Cuscuta compacta*) has been reported to approximately 5 per cent of sweet orange (*Citrus sinensis*) seedlings tested (Price, 1963). Psorosis virus is a type member of ophioviruses. Other ophioviruses *(Mirafiori lettuce big vein virus, Ranunculus white mottle virus)* have been shown to be transmitted by *Olpidium brassicae.* Psorosis virus was detected in zoospores released from sporangia by RT-PCR suggesting that psorosis virus may also be transmitted by *O. brassicae* (Palle *et al.*, 2005).

NATURAL HOSTS

Grapefruit, mandarins, sweet oranges, lemons, limes and Dweet tangar are the natural host of psorosis virus. Citranges, Lemons, Mandarin (Comun de Concordia), *P. trifoliata*, Rough lemon, Sour orange, *Citrus daidai*, and *Clausena* sp. are symptomless hosts of psorosis-A virus in field.

EXPERIMENTAL HOSTS

Under experimental condition, psorosis virus has been transmitted to several herbaceous plant species including *Chenopodium amaranticolor* (Storm and Streets, 1963; Timmer *et al.*, 1978). Bark scaling symptoms are more severe in grapefruits and sweet oranges but less severe in Orlando tangelo. Leaf symptoms are more severe in sour orange, sweet orange, Dweet tangor and West Indian lime but less severe in citron, grapefruits, lemons, mandarins, pummelo, rough lemon and tangelos. Hosts like Etrog citron, *C. excelsa*, and sweet oranges show shock reaction, young and mature leaf symptoms are produced in sweet oranges and Clementine. Dweet tangor produces only mature leaf symptoms. Some cultivars of Citranges (Carrizo, Rusk,Troyer), *Citrus daidai, Clausena* sp., *Lemons, Poncirus trifoliata, Rough lemon* and *Sweet lime* are tolerant to psorosis-A infection.

DIAGNOSTIC AND ASSAY HOSTS

Sweet orange seedlings are used as diagnostic host. Inoculated seedlings will show symptoms of shock reaction, young leaf flecking along veinlets and oak-leaf pattern in young leaves and ring like pattern on mature leaves within a few weeks at cool temperature. Such plants will show bark scaling and flaking within six years or more. Seedlings of mandarin, Dweet tangor, citron and Mexican lime may also be used as indicator hosts of psorosis-A virus. *Chenopodium amaranticolor* may be used local lesion hosts of psorosis virus. Virus is maintained and propagated in *Citrus* spp.

PHYSICAL PROPERTIES

The virus has longevity *in vitro* (LIV) of 2 days (at 4°C), dilution end point is usually around 2 and Thermal inactivation point is 50°C in crude sap.

THE PATHOGEN

The causal agent of psorosis is presumed to be *Citrus psorosis virus* (CPsV), the type member of the genus *Ophiovirus* (Milne *et al.*, 2000). CPsV virions are highly kinked filaments 3–4 nm in diameter of at least two sizes (García *et al.*, 1994), which contain three single-stranded RNAs of negative polarity and a 48- to 50-kDa coat protein (Sánchez de la Torre *et al.*, 1998; Milne *et al.*, 2000). Psorosis virus particles are elongated with helical symmetry. Virus preparations contain more than one particle components. Citrus psorosis disease agent behaves as a two component ssRNA Virus (Garcia *et al.*, 1989). Purified preparations from psorosis inoculated citrus and *Chenopodium quinoa* showed filamentous particles in at least two sizes (contour lengths) and occur in an open circular form and a linear form, plus intermediates. The linear (or *L*) form may appear as a sinuous structure about 10 nm in diameter and up to 2,500 nm long; it is in reality a two-stranded structure with loops at the ends, formed from a collapsed circle. The open circular form (*0* form) is a highly kinked filament 3 to 4 nm in diameter. The morphology suggests that this type of virus represents a new genus of viruses similar to the Tenuiviruses (Milne *et al.*, 1996). They proposed the name 'Ophiovirus' (from the Greek *ophis,* meaning 'serpent') for this new genus, because the particles have a snaky appearance. Virus particles have a length of 1500–2500 nm; 760 nm and a width of 3 nm, or 9 nm. Psorosis virus is a bipartite single stranded, negative strand RNA virus classified as genus 00.094.0.01 Ophiovirus in family Ophioviridae (Martin *et al.*, 2005).

GENOME

Genome of psorosis virus is segmented and consists of three segments of circular, negative-sense, single-stranded and double-stranded RNAs; that form a panhandle, a structure formed of complementary sequences. Virions may also contain dsRNA replicative intermediates. RNA-2 is fully or partially sequenced; complete sequence is 1100-1200 nucleotides long and sequenced region is 1644 nucleotides long and encodes unknown product (accession number [AF218572]). RNA-3 is genomic and partially sequenced, sequenced region is 1377 nucleotides long and encodes coat protein (accession number [AF036338]). Nucleotide sequences at the 3'-terminus are thought to be complementary to the 5' end and form a panhandle. An isolate of citrus psorosis virus called citrus ringspot virus (CtRSV-4) contains at least three genomic RNAs; two of them (RNA 2 and RNA 3) are found in the top component and the largest (RNA 1) in the bottom component. RNA 3, the smallest RNA, was cloned and a partial sequence of 1,454 nucleotides was determined. One open reading frame (ORF), which encodes a 439 amino acid protein with a MW of 48,654 was found. This was identified as the coat protein. The nucleotide sequence of RNA 2 was determined and found to consist of about 1,620 nucleotides. In the viral-complementary RNA, one open reading frame (ORF) is present coding a 476 amino acid protein of MW 53,694 with unknown function. No subgenomic RNAs were found (Sánchez de la Torre *et al.*, 2000).

Viral genome encodes structural proteins of 48700 Da which has been sequenced and a function assigned; sequence with accession number [O89526]; is coat protein. A full-length clone of the CPG was transcribed and used in Northern hybridization assays to establish that short particle RNA of CPV is negative sense and contains the CPG. Moreover, the CPG was not found on RNA extracted from long particles or on the sedimentable dsRNA from CPV infected tissue. RT-PCR assays were developed for the amplification of a 600 bp fragment of CPG and for the complete CPG (1317 bp). The 600 bp fragment from a biologically and serologically different isolate, CPV-6, was cloned, sequenced and found to share 86% (nucleotide) and 96% (amino acid) identity with CPV-4. BLAST analysis of sequences from CPV-4 and CPV-6 detected no significant nucleic acid or protein similarity with any known viral sequences (Barthe *et al.*, 1998). Genetic variation in populations of Citrus psorosis virus has been observed (Martin *et al.*, 2006).

DIAGNOSIS AND MANAGEMENT

Field diagnosis of CPV is done on the basis of bark scaling symptoms on tree trunk or limbs. Commonly, laboratory diagnosis is done by grafting samples on indicator seedlings (sweet orange or mandarin). Inoculated plants will show a shock reaction, vein clearing/vein banding of young leaves and mottling or ringspot on mature leaves. Mechanically inoculated *C. quinoa* leaves will show local lesions diagnostic of CPV. Biological indexing is one of the best methods of CPV diagnosis (Roistacher *et al.*, 2000; Garnsey, 1993). In ultrathin sections inclusions resembling P2 like protein in sieve elements of phloem tissues are observed in infected plants but not in healthy (Ahlawat, 1989). Antibodies have been prepared and used in diagnosis of CPV by serological methods. Monoclonal antibodies have also been developed and used for virus specific detection in DAS-ELISA. Different epitopes and serogroups have been identified in CPV. Direct tissue blot immunoassay (DTBIA), a sensitive and simple technique that requires very little sample manipulation, has been applied for the rapid detection of *Citrus tristeza* virus (Garnsey *et al.*, 1993; Cambra *et al.*, 2000; Djelouah *et al.*, 2000) has also been used for detection of CPV (Djelouah *et al.*, 2000; D'Onghia *et al.*, 2001). This technique is more sensitive and cheaper than ELISA. Recently, TAS-ELISA using horseradish peroxidase and alkaline phosphatase enzymes has been developed for psorosis detection which is rapid, reliable, faster and more sensitive technique than the currently used for psorosis virus (Maria *et al.*, 2006). Psorosis virus has also been detected in immunoblotting using citrus ringspot antiserum (da Graca *et al.*, 1993). Diagnosis of CPV is more reliably done in RT-PCR (Garcia *et al.*, 1997), nucleic acid hybridization and immunosorbent electron microscopy (Martin *et al.*, 2004). Psorosis in California is also detected by PAGE and double stranded RNA technology.

The first budwood certification programme was started against citrus psorosis in California. To initiate certification programme, an apparently healthy mother tree is identified which is continuously observed by visual inspection for two years to see if any leaf symptom appears. Four trees around the mother tree are also inspected to avoid the possibility of transmission by root grafting. The identified tree is indexed biologically on indicator hosts. Since the virus has been suspected to be seed transmitted rootstock seedlings should also be indexed. At the time of inception of budwood certification programme, no serological or molecular techniques were available for virus testing and biological indexing was the only method of testing. In spite of this, the programme was very successful to eradicate citrus psorosis in California.

Another option to get virus free planting material is the use of nucellar clones. If no healthy source of a citrus variety is known, nucellar seedlings of that variety can be obtained which will preserve the varietal characteristic and at the same time will be free from the virus. Psorosis virus is also eliminated through heat treatment of infected citrus plants by keeping them for 2–3 months in hot air at alternating temperature of 40°C for 16h during day and 30°C for 8 h during night (Calavan *et al.*, 1972). Field trees of citrus are made free from bark lesions by scrapping them by hand. This treatment is described "hand-scraping". Bark lesions can also be removed by treatment with a solution of 1% dinitro-o-cyclohexilphenol in kerosene which will also stop spreading of bark lesions. These are some of the management practices used earlier to prevent psorosis infection. Recently, more advances have been made in diagnostic technology which help in developing assured healthy planting material with the result now none or only a few psorosis infected trees are planted. Earlier approaches like treatment of affected trees etc. are now abandoned and psorosis disease is now being eliminated by the use of virus-free healthy planting material which may also be obtained through shoot tip grafting (Navarro *et al.*, 1980).

In Brazil, a bark scaling disease of citrus (BBSC) is found in Bahia state commonly known as Bahia psorosis. It affects most of citrus species and varieties, but especally severe on grapefruit. BBSC symptoms include a darkening and thickening of the bark leading to scaly lesions on trunk and branches; young branches show dieback; and significant gum extrusion. Symptoms resemble to psorosis-A but data revealed that BBSC is distinct disease, although its aetiology remains unknown (deSimone *et al.*, 2010). A combined strategy of strong sanitation effort, lower planting densities and rouging has been recommended to manage Bahia bark scaling disease.

Citrus Tristeza

There are very few diseases that caused huge crop losses as tristeza did in citrus. Phytophthora epidemic caused worldwide destruction of citrus in the 19th century. Sour orange was then identified a rootstock which was resistant to *Phytophthora* and was used as acceptable rootstock to raise sweet orange and mandarin trees in many countries. However, in South Africa, the trees raised on sour orange rootstock declined/died within two to three years (Oberholzer *et al.*, 1949). Subsequently, it was found that Satsuma mandarin is also a failure on sour orange in Florida and Gulf States (Swingle, 1909). A similar condition was also noticed in Java, Argentina, Brazil and several other countries where sour orange was used as root stock. The term "tristeza" was coined by Moreira (1942) to this decline malady which means "sadness" in Portuguese language. Reason to this decline was attributed as an incompatibility between sweet orange and mandarin scions and sour orange rootstock (Bioletti *et al.*, 1904). But it was later described as a disease condition (Webber, 1925). This combination of sweet orange, mandarin or grapefruit on the sour orange rootstock was susceptible to a new highly destructive vector-transmitted disease now known as tristeza. First evidence that this disease was graft transmissible and a virus was involved in this malady was published in 1946 (Fawcett and Wallace, 1946). At the same time, it was reported from Brazil that tristeza disease was transmissible by brown citrus aphid, *Toxoptera citricidus* Kirk (Meneghini, 1946). Finally, electron micrographs of thread like particles were obtained from purified preparations from tristeza infected lime trees under electron microscope (Kitazima *et al.*, 1964).

Citrus tristeza virus (CTV) had ruined citrus industry in many countries where sour orange was used as rootstock although no accurate statistics for the loss is available. Tristeza destroyed sweet orange, mandarin and grapefruit trees on sour orange roots in a large proportion, 15 million trees in Argentina, 6 million or more trees in Tucuman, 6 million trees in Brazil, 3 million trees in California and an unestimated number in many citrus growing countries. However, in countries where sour orange was not used as a root stock like Japan, India, the Philippines, Taiwan etc. the trees were not so severely damaged (Reuther *et al.*, 1978). In India, decline symptoms were observed on either sweet orange grafted on rough lemon or sweet orange rootstocks (Vasudeva and Capoor, 1958; Vasudeva *et al.*, 1959). In India, Tristeza was first observed in Bombay state (Nagpal, 1959) and later reported from several parts of India (Capoor, 1963). Tristeza in India is widely distributed in most citrus growing regions of the country (Capoor and Rao, 1967; Nariani and Raychaudhuri, 1968; Nariani *et al.*, 1965, 1966, 1970; Raychaudhuri *et al.*, 1977; Reddy, 1965; Reddy and Rao,

1961). Estimates indicate that tristeza destroyed about a million trees in India (Ahlawat, 1997b). CTV is more damaging to Indian sweet oranges (Mosambi) than to other citrus cultivars (Ahlawat *et al.*, 1995) and about 70% incidence of tristeza (stem pitting) has been recorded in Mosambi sweetorange trees on different rootstocks (Mali and Chaudhuri, 1993). In places where aphid vector, *Toxoptera citricida* is present, severe strains of CTV are commonly available. These severe strains have been defined based on their reaction in Kagzi (Mexican) lime indicator seedlings as vein clearing and stem pitting (Ahlawat and Raychaudhuri, 1988). Mild strains cause only vein clearing. Some strains produce stunting and chlorosis in seedlings of sour orange, acid lime or grapefruit. In certain citrus growing regions of India CTV strains that cause seedling yellows are also found (Ahlawat and Raychaudhuri, 1988).

Tristeza and its vector, *Toxoptera citricida* might be originated in Asia which is also the original home of citrus. CTV along with its vector, *T. citricida* is widely distributed in seedling trees of Darjeeling orange in Darjeeling and Sikkim hills suggesting the spread of the disease by vector in this region of citrus origin (Ahlawat and Sardar, 1976). Tristeza disseminated to most parts of the world through the movement of infected planting material probably through traders and tourists and within countries by aphid vectors. Now tristeza is known to occur in all citrus growing countries. Some other diseases are now known to be caused by strains of tristeza virus such as grapefruit stem pitting, gold coast lime disease on sour lime, pera orange stem pitting in Brazil and hassaku dwarf in Japan (Tanaka *et al.*, 1969).

Tristeza disease is worldwide in distribution wherever citrus is grown (Price, 1970; CABI, 2010). This disease is widespread in several countries of EPPO region (Israel, Spain, Turkey, France, Algeria, Cyprus, Morocco and Tunisia), Asia (Bhutan,Breuni, China, Cyprus, Georgia, India, Indonesia, Iran, Japan, Jordan, Korea, Malaysia, Nepal, Pakistan, Philippines, Saudi Arabia, Sri Lanka, Taiwan, Thailand, Turkey, Viet Nam and Yemen), Africa (Nigeria, Cameroon, Centrl African Republic, Chad, Egypt, Ethiopia, Gabon, Ghana, Kenya, Mauritius, Morocco, Mozambique, Nigeria, Reunion, South Africa, Tanzania, Tunisia, Zaire, Zambia, and Zimbabwe), North America, Central America and Carribbean (Antigua and Barbuda, Bahamas, Belize, Costa Rica, Dominian Republic, El Salvador, Gautemala, Honduras, Jamaica, Nicaragua, Panama, Peurto Rica, St. Lucia, Trinidad and Tobago, South America (Argentina, Bolivia, Brazil, Chile, Colombia, Ecuador, Guyana, Peru, Paraguay, Surinam and Uruguay) and Oceania (Australia, Fiji, French Polynesia, New Caledonia, New Zealand and Samoa (CABI and EPPO data sheets on quarantine pests). Bove and Vogel (1981) described the distribution of tristeza disease in Individual countries. Tristeza disease has also been recorded even in countries where sour orange rootstock was never used. Tristeza disease moved from one country to another by man in propagative material and within countries by aphid vectors. In India, a mixed infection of tristeza and greening bacterium is common in declining sweet oranges in Maharashtra (Ghosh *et al.*, 2003). Synonym(s) of tristeza virus are citrus quick decline virus (Fawcett and Wallace, 1946; Capoor, 1963), grapevine A virus, hassaku dwarf virus, and lime dieback virus.

SYMPTOMS

Tristeza affected trees look wilted as if they are affected by some root pathogen. Leaves of affected trees drop, twigs become defoliated, die back starts from tip and finally the tree is declined (Plate 3). These symptoms are caused by the starvation of roots as a consequence of sieve tube necrosis induced by the virus below the bud union. CTV causes different symptoms in different *Citrus* species depending on the virus strain, the species or variety of citrus, and the scion-rootstock combination. In Asia, various strains of CTV, generally referred to as seedling yellows (CTV-SY), tristeza (CTV-T), stem pitting (CTV-SP), and a mild type, have been widespread for many years. Any of these strains may exist in a citrus plant, or they may occur together, as a complex. As described earlier the most important symptom of the disease is decline, sudden wilting and drying of leaves in certain stock-scion combinations. In fruiting citrus trees, leaves of wilting trees fall down but the fruits remain hanging till they become dried or mummified. Rotting of roots and absence of functional feeder roots is a common phenomenon in CTV infected trees. Sometimes, affected trees may recover but this condition is rare. Absence of new growth and bronzing of leaves in CTV infected field trees is also common. CTV infected trees flower heavily with comparatively more fruit setting. In highly susceptible hosts like Mexican lime and *Citrus excelsa*, CTV produces vein clearing of leaves (Plate 5) and stem pitting (Plate 4). Decline, stem pitting and seedling yellows are also the reaction of tristeza and are being described below.

Decline on sour orange

Certain strains of CTV are responsible for decline of sweet orange, grapefruit or tangerine trees on sour orange roots. The reaction is very severe showing dwarfing or often death of trees. Tree decline is caused due to degeneration of phloem tissues of sour orange just below the union. Decline due to tristeza infection is of two types, slow decline and quick decline. Trees that decline slowly have a budge above the bud union and honey combing is present on the wood below the bark in the sour orange rootstock. In quick declining trees, honey combing is absent but a brown line appears at the bud union which can be seen by removing the bark.

Stem pitting (SP)

Stem pitting is a term used to describe depressions of different shapes and sizes in the wood of infected plant. The shape, size and number of pittings depend on the strain of CTV and the host variety. For example, in CTV inoculated Mexican lime seedlings, longitudinal depressions of various lengths develop on main shoot and twigs which can be seen by removing the bark. However, pitting should not consider as a definite symptom of CTV as other maladies like cristacortis and xyloporosis commonly show such symptoms. Stem pitting affected trees of limes, grapefruit and sweet orange are stunted and have a bushy appearance. Trees with mild pitting may not show visible canopy symptoms. Leaves are chlorotic, and the twigs are brittle and break easily when bent. Gum is often associated with fine pits. Hassaku is a natural pummelo hybrid in Japan and a disease known as hassaku dwarf is common there. Hassaku dwaf disease is also a form of CTV stem-pitting virus. A pummelo dwarf disease is known to occur in China and other Asian countries. This disease is also caused by CTV stem pitting virus. In India, Stem pitting in Coorg mandarin is known to occur due to CTV infection (Manjunath *et al*., 1989). Sem pitting CTV isolates may cause pitting in one *Citrus* spp. but not in others. Garnsey *et al.* (2005) recognised 11 distinct general reaction patterns in CTV using five indicators (Mexican lime, sour orange, Madam vinous sweet orange, Duncan grapefruit seedlings, and sweet orange plants budded on sour orange rootstock) inoculated by grafting and a screening method designed to measure only fairly large differences in symptoms do not fully reflect the complexity of CTV symptom responses. The different patterns of symptoms observed in the five indicators suggested that seedling yellows, the decline syndrome in sweet orange grafted on sour orange, stem pitting in sweet orange, and stem pitting in grapefruit are independent expressions of CTV pathogenicity and may occur in various combinations (Garnsey *et al*., 2005).

Seedling yellows (SY)

Samples of sweet orange and mandarin trees infected with some isolates of tristeza virus when inoculated by tissue graft to seedling of Eureka lemon, Marsh grapefruit, Seville sour orange and some citron (*C. medica*), a reaction of severe stunting and yellowing of leaves occurs in these hosts. This reaction was named as seedling yellows (Fraser, 1952). Seedling yellows was initially thought to be a distinct virus than CTV but studies revealed that this is an isolate of tristeza virus (Wallace and Drake, 1972). Seedling yellows spreads slowly in field trees but pose a problem when trees infected with this strain are topworked with grapefruit or lemon. In India, seedling yellows strain of tristeza has been observed earlier but is restricted in its distribution (Vasudeva *et al*., 1959; Capoor, 1965).

Tristeza infected trees also show some microscopic symptoms before the visual symptoms appear. Necrosis of sieve tubes and some anatomical changes have been noticed in phloem tissues of sweet orange trees on sour orange rootstock (Schneider, 1959). After infection, the amount of phloem is reduced resulting in a girdling effect. Girdling prevents movement of photosynthetic food to the roots resulting utilization of reserve starch and subsequent decay of roots and appearance of top symptoms (Schneider, 1959).

TRANSMISSION

CTV is readily transmitted by bud or scion grafts. It is also transmitted by dodder, *Cuscuta americana, C. subinclusa* and *C. reflexa.* CTV is transmitted mechanically by bark wounding or by slashing stem or leaf abrasion (Garnsey and Muller, 1988). In India and many other countries, CTV is transmitted by seven species of aphids colonizing on citrus plants.These aphid species are: *Toxoptera citricida* Kirk., *Aphis gossypii* Glover, *A. citricola* van der Goot (= *A. spiraecola* Patch), *A. craccivora* Koch, *M. persicae* Sulz., *Toxoptera aurantii* Bsy and *Dactynotus jaceae* L. (Vasudeva *et al.*, 1959; Varma *et al.*, 1960, Rao and Capoor, 1965; Meneghini, 1946; Dickson *et al.*, 1951; Norman and Grant, 1956). A brown citrus aphid (BCA), *T. citricida* is the most effecient vector of CTV throughout the world. *A. gossypii* is an important vector, especially in Spain, Israel, and California where *T. citricida* is not established as yet. Brown citrus aphid (*T. citricida*) introduced in Florida in 1990s (Lee *et al.*, 1992) and due to this aphid introduction, CTV rate of spread has increased dramatically within Florida. *Aphis citricola* and *Toxoptera aurantii* (Capoor and Rao, 1967) have also been shown to be potential vectors of CTV but their efficiency is very low. All these aphid vectors transmitted CTV in semi-persistent manner. Aphids can acquire CTV within few minutes while feeding on infected plant and can transmit it by a transmission feeding of 24 h. Severe strains of CTV and seedling yellows strains are restricted to places where *T. citricida* occurs. In places where *T. citricida* is not present, *Aphis gossypii, A. craccivora, Myzus persicae* and *Dactynotus jaceae* play the role of vectors, but they are less efficient (Ahlawat and Raychaudhuri, 1988).

HOSTRANGE

Most of the citrus species/cultivars and citrus relatives are host of citrus tristeza virus including SP and SY strains. CTV exhibits visible symptoms only in a few hosts and in most of the species of citrus it remains symptomless. Trifoliate orange has been reported to be immune to tristeza (Costa *et al.*, 1949) but later it was shown to be infected by some strains of seedling yellows (Wallace and Drake, 1972). Rootstock species, *Poncirus trifoliata* is resistant to most isolates of CTV, but not to members of the CTV resistance-breaking (RB) strain presently found in New Zealand (Harper *et al.*, 2010). Hosts like Rusk citrange, Grapefruit, Hassaku, Kumquats, Limequats, Mandarins (Stem pitting in India, Japan and Indonesia), Mediterranean Pera, Tahiti lime and Tangor show stem pitting symptoms. Hosts like Limes (*Citrus aurantifolia)* cultivars Kagzi, Key, West Indian, and Mexican lime, and sweet oranges show vein clearing and stem pitting symptoms respectively. Pummelo shows both stem pitting and seedling yellows symptoms. Following hosts show variable symptoms: Uvalve, Citron, Etrog citron, Citrumelo 1452, *Citrus amblicarpa, Citrus daidai, Citrus excelsa, Citrus hystrix, Citrus junos, Citrus macrophylla,* Sweet lime, *Aegle marmelos, Aeglopsis chevalieri, Afraegle paniculata, Citropsis gelatiana, Citrus aurantium, Citrus excelsa, Citrus limon, Citrus medica, Citrus paradisi, Citrus reticulata, Citrus sinensis, Microcitrus australis, Pamburus missionis, Passiflora gracilis, P. caerulea, P. incense and P. incarnate.*

Symtomatic host plants of CTV

Strains of tristeza show stem pitting symptoms in certain hosts whether used as seedlings, scions or rootstocks. The hosts of CTV stem pitting in different countries are given below:

(a) ***Strains of California, Florida, Spain*:** Walve citrange, Etrog citron, *Citrus excelsa, C. hystrix, C. junos, C. macrophylla,* Sweet lime, West Indian lime.

(b) ***Strains of South Africa, South America, Japan*:** Species and varieties in (a) and: Citranges (Morton, Rusk, Savage), Grapefruits, Hassaku, Kumquats, Limequats, Pummelo, Sweet orange (Mediterranean, Pera), Tahiti lime, Tangor.

(c) ***Strains of certain areas of Brazil, Japan, Southeast Asia, Cameroon, Mauritius, Reunion*:** Species and varieties in (a) and (b) and Citrumelo 1452, Rangpur lime, Tangors.

Species or varieties except lemons and sour orange show decline when grafted on sour orange or on lemons. However, Mandarins on Rangpur lime (Brazil), Sweet orange on Rangpur lime (Brazil), Sweet orange on *P. trifoliata* (South America), Sweet orange on Troyer citrange (California) decline under certain conditions.

Natural host and symptoms

Citrus spp. grafted onto *Citrus aurantium* (sour orange) rootstock show following symptoms:

Citrus paradisi (Grapefruit) — stunting.

C. aurantifolia (Lime) — die-back.

C. aurantifolia (Seville orange) — seedling yellows.

C. reticulata (Mandarin) — decline.

In Northeastern region of India, natural infection of CTV was observed in seedling trees of Khasi mandarin, Rough lemon, Assam lemon, and kapura tenga. However, wildly distributed *Citrus indica* neither showed any CTV symptoms in field nor showed CTV symptoms when inoculated to indicator host, Kagzi lime (Bhagabati *et al.*, 1989).

Symtomless hosts of CTV

Citranges (Carrizo, Troyer), Citrumelo 4475, Lemons, Mandarins, *P. trifoliata*, Rangpur lime, Rough lemon, Satsuma, Sour orange, Sweet orange and Tangelo do not show stem pitting whether used as seedling, scion or rootstock. Some tristeza isolates show stem pitting on some of the following hosts - Lemons, Mandarins, Rangpur lime, Rough lemon and sweet orange. There is no decline when grafted on Mandarins, Sweet oranges on Citranges (Carrizo, Troyer), Citrumelo 4475, Cleopatra mandarin, *P. trifoliata*, Rangpur lime, Rough lemon

Tolerant hosts of CTV

Following hosts are tolerant to CTV: Citranges (Troyer, Carrizo), Citrumelo 4475 (Swingle), Clementine, Lemons, *Poncirus trifoliata, Rangpur lime,* Rough lemon, Satsuma, Sweet orange and Orlando Tangelo.

Diagnostic hosts

Diagnostic host species and symptoms on them described by Wallace and Drake, (1951) and Lee and Bar-Joseph (2000) are described below:

- ***Citrus aurantifolia*** cvs West Indian lime, Key lime, Mexican lime, Galego lime, Kagzi lime — chlorotic vein flecks, vein corking, leaves cupped or canoe-shaped; stem pitting.
- ***C. sinensis*** on *C. aurantium* roorstock — sudden wilting, decline and death, overgrowth on bud union and 'honeycombing' just below union.
- ***C. sinensis, C. paradisi*** — stem pitting.

Citrus aurantifolia, C. excelsa and *C. sinensis* are used for maintenance and propagation of CTV isolates.

ASSAY/INDICATOR HOSTS

Best assay host for CTV is small fruited acid lime seedlings (*Citrus aurantifolia)* West Indian lime, Key lime, Mexican lime, Galego lime, Kagzi lime (Capoor, 1961). Symptoms on inoculated plants develop within a few weeks under cool climate. Main symptoms are vein clearing, leaf cupping, stem pitting and stunting of plants. Other indicator hosts are *Citrus macrophylla* and *C. excelsa*. Symptoms in these hosts will develop as veinlet clearing in young leaves and pitting in stem. *Citrus amblycarpa* (*Nasnaran* Mandarin) is also a good indicator host for CTV showing symptoms similar to Kagzi lime (Pandey *et al.*, 1987). A strain of Tristeza virus (K) has been reported which does not induce any symptoms in Mexican lime (Djelouah and d'Onghia, 1986). Indicator hosts for seedling yellows are grapefruit, lemon, and sour orange seedlings. Symptoms on these hosts develop within a few months as stunting and leaf yellowing on top of inoculated seedlings.

THE PATHOGEN

Citrus tristeza virus particles are highly flexuous and filamentous measuring 10–12 nm in width and 2000 nm in length (Fig. 2A). CTV is a member of the genus closterovirus within the family Closteroviridae, is one of the more complex single stranded RNA plant viruses (Bar-Joseph *et al.*, 1970, 1979; Bar-Joseph and Lee, 1989). The virus is located in phloem tissues of infected host. Density of CTV virions is 1.257 g/ml. The sedimentation coefficient is 140 S_{20w} (~10). Tristeza virus is an ICTV approved species of the genus 00.017.0.01. *Closterovirus*; family 00.017. *Closteroviridae.*

Several strains have been described in CTV based on differences in symptoms induced in several hosts (Bove and Vogel, 1981), aphid transmissibility (Bar-Joseph *et al.*, 1983), dsRNA patterns (Dodds *et al.*, 1987), peptide maps (Guerri *et al.*, 1990) and serological reaction with monoclonal antibodies (Permar *et al.*, 1990).

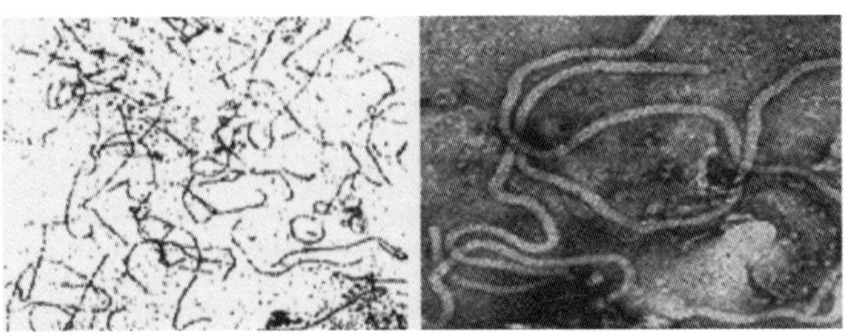

Fig. 2A: Highly flexuous particles of *Citrus tristeza* virus. Left shows CTV practicles in partially purified preparation

GENOME

Tristeza virus genome is monopartite. The nucleic acid is linear, positive-sense and single-stranded RNA of approximately 19.3 kb (Bar-Joseph *et al.*, 1989) with two proteins (CP) 25 and 27kDa covering 95 and 5% of particle length respectively. Complete CTV genome has been sequenced and is 19,226 to 19302 nt long. Several workers have reported complete sequences of CTV genome of several isolates including the coat protein gene and deposited in GenBank (Bar-Joseph and Lee, 1989; Harper *et al.*, 2010; Karasev *et al.*, 1995; Pappu *et al.*, 1993; Roy and Brlansky, 2010). Partial genome of CTV has also been sequenced from a CTV isolate in a mixed infection (Manjunath *et al.*, 2005). Single stranded positive sense genomic RNA encodes 12 open reading frames (ORFs). CTV consists of two capsid proteins: 25 kD protein covering about 95% of particle length and a major 27 kD coat protein that covers one end of the particle forming a rattle snake structure (Bar-Joseph *et al.*, 1989; Karasev *et al.*, 1995). Coat protein genes of several biologically distinct CTV isolates have also been sequenced worldover (Pappu *et al.*, 1993; Sekiya, 1991). Coat protein genes of several Indian isolates of CTV have also been cloned and sequenced (Manjunath *et al.*, 1993; Ghosh *et al.*, 2009). CTV coat protein genes (CPG) of four Indian isolates were amplified by RT-PCR using CPG specific primers, yielded an amplicon of 672 bp for all the isolates (Roy *et al.*, 2003). The complete nucleotide sequence of Indian isolates have been deposited in the GenBank database under accession numbers AF501867 to AF501870. Coat protein genes of Bangalore and Pune isolates have also been sequenced (Accession No. AF336121, AY936477). Sequence analysis of coat protein gene of an isolate from central India showed 97% homology with Chinese isolate CTV-0002 and S4. In Phylogenetic analysis, this isolate was displayed in different clade than other known Indian CTV isolates, but shared a separate clade with isolates from China, Israel, Jordan and New Zealand (Ghosh *et al.*, 2009). Based on genomic sequences, four different clusters *viz* Kpg1, Kpg2, Kpg3 and CTK 27 have been identified from Northeast Himalayan region which is considered as home of citrus. These isolates were found to be genetically related to exotic isolates like T-30 (mild isolate from Florida), T-385 (severe strain from Spain), SY 568 (Severe isolate from California), VT (Israel Severe) and PB 61 from Australia indicating that diversified CTV population is present in the NE Himalayan region of India from where CTV originated (Biswas *et al.*, 2008). Information is now available on the population dynamics in CTV mixtures and the generation of virus recombinants after aphid transmission (Roy and Brlansky, 2009). The sequence variation of the Indian CTV-B165 isolate was compared to the three well known CTV genotypes, T36, T30, and VT. The genome of the predominant component of CTV-B165 (isolate from Bangalore) was 19,247 nt in length with 12 open reading frames (ORFs) and was structurally identical to other CTV isolates. To determine the presence of the CTV-B165 genotype, a pair of genome specific primers has been designed and standardized for reliable detection of the novel CTV genotype by reverse-transcription polymerase chain reaction (Roy and Brlansky, 2010). Molecular characterization of CTV isolates has been done by several workers throughout world and coat protein gene sequences selected from genbank were used in sequence and phylogenetic analysis to determine mild or severe isolates (Barbarossa and Vovlas, 2010). It has been suggested that there is extreme genome stability and low mutation rates in CTV. These studies have significant implication in evolution and emergence of new CTV strains, and in the use of CTV as a stable expression vector (Weng *et al.*, 2010).

Strains in CTV

Several genotypes are known to exist in CTV which have been studied including its epitope diversity (Garnsey *et al.*, 1989; Cambra *et al.*, 1993). CTV isolates have been grouped into six genotypes; T3, T30, T36, VT, trifoliate resistance breaking (RB) and B165 (Harper *et al.*, 2010; Roy *et al.*, 2010). CTV strains cause symptoms such as stem pitting, decline and death of sweet orange and grapefruit on sour orange rootstock, seedling yellows, vein clearing and vein corking. Most of the strains in CTV have been identified under experimental conditions. CTV strains have been detected by way of differential transmission, differential hosts, serological or through molecular techniques. (Balaraman and Ramakrishnan, 1977; Chakraborty *et al.*, 1993; Lee *et al.*, 1997). Although polyclonal antibodies have been developed against several isolates of CTV world wide, a monoclonal antibody (MCA-13) has been produced to discriminate mild and severe strains (Permar *et al.*, 1990). This antibody recognizes only the severe strain in Enyme linked immunosorbent assay (ELISA). Severe and mild strains are also recognized by polyclonal antibody in ELISA. However, if a strain is recognized by polyclonal antibodies but not by monoclonal (MCA-13), then it is considered to be a mild strain (Chakraborty and Ahlawat, 2001; Chakraborty *et al.*, 1993; Cambra *et al.*, 1991). The isolates in mixture may be positive and negative to MCA-13 monoclonal antibody (Brlansky *et al.*, 2003). Field isolates have complex of CTV isolates comprising both mild and severe strains. These isolates can be separated by aphid transmission or reaction on Kagzi lime. CTV strains can also be identified by Bi-directional PCR (Roy and Ramachandran, 2002). Recently developed genotype specific primers for 5 known CTV genotypes revealed the presence of all CTV genotype in India by RT-PCR (Roy *et al.*, 2010).

Severe strains in India

In India, severe strain was originally isolated from an infected sweet orange tree in Pune and transferred to key lime seedlings by *Toxoptera citricida*

(Capoor and Rao, 1967). This strain produced severe leaf reaction in Key lime and *Aeglopsis chevalieri* and showed severe stem pitting in Key lime. Howevere, it did not induce seedling yellows reaction in Eureka lemon, sour orange or grapefruit. In India, severe stem pitting is rarely seen in Northern India but present in some of the endemic areas in the country like Karnataka and Maharashtra where BCA is an active vector.

Mild strain in India

Mild strain of CTV shows mild vein clearing or flecking on leaves of Kagzi lime. It produces either very mild pits in stem of infected Kagzi lime or pits are altogether absent. Trees having mild strains behave like healthy trees with mild or no effect on yield for a long time even on highly susceptible combinations. This strain appears to be more prevalent in Indian citrus especially in North India and probably protecting severe strain under natural condition (Ahlawat and Raychaudhuri, 1988). Presence of mild strain was confirmed by Bi-directional and multiplex genotype specific RT-PCR (Roy and Ramachandran, 2002; Roy *et al.*, 2010).

Seedling yellows strain in India

In India, seedling yellows strain in CTV was originally isolated from declining mandarin trees in Coorg region of Karnataka. Mandarin trees in this region also showed stem pitting symptoms. This strain produced shock reaction in Kagzi lime seedlings and typical yellows reaction in seedlings of Eureka lemon, sour orange and grapefruit. In addition, it induces typical tristeza reaction on prepared plants of sweet orange on sour orange roots, and pin-hole type stem pitting in Eureka lemon. This strain was identified as seedling yellows strain (Capoor, 1965).

DIAGNOSIS AND MANAGEMENT

CTV spread in nature is either by infected planting material or through insect vectors. In most *Citrus* spp. CTV does not produce any symptoms. Isolates of CTV usually falls into one or more of the biological symptom categories and noticeably differ in their pathogenicity on various host species and cultivars (Garnsey *et al.*, 2005). Therefore, efforts by the scientists were concentrated to develop reliable diagnostics to detect this highly damaging virus in planting material so that losses caused by this virus can be minimized. CTV is easily detected by graft inoculation of samples to Mexican lime (Kagzi lime) indicator host where the leaves of inoculated plants will develop vein clearing or flecking symptoms within a few weeks under low temperature conditions (18–25°C). Traditionally,

CTV has been detected and categorized by different biological indexing methods inoculating the appropriate indicator plants and observing the symptoms (Garnsey *et al.*, 2005) but, the major limitations of this technique are time consuming to confirm the detection, expensive and laborious and often imprecise. Polyclonal and monoclonal antibodies against CTV have been prepared against various isolates/strains in several laboratories world over including from the recombinant coat protein gene of citrus tristeza virus. In practice, CTV is detected by electron microscopy, enzyme linked immunosorbent assay (ELISA), dot immunobinding assay (DIBA), Immuno electron microscopy (IEM) with polyclonal or monoclonal antibodies, cDNA probes, light microscopy after azure-A staining of sections (Aggregates of CTV are seen in light microscope) and by PAGE observing dsRNAs (Dodds *et al.*, 1984). Antibodies raised by the expression of coat protein gene of CTV are efficient at trapping virus particles in double antibody sandwich ELISA assays. Serological detection techniques like ELISA, immunoprinting ELISA or tissue blot assay are more commonly used for CTV diagnosis in indexing programmes all over the world (Ahlawat and Raychaudhuri, 1988; Ahlawat *et al.*, 1997; Ahlawat *et al.*, 1992; Ahlawat and Pant, 2003; Rocha Pena and Lee, 1991; Bar-Joseph *et al.*, 1979; Garnsey *et al.*, 1993; Cambra *et al.*, 2000; Biswas *et al.*, 2009; Ghosh *et al.*, 2009). Antibodies specific for the recombinant coat protein (rCP) of the *p25* gene of *Citrus tristeza virus* (CTV) were developed in goats and rabbits and further evaluated as a complete kit for the detection of the virus using healthy and CTV-infected tissue. The combination of goat T1 used as primary (coating) and rabbit C3 as intermediate (detecting) rCP antibodies reacted efficiently, with optical density at 405 nm (OD_{405}) values between 0.250 and 2.000 with samples values between 0.250 and 2.000 with samples from an international collection of diverse CTV isolates (Iracheta-Cárdenas *et al.*, 2009).

Use of polymerase chain reaction (PCR), not yet a competitor of immunological methods of CTV in field diagnosis and indexing because of their reliability, rapidity and low relative cost, offers a quick detection and alternative technique (Karasev *et al.*, 1995; Pappu *et al.*, 1993; Biswas, 2008). Other methods involved probing the amplified product obtained by PCR with 'strain-specific probes"or single stranded conformational polymorphism of the CTV gene product obtained by PCR. A multiplex real-time PCR assay has also been developed to detect and quantify CTV genotypic mixture present in infected plants. Multiplex real-time quantitative assays are sensitive, efficient, and take less time than other existing methods. This relative quantitative assay will be a reliable tool for diagnosis, detection and genetic diversity studies on CTV (Ananthakrishnan *et al.*, 2010). To simplify the identification and differentiation of CTV genotypes, a multiplex RT-PCR technique for the screening of CTV isolates has also been developed (Roy *et al.*, 2010). A multiplex Taqman-based real-time reverse transcription (RT) polymerase chain reaction (PCR) assay is developed to identify potential severe strains of *Citrus tristeza virus* (CTV) and separate genotypes that react with the monoclonal antibody MCA13. The utility of the assay has been validated using field isolates collected in California and Florida (Yokomi *et al.*, 2010). More recently, quantitative reverse transcriptase polymerase chain reaction (Q-RT-PCR) has been developed for detection and quantification of CTV from infected tissue and aphids (Saponari *et al.*, 2008).

CTV is detected both in plant tissue and in single aphid by 'print capture' and 'squash capture' PCR techniques (Cambra *et al.*, 2000). Sometimes, citrus trees are also infected with a mixed population of citrus yellow mosaic badnavirus-a dsDNA virus and tristeza virus (Fig. 2B). An easy approach has been developed to detect both these RNA and DNA viruses by duplex PCR technique (Ghosh *et al.*, 2008; Roy *et al.*, 2005). Quantitative detection of CTV in citrus and aphids can be done by real-time reverse transcription- PCR (Saponari *et al.*, 2008; Yokomi *et al.*, 2010). RT-PCR, multiple molecular marker assay, coat protein (CP) gene analysis and Single strand conformation polymorphism analysis are also used for CTV detection in field trees (Yokomi *et al.*, 2010). Identification of diversified citrus tristeza virus population isolates from eastern Himalayan region of India has been determined by heteroduplex mobility assay and nucleotide sequence analysis (Biswas *et al.*, 2004). CTV is commonly found in mixed population in infected trees. Single strand conformation polymorphism (SSCP) can detect point mutation in DNA fragments and determine the structure of viral population. Based on SSCP profile, typing of CTV isolates of unknown biological activities is possible. SSCP protocol could separate phylogenetically and biologically different strains of CTV (Saponari *et al.*, 2007). Due to the wide biological diversity, knowledge to identify the actual CTV strains present in a certain region is essential before to design any strategy to control the damage caused by mixture of tristeza

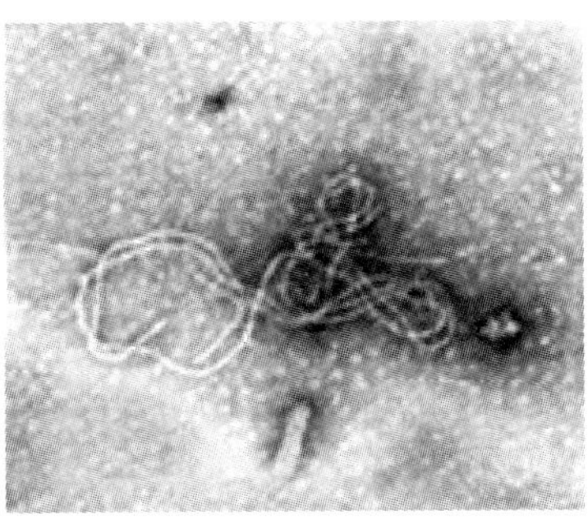

Fig. 2B: Mixed infection of CTV and citrus yellow mosaic badnavirus

virus strains. CTV control depends on prevalence and severity of citrus tristeza disease. Several methods have been suggested to contain CTV which are briefly described below:

- Introduction and prevention of CTV infected material by quarantine. Eradication of infected trees in the field must continue as far as possible to avoid spread of CTV through insect vectors. For example, MCA-13 positive trees are being removed to avoid spread of severe strains of CTV in central California (Yokomi *et al.*, 2010).
- To identify CTV or vector- free areas (many places of Northern plains of India) for development of nurseries and Foundation blocks (Ahlawat and Sardar, 1976).
- To develop and supply healthy planting material through budwood certification programmes. The certification programmes include provision for procurement, indexing, protective maintenance, inspection, increase and distribution of clean material (Reuther, 1978; Calavan, 1974). Detailed of budwood certification are mentioned in Chapter 8. Tristeza virus-free planting material is now being obtained by shoot tip grafting. In India, shoot tip grafting (STG) is used to eliminate CTV from exotic germplasm and also for the production of disease-free planting material (Vijayakumari *et al.*, 2006a, 2006b). CTV has also been managed by *in vitro* propagation of *Citrus* spp. The protocol has been developed and used for regeneration of citrus plants (Mandal *et al.*, 2001).
- CTV tolerant rootstocks should be used. Rangpur lime, rough lemon and volkamer lemon are normally considered as tolerant to CTV but they may show stem pitting when infected by some isolates. Trifoliate orange and its hybrids are tolerant to CTV but are susceptible to viroids and citrus tatter leaf virus. Gou Tou orange root stock in China showed tolerance to CTV but needs more experimentation.
- In places where severe strains of CTV and its effective aphid vector are endemic, cross protection technology using mild strain is the only practical method to reduce the severe effect of the virus. In this technology, planting material is immunized with mild strains. Such immunized trees are protected against severe strains. Cross protection has been successfully used in Brazil, South Africa, Australia and in several other countries. In India, strainal variation observed in CTV and protective strains were used for cross protection (Balaraman and Ramakrishnan, 1980). However, this technology could not be as successful in India as elsewhere

(Chakraborty and Ahlawat, 2001). Similar results of failure of cross protection have been reported from California (Roistacher and Dodds, 1993). Sour orange is one of the best rootstocks but all attempts at cross protection to salvage sour orange as a rootstock have failed. Protective isolates of stem pitting/small fruit CTV were found in Brazil, Argentina, Australia, South Africa, India, and Japan and recently in Peru. It may take 10 to 15 yr or longer to find and develop effective protection against stem pitting tristeza as was done by Costa and Müller in Brazil, van Vuuren in South Africa and by Bederski in Peru. The protective CTV isolates developed by Müller and Costa have proven highly effective for over 40 years in Brazil. Protective CTV isolates can be obtained by attenuating severe isolates by passage through *Passiflora gracilis* or *P. caerulea* via aphid transmission. Some of these isolates have proven highly effective in the revival of the citrus industry in Peru. Cross protection details can be seen on EcoPort slide show #103 (Roistacher *et al.*, 2010). Mild and severe isolates can be differentiated by (i) multiplex one-step real-time RT-PCR; (ii) restriction fragment length polymorphism (RFLP) analysis. The multiplex real-time assay simultaneously detected of all CTV isolates and differentiated VT and T3 genotypes. RFLP assay is based on primers which amplified a target sequence containing unique restriction D*de*I or S*sp*I sites. The D*de*I restriction site was conserved among VT and T3 genotypes and absent in T30 (mild) and T36 (severe) genotypes. In contrast, the S*sp*I restriction site was conserved in mild isolates and absent in the severe strains (Saponari and Yokomi, 2010).

- Aphid vector(s) control is an important method to minimize CTV incidence. It can be done by spraying insecticides during the time of aphid population build up. Aphids can be controlled by biological methods. However, it is practically difficult to prevent tristeza spread by aphid control. Sprays at early stages of infection or during period with high aphid populations, reduces the rate of spread.
- A long term solution to CTV management is to produce transgenic citrus genetically resistant or immune to CTV. Pathogen-derived transgenic *Citrus* for resistance against *Citrus tristeza virus* has been developed (Febres *et al.*, 2005). Many different virus sequences have been used to confer resistance to CTV. Genes used for this purpose were native and untranslatable versions of CP genes, (Roman *et al.*, 2005), replicase genes, antisense sequences (Baulcombe, 1996) and RNA interference (Lee *et al.*, 2007). Genetic transformation of *Citrus paradisi* was obtained with antisense and

untranslatable RNA-dependent RNA polymerase genes (Cevik *et al.*, 2006; Moore *et al.*, 2000). Sour orange (*C. aurantium*), Mexican lime/Kagzi lime (*C. aurantifolia*), and grapefruit (*C. paradisi*) were transformed with CTV genes (Gutierrez *et al.*, 1997; Dominguez *et al.*, 2000; Febres *et al.*, 2002; Roman *et al.*, 2005; Moore *et al.*, 2000; Pérez *et al.*, 1998; Tarafdar *et al.*, 2009). Transgenic citrus has been reviewed by Peña and Navarro (1999).

Note: Hassaku (*Citrus hassaku*) is a popular variety in Japan. It is monoembryonic and most of Hassaku trees are infected with a dwarfing disease known as Hassaku dwarf. This disease is of concern in Japan because the losses caused by it are heavy. Characteristic symptoms of the disease are small, pale, upward rolling of leaves, concavities on the trunk and main shoots and overall stunting of the affected trees. Affected trees bloom heavily, setting of fruits reduced and very few fruits mature. Fruits are small with thickened rind. On graft inoculation to Mexican lime, typical CTV symptoms of vein clearing, stem pitting and vein corking (Plate 5) develop on inoculated plants. Hassaku dwarf is transmitted by *Toxoptera citricida* and the virus particles associated with hassaku dwarf are similar to tristeza. These evidences clearly indicate that hassaku dwarf disease in Japan is caused by a strain of citrus tristeza virus (Tanaka *et al.*, 1969).

Citrus Crinkly Leaf

Crinkly leaf disease of citrus was first described by Fawcett in 1936 followed by Fawcett and Bitancourt in 1943. Detailed description of the crinkly leaf disease was given by Wallace, 1959. In India, crinkly leaf disease was reported on Lisbon and Nepali round lemon from Darjeeling hills (Ahlawat and Sardar, 1976). Crinkly leaf disease has also been reported from Florida (Garnsey, 1968) and also from Australia (Fraser, 1961). Lemon is the main host of crinkly leaf virus and the disease popularly known as lemon crinkly leaf disease. This disease was initially considered as a strain of psorosis disease complex but later studies revealed that it is distinct from psorosis viruses.

SYMPTOMS

Symptoms of crinkly leaf disease appear both on young leaf and mature leaf. The young leaves of infected lemon plants show vein flecking symptoms but sometimes Eureka and Lisbon lemon show numerous necrotic pin-point spots on newly developing leaves (Fraser, 1961). The spots were circular clear areas 2 to 4 times larger than the oil glands and were distributed without relation to veins, their margins were usually fairly sharply defined and their numbers often so numerous that the leaves were noticeably paler in colour than unaffected leaves. The numbers of such spots were greatest on the growth which appears immediately following inoculation and decreased in later growth. As the spotted leaves grew and reached maturity, they develop crinkling. Mature leaves of infected lemon show leaf puckering or crinkling and blistering (Plate 6). Affected leaves also show warping and pocketing appearance due to irregularities of growth in different parts of the leaf blade (Fawcett, 1936; Klotz and Fawcett, 1941, Fawcett and Bitancourt, 1943). Leaf flecking symptoms of crinkly leaf disease resemble severe feeding by citrus red mite (*Panonychus citri*). The intensity and distribution of symptoms are variable. Some trees showed symptoms on one or two branches but in other trees most branches developed noticeably affected leaves. Seasonal variations also occur. Affected trees show an upright, restricted type of growth. Fruits on infected lemon trees remain small, coarse and bumpy. When sweet orange trees are inoculated with crinkly leaf infected lemon scions, typical veinal flecking symptoms of psorosis complex are developed (Fawcett and Bitancourt, 1943; Reuther *et al.*, 1978). Hence, at one time this disease was considered to be a component of psorosis complex. Pin point spotting on young leaves of inoculated Eureka lemon plants has been observed in Australia and India (Fraser, 1961; Ahlawat and Sardar, 1976). It was suggested by Fraser that probably California isolates had some other virus with crinkly leaf infection. Observations of Fraser were confirmed later (Wallace, 1968). However, Dweet tangor consistently developed psorosis young leaf symptoms when infected with pure culture of crinkly leaf virus.

TRANSMISSION

Citrus crinkly leaf virus is transmitted by grafting including tissue grafting to Mexican or Kagzi lime (*Citrus aurantifolia)*, Eureka lemon (*C. limon*), Lisbon lemon, and Dunken grapefruit (*C. paradisi*). Crinkly leaf virus is also transmitted by mechanical inoculation from citrus to citrus (seedlings of Dunken grapefruit and Eureka lemon) and herbaceous hosts like beans, cucumber, petunia, tobacco and tomato. The virus was transmitted by back inoculation from Petunia and tobacco plants to citrus. Local lesion symptoms are developed on mechanically inoculated plants of cowpea (*Vigna sinensis*) and bean (*Phaseolus vulgaris*) after 2–4 days of inoculation. No vector of this virus is known but natural spread is sometimes observed probably by contamination of operational tools. Seed transmission of the virus has also been observed. Seedlings of sweet lime, sour orange, rough lemon, *Poncirus trifoliata,* Emperor and Cleopatra mandarin, and Orlando tangelo showed no symptoms. Strains in Crinkly Leaf Virus have also been suggested (Fraser, 1961).

HOST RANGE

Susceptible host species are found in the Family Apocynaceae, Leguminosae-Papilionoideae, and Rutaceae. Following *Citrus* species are susceptible to experimental inoculation: Citron, Etrog citron, Clementine, Grapefruit, Lemons, Lime (kagzi, Key, Mexican and West Indian), Cleopatra mandarin, Pummelo, Satsuma, Sour orange, Sweet lime and Sweet orange. Following species of herbaceous hosts are susceptible by mechanical inoculation: *Chenopodium amaranticolor, C. quinoa, Crotalaria spectabilis, Cucumis sativus, Gomphrena globosa, Momordica balsamina, Phaseolus vulgaris cv. Red Kidney, P. aureus, Vigna unguiculata* and *Catharanthus roseus. Citrus aurantifolia, Citrus limon, Citrus paradisi, Citrus reticulata, Citrus sinensis,* Lemons and limes (Maxican lime) are highly susceptible. Puckering symptoms in Mexican lime are comparatively severe. Puckering has also been observed on leaves of Etrog citron (*C. medica* L.) and *C. excelsa*. However, symptoms on sweet orange and mandarins are mild and may be visible to some young leaves only. Indian isolates of crinkly leaf virus showed symptoms on inoculated Washington navel, Rangpur lime and Kagzi lime in addition to Eureka and Lisbon lemons. The virus is transmitted mechanically to *Sesamum indicum* in addition of some other hosts listed above. Some of the diagnostic host species and symptoms on them after inoculation are given below.

- ***Lisbon lemon*:** Lisbon lemon seedlings were somewhat more sensitive than those of Eureka lemon and the spots produced on early formed leaves were in some cases so numerous that they became confluent, and some distortion of the young leaves resulted.
- ***Eureka lemon*:** Asteroid local lesions developed.
- ***Meyer lemon*:** Few or more spots and slight crinkling developed.
- ***Pear-shaped and knight citrons*:** The spots were few and not well defined. In a few cases very slight crinkling occurred in one or two leaves.
- ***Sweet orange*:** Few spots and slight crinkling developed.
- ***Vigna unguiculata*:** *V*ein clearing and malformation developed.
- ***Crotalaria spectabilis*:** Systemic chlorotic veinbanding developed.

ASSAY/INDICATOR HOSTS

Citrus crinkly leaf is assayed by mechanical inoculation on local lesion hosts, cowpea and bean but better local lesion assay host is *Vigna unguiculata* cv. 'Blackeye'. Best indicator hosts in *Citrus* spp. are: lemons, limes, Etrog citron and sour orange. Symptoms on these hosts are crinkling of mature leaves and vein flecking or pin point spotting of young leaves. Main indicator hosts and symptoms on them are given below.

- ***Lemon seedlings, Etrog citron*:** Leaf crinkling develops after a few weeks of inoculation.
- ***Cowpea cultivar Blackeye*:** Chlorotic lesions develop on mechanically infected leaves within a few days.

THE PATHOGEN

A polyhedral virus approximately 27 nm has been reported from California isolates of crinkly leaf (Yot-Dauthy and Bove, 1968). Crinkly leaf virus is now classified as ilarvirus of subgroup 2 of ilarviruses with four nucleoprotein components. The size of virus particles ranges from 26 to 32 nm. Crinkly leaf virus is serologically related to other citrus ilarviruses causing citrus variegation and citrus leaf rugose diseases. Crinkly leaf virus is considered to be a strain of infectious variegation ilarvirus. Genome of ilarviruses will be discussed in the chapter of leaf rugose and infectious variegation.

DIAGNOSIS AND MANAGEMENT

Citrus crinkly leaf ilarvirus is detected by inoculation to lemon seedlings and by sap inoculation to local lesion hosts, cowpea and bean. The virus is also detected by ELISA using polyclonal and monoclonal antibodies. Since citrus ilarviruses have been sequenced (Scott and Ge, 1995 a, b), crinkly leaf virus is successfully detected by molecular techniques like Polymerase chain reaction (PCR) and nucleic acid hybridization. Crinkly leaf virus can be managed by the use of healthy planting material. *Poncirus trifoliata* is tolerant to crinkly leaf virus and hence can be used as rootstock in endemic areas.

Citrus Infectious Variegation

Infectious variegation in citrus was described as a possible virus disease in 1931 by Petri. His description of symptoms and his illustrations are similar to those reported by Fawcett and Klotz (1939), who transmitted the causal agent by bud-grafting from variegated lemon to sour orange. They noticed that some of the symptoms were similar to those associated with psorosis and suggested that the disease might possibly be related to psorosis. Klotz and Fawcett (1941) noted that crinkly leaf of lemon is sometimes associated with infectious variegation and it was not certain whether infectious variegation differed sufficiently from crinkly leaf to be considered a separate disease. From cross-protection studies, it was concluded that infectious variegation is caused by a strain of the psorosis virus. Grant and Smith (1960) reported infectious variegation in Florida and transmitted the causal agent by bud and leaf-piece grafts to several different species of citrus.

Initially, infectious variegation disease was reported to infect lemons and Sour orange in California but now the disease is probably distributed worldwide (occurs in most citrus growing countries, but is not common). This disease was observed on Marsh grapefruit trees in Florida, Cyprus, Turkey, Australia and Mediterranean countries, Italy and Portugal (Grant and Smith, 1960; Bahia Bennoni *et al.*, 2004; Gerola *et al.*, 2005). In India, infectious variegation was recorded on *Citrus decumana* and lemon plantations in Darjeeling and Sikkim hills by the author (Yora *et al.*, 1977). This disease was initially studied as a strain of citrus psorosis (Wallace, 1957) but later investigations revealed it to be a distinct disease. Strains in infectious variegation virus do exist. For example, California and Florida strains of infectious variegation virus differ in reaction on *Crotalaria spectabilis* (Desjardins and Wallace, 1966).

SYMPTOMS

Infectious variegation disease is commonly found in lemon (*Citrus limon*) and Sour orange (*C. aurantifolia*) trees. In these hosts, common symptoms are malformation of mature leaves and psorosis-like flecking of young-leaves. Infectious variegation virus causes mosaic-like symptoms including crinkling, distortion, flecking of young leaves, pin point spotting and shoe-string effect on leaves. Affected trees remain stunted.

HOSTRANGE

Citrus cultivars susceptible to infectious variegation are; Bergamot, Citrons, Etrog citron, Clementine, Grapefruit, lemons, Lime (Kagzi, Key, Mexican,

West Indian), Mandarin (Cleopatra), Pummelo, Satsuma, Sour orange, Sweet lime and sweet orange. Lemons, Cleopatra Mandarin and Satsumas produce better symptoms of infectious variegation than Etrog citron, Clementine, Grapefruits, Pummelo, Sour orange, Sweet lime, and West Indian limes. *Chenopodium quinoa, C. amaranticolor, Crotalaria spectabilis, Cucumis sativus, Gomphrena globosa, Momordica balsamina, Phaseolus aureus, P. vulgaris,* and *Vigna sinensis* (cowpea) are herbaceous hosts that are susceptible to infectious variegation virus. Willowleaf mandarin and *P. trifoliata* are symptomless hosts of the virus. *Poncirus trifoliata* appears to be tolerant to infectious variegation virus. Symptoms on common hosts are given below:

- ***Eureka lemon, Etrog citron, C. decumana and Grapefruit*:** Chlorotic mottle and puckering.
- ***Lemon, Sweet and Sour Orange*:** Leaf distortion, epinasty, chlorotic mottle and leaf pattern (Plate 7).

TRANSMISSION

Infectious variegation virus is transmitted by mechanical inoculation (Grant and Corbett, 1960); transmitted by grafting (Fawcett and Klotz, 1939); and possibly transmitted by seeds. Following plant species are susceptible to experimental virus inoculation: *Capsicum frutescens, Capsicum frutescens* var. *grossum, Citrus aurantifolia, Citrus aurantium, Citrus limon, Crotalaria spectabilis, Sesamum indicum* (Desjardins, 1969), *Vigna unguiculata*. Symptoms produced by infectious variegation virus upon mechanical inoculation in some major hosts are described below (Desjardins and Wallace, 1966; Garnsey, 1968):

- ***Cucumber (Cucumis sativus* L.):** Diffuse chlorotic rings on inoculated cotyledons.
- ***Kidney bean (Phaseolus vulgaris* L.):** On variety Bountiful, the symptoms appear as distinct yellow vein banding in systemic manner.
- ***Gomphrena (Gomphrena globosa* L.):** Diffuse spots on inoculated leaves and interveinal chlorosis on systemically infected leaves.

Blackeye cowpea {*Vigna sinensis* (Torner) Savi} inoculated from citrus crinckly leaf virus isolate of California and infectious variegation isolate of Florida develop similar symptoms as chlorotic local lesions on inoculated leaves followed by systemic infection. The only difference between two viruses was that the symptoms were developed earlier on plants inoculated by infectious variegation virus (Dauthy and Bove, 1965).

ASSAY HOSTS

Although several varieties of cucumber are susceptible to infectious variegation but variety National Pickling was found to carry the stable symptoms as mentioned above and is used as an assay host. *Vigna unguiculata* (W) and *Capsicum frutescens* var. *grossum* (W) are also used as assay hosts. Following hosts are used as indicator hosts:

- ***Lemon seedlings***: Leaf variegation develops within a few weeks after inoculation.
- ***Cowpea cultvar blackeye***: Chlorotic lesions develop on mechanically inoculated leaves within a few days.

THE PATHOGEN

The virus associated with infectious variegation disease was initially purified and polyhedral particles of approximately 30 nm were obtained (Corbett and Grant, 1967; Garnsey, 1974; Rana *et al.*, 1974; Desjardins and French, 1970). Spherical virus particles of 30 nm were also observed from tissues of sour orange infected with Indian isolates of infectious variegation in leaf dip preparations (Fig. 3) (Yora *et al.*, 1977). Virions have icosahedral symmetry. Capsid is quasi-isometric or irregularly isometric and has a diameter of 26–35 nm. Virus preparations contain more than one particle component. Infectious variegation virus is an ICTV approved species of the genus 00.010.0.02. *Ilarvirus* in the family 00.010. *Bromoviridae.*

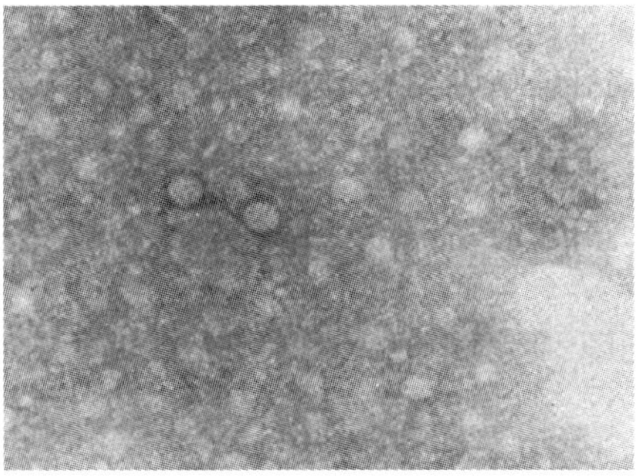

Fig. 3: Virusus particles associated with Infectious variegation in dip Preparation

GENOME

The genome of infectious variegation ilarvirus is segmented; tripartite, segements are distribute among 3 particle types of different size; consists of three segments of linear, positive-sense, single-stranded RNA. Minor species of non-genomic nucleic acid are also found in virions. Encapsidated nucleic acid is mainly of genomic origin, but virions may also contain subgenomic RNA. Sequence can be accessed from EBI-EMBL and GenBank. RNA-3 is fully sequenced and encodes movement protein and coat protein genes (Scott and Ge, 1995). The RNA species of citrus variegation virus have been characterized through complementary DNA clones (Calvert *et al.*, 1988).

Coat protein gene and part of the intergenic region of the RNA 3 of several isolates of Citrus variegation virus (CVV) producing either infectious variegation or crinkly leaf symptoms were amplified by RT-PCR, cloned and sequenced. Some isolates were composed of a mixture of sequence variants. Coat protein gene appeared to be highly conserved (lowest similarity among all CVV sequences 93%), especially at the N-terminal, indicating at the molecular level that both types of symptoms are indeed produced by the same virus species. Sequence analysis showed low genetic variability of CIVV isolates in the CPg confirming the high degree of coat protein gene conservation (Kubaa *et al.*, 2010). No relationship could be found with the geographic origin. Sequences obtained from isolates producing infectious variegation clustered in a different branch of a dendrogram than those originating from crinkly leaf symptoms. Both clusters could further be distinguished by two parsimonious sites in the coat protein gene. In the short stretch of the intergenic RNA 3 region analysed, a stable hairpin exists in addition to the previously reported hairpin that constitutes the core promoter for the RNA 4 transcription. This second hairpin could also be recognised in the other subgroup 2 Ilarviruses. Surprisingly, at the nucleotide and amino acid levels and in the secondary features of the intergenic region, CVV appeared closer to the other subgroup 2 Ilarviruses than to Citrus leaf rugose virus to which it is serologically related and has been considered to be evolutionary related (Bahia Bennani, 2004; Gonsalves and Garnsey, 1975, 1975a, b). Viral genome encodes structural proteins which are involved in viral encapsidation and in protein movement. A non-structural protein is also located in the capsid. The virus is serologically related to citrus leaf rugose virus but distantly.

HISTOPATHOLOGY

The sieve tubes and companion cells in ultrathin sections of infectious variegation infected *Phaseolus vulgaris* leaves when viewed in electron microscope contained a slime-like substance, more or less organized into compact systems. In the mature sieve elements of virus-infected plants systems of parallel membranes were seen along which spherical particles, of about 30 nm diameter, were aligned in simple rows (Gerola *et al.*, 2005).

DIAGNOSIS AND MANAGEMENT

Citrus infectious variegation ilarvirus can be detected by graft inoculation to lemon seedlings and by sap inoculation to susceptible herbaceous hosts as described above. *Citrus aurantifolia, C. limon* cv. Eureka, Citron, Sweet orange and *Cleopatra mandarin* are used as hosts for disease diagnosis. Symptoms in these hosts are leaf flecking, crinkling, variegation and malformation. Polyclonal and monoclonal antibodies have been produced against this virus (Davino and Garnsey, 1984) and used for serological detection in techniques like ELISA. Other techniques like nucleic acid hybridization assays and PCR are also being used to detect the infectious variegation virus (Bahia Bennani *et al.*, 2004). Using primers designed for the conserved region, the virus was detected with a prevalence of 25% and 13% in Portuguese and Moroccan citrus collections. A group of RT-PCR positives was further confirmed by ELISA and biological indexing (Bahia Bennani, 2004).

Since infectious variegation virus is not vector transmitted, chances of field spread are very rare. This virus is being disseminated to new plantations by planting infected trees obtained from infected mother trees. This disease can be effectively controlled by propagating young plants from virus-free mother trees. The virus free mother trees can be obtained by testing them against the virus using the diagnostic techniques described above. Infectious variegation virus can also be inactivated by keeping infected plants for two-three months for hot air treatment at alternating temperature of 40°C for 16h/day and 30°C for 8 hr/night (Calavan *et al.*, 1972). An antiviral principle obtained from seeds of *celosia cristata* inhibited tobacco streak ilarvirus and may be effective to infectious variegation ilarvirus (Sumiya *et al.*, 2004).

Citrus Leaf Rugose

Citrus leaf rugose virus has earlier been reported by the name of *Citrus psorosis* virus complex and Citrus crinkly-leaf-type virus.The name "citrus-leaf-rugose virus" (CLRV) was proposed for a mechanically transmitted citrus virus which was discovered in Florida and has been described earlier as "crinkly-leaf-type virus"(Garnsey, 1974, 1975). The disease has been described in detail by Garnsey and Gonsalves, 1976). This disease is also found in Australia but no spread was observed (Brunt *et al.*, 1996).

SYMPTOMS

Trees of Indian Key lime or Mexican lime *(Citrus aurantifolia),* Lemon (*C. limon),* and grapefruit (*C. paradisi)* are found to be naturally infected with leaf rugose virus in Florida. Symptoms on these hosts appear as flecking of young leaves and malformation or puckering of old leaves. These symptoms are clearly seen in outdoor Eureka lemon trees. Fruits on affected trees are small and lumpy. Citrus leaf rugose virus (CLRV) does not produce symptoms in many citrus species. CLRV causes leaf flecking in Eureka lemon, leaf rugosity in Mexican lime, and severe stunting of young Duncan grapefruit (*C. paradisi*) plants (Garnsey, 1968). The virus is present in most plant tissues.

TRANSMISSION

Leaf rugose is not transmitted by a vector but field spread is suspected (Garnsey, 1968). CLRV is experimentally transmitted through contamination of cutting tools. It is transmitted by mechanical inoculation and by grafting. It is possibly transmitted by seeds but there is no authentic report.

HOST RANGE

Host range of citrus leaf rugose virus includes both citrus and herbaceous hosts. Susceptible host species are found in the Family Apocynaceae, Leguminosae-Papilionoideae, and Rutaceae. Among citrus species, Lime (Kagzi, Key, Mexican, West Indian), Dunken grapefruit, Eureka lemon, sweet orange are important hosts. Herbaceous hosts like *Chenopodium amaranticolor, C. quinoa, Crotalaria spectabilis, Cucumis sativus, Gomphrena globosa, Momordica balsamina, Nicotiana tabacum, Petunia hybrida, Phaseolus vulgaris cv. Red Kidney and Vigna unguiculata* are susceptible to leaf rugose infection (Garnsey, 1975). The following species are susceptible to mechanical inoculation: *Catharanthus roseus, Citrus aurantifolia, Citrus limon, Citrus paradisi, Citrus reticulata, Citrus sinensis, Crotalaria spectabilis, Phaseolus vulgaris* (cv. Red Kidney) and *Vigna unguiculata.* Symptoms develop upon inoculation on some important hosts are given below:

- ***Mexican lime or Indian key lime*:** Variable rugosity of leaves.
- ***C. limon (Eureka lemon)*:** Pinpoint chlorotic flecks on expanding leaves, persistent symptoms without leaf distortion
- ***C. paradisi (Duncan grapefruit)*:** Severe stunting and chlorosis in some isolates.
- ***Citrus sinensis, C. limon, C. reticulata*** — crinkling and flecking
- ***Phaseolus vulgaris (bean) cv. Red Kidney*:** Small, necrotic local lesions develop on inoculated leaves. No systemic infection.
- ***Vigna unguiculata*:** Vein clearing and malformation of leaves.
- ***Crotalaria spectabilis*:** Systemic chlorotic veinbanding.

ASSAY/INDICATOR HOSTS

Phaseolus vulgaris cv. 'Red Kidney', *Vigna unguiculata* (Early Ramshorn cowpea), and *Crotalaria spectabilis* are local lesion hosts (Garnsey, 1975) and can be used as assay host of the virus. Following citrus and herbaceous hosts are used as indicator hosts of virus:

- ***Mexican lime seedlings*:** Symptoms develop as leaf puckering within a few weeks.
- ***Red Kidney bush bean*:** Local lesions develop on inoculated leaves within a few days.

Leaf rugose virus is propagated on *C. aurantifolia,* Etrog citron (*C. medica*) and most other citrus varieties which are susceptible. Cultures may also be maintained on *Gomphrena globosa, Nicotiana tabacum* and *Petunia hybrida.*

PHYSICAL PROPERTIES

Purified virus had an absorption minimum at 244 nm, a maximum at 260 nm, and 260/280 ratio is 1.4 to 1.45. Purified CLRV was stable for several months at 4°C. Thermal inactivation point of the virus is near 60°C., longevity *in vitro* was 1 to 2 days at room temperature in crude extracts.

THE PATHOGEN

Citrus leaf rugose virus (CLRV) has been purified (Garnsey, 1974). Virions are round with icosahedral symmetry. The capsid is quasi-isometric and has a diameter of 25–32 nm (Yot-Dauthy and Bove, 1968; Garnsey, 1975). Virus preparations contain more than one particle component. Virions have isometric or irregularly shaped capsid. CLRV belongs to group IV positive sense, single stranded RNA viruses. The virus is an ICTV approved species of the genus 00.010.0.02. *Ilarvirus* in family 00.010. *Bromoviridae*.

GENOME

The genome of CLRV is tripartite, segements are distributed among 3 particle types of different size; consists of four segments of linear positive-sense, single-stranded RNA. Minor species of non-genomic nucleic acid are also found in virions. The encapsidated nucleic acid is mainly of genomic origin, but virions may also contain subgenomic RNA, that is mRNA derived from genomic RNA-3. Based on A_{260}/A_{280} ratios amount of RNA in the particles is estimated to be similar to that of Tulare apple mosaic and tobacco streak ilarviruses (12–14%). The RNAs in NPs 1, 2, 3 and 4 are designated as RNAs 1, 2, 3 and 4, respectively (Gonsalves and Garnsey, 1974). Their Molecular Weights are: *c.* 1.1×10^6, *c.* 1.0×10^6, *c.* 0.7×10^6, and *c.* 0.3×10^6, respectively. Sedimentation data indicate that two molecules of RNA 4 are probably enclosed in one NP 4 particle. Very small amounts of a fifth RNA (RNA 4a) are usually observed in gels. The RNAs are not infective singly. However, non-infective mixture of RNAs 1+2+3 became highly infective with the addition of either RNA 4 or coat protein (Gonsalves and Garney, 1975*a*). Coat protein and RNA-4 of citrus variegation virus or alfalfa mosaic virus, and coat protein from tobacco streak virus are also effective (Gonsalves and Garnsey, 1976). RNA 4 or coat protein from citrus leaf rugose virus will activate non-infective mixtures of RNAs 1+2+3 from alfalfa mosaic virus, and from citrus variegation virus. A single polypeptide species with M.Wt of *c.* 26,000 based on SDS-polyacrylamide gel-electrophoresis is present.

The complete genome of CLRV has been sequenced and is 7888 nucleotides long. RNA-1 has 3099 nucleotides, RNA-2 is about 2817 nucleotides long. RNA-3 has also been sequenced, but only an estimate is given, complete sequence is 1972 nucleotides long. Subgenomic RNA-4 is an mRNA derived from RNA-3 negative strand template. The multipartite genome is divided among more than one type of particle and the segments are distributed between 3 different types of particles. The largest particles contain each one molecule of RNA-1 (sedimenting component B). The medium sized particles contain each one molecule of RNA-2 (sedimenting component M). The smallest particles contain one molecule each of RNA-3 and RNA-4 (sedimenting component T). Sub-genomic RNA is present in infected cells; encoding the coat protein. The genomic organization of ilarviruses is identical to that of other genera in the family *Bromoviridae* (Scot and Ge, 1995). The viral genome encodes structural proteins and non-structural proteins. Virions consist of 1 structural protein(s) located in the capsid which is involved in viral encapsidation and involved in protein movement.

RELATIONSHIPS

Two strains, which differ in their reaction on Duncan grapefruit, have been described (Garnsey, 1975). This virus is serologically related but not identical to citrus variegation virus, citrus crinkly leaf virus and Tulare apple mosaic virus (Garnsey, 1974, 1975). The virus has no serological relationships to tobacco streak, cowpea mosaic and alfalfa mosaic viruses (Garnsey, 1975; Gonsalves and Garnsey, 1976). All the three viruses (citrus crinkly leaf, citrus infectious variegation and citrus leaf rugose) are ilarviruses and have many properties in common. CLRV has properties similar to Tulare apple mosaic virus. It is therefore, important to know how to distinguish individual viruses. although these viruses are related, they are easily distinguished by biological properties and serological differences. Citrus variegation virus does not produce rugose symptoms in Mexican lime, and citrus leaf rugose virus does not cause severe distortion and variegation patterns in Etrog citron. Citus leaf rugose virus causes only local necrotic symptoms in the bean and cowpea varieties tested, whereas citrus variegation virus commonly causes systemic symptoms. Sedimentation profiles for the nucleoprotein components of the two viruses are different, with citrus leaf rugose virus showing more NP 4 and poorer separation of NP 1 and NP 2 (Gonsalves and Garnsey, 1975b). Recognizable differences in the nucleic acid species of these two viruses also exist (Gonsalves and Garnsey, 1975b). Similarly, citrus leaf rugose virus differs in biological and physical properties from Tulare apple mosaic virus. The latter causes chlorotic local lesions in cucumber (*Cucumis sativus*), vein banding in *Gomphrena globosa* and necrotic local and shock symptoms in tobacco, none of which is caused by citrus leaf rugose virus. Tulare apple mosaic virus is less stable, and the sedimentation profile of its nucleoprotein components is different.

DIAGNOSIS AND MANAGEMENT

Citrus leaf rugose virus is detected by graft inoculation to lemon seedlings and by mechanical inoculation to several herbaceous hosts like Red kidney bean. Antisera have been prepared against CLRV, which reacted to purified virus and to extracts from CLRV-infected citrus and herbaceous hosts. Antisera with titres of 1/256 to 1/512 can be used to detect the virus in gel diffusion (Garnsey and Purcifull, 1969). CLRV antiserum reacted heterologously to citrus variegation virus, but not to tobacco streak or cowpea mosaic viruses. Monoclonal antibodies have also been prepared against CLRV and have been used to detect CLRV in ELISA and gel diffusion tests. Since the virus genome has been sequenced, it is possible to detect the virus in PCR and nucleic acid hybridization tests (Ge and Scott, 1994; Scott and Ge, 1995a, b). Citrus leaf rugose disease can be managed by similar methods as have been mentioned earlier with other ilarviruses.

Citrus Leprosis and Zonate Chlorosis

Citrus leprosis disease was originally referred to as scaly bark and neil-head rust disease in Florida (Fawcett, 1911). Leprosis disease was first reported from Florida on sweet orange (Knorr, 1968; Hume, 1901). However, this disease has not been reported in USA since 1960 (Childers *et al.*, 2003). Leprosis comes from the word lepra = Ulcers (Knorr and DuCharme, 1951). This is probably the most important virus disease of citrus in Brazil at the present time (Rodrigues, 2000). The Brazilian citrus industry spends over US$100 million annually (about 21% of citrus production cost) on miticides to control the *Brevipalpus* mite vector of leprosis virus (Omoto, 1998; Rodrigues *et al.*, 2003). Without control of the mite vector and reduction of inoculum in the field by pruning, the disease will kill citrus trees in three years. Presently, this disease is present in South America (Argentina, Paraguay, Brazil, Bolivia, Colombia and Venezuela) (Bitancourt, 1937; Childers *et al.*, 2003, 2001; Fawcett and Lee, 1926; Knorr *et al.*, 1968; Rodrigues *et al.*, 2003), Central America (Panama, Costa Rica, Guatemala (Palmieri *et al.*, 2005; Dominguez *et al.*, 2001) and Honduras. It has also been reported from Mexico (Rodrigues *et al.*, 2003; Bastianel *et al.*, 2006). There are unconfirmed reports of leprosis from Costa Rica (Araya-González, 2000), Guatemala (Palmieri *et al.*, 2003), Honduras, and Nicaragua (Guerra-Moreno *et al.*, 2005). Symptoms of leprosis-like diseases have been reported from citrus-producing areas of Asia and Africa (Fawcett and Lee, 1926; Knorr and Price, 1958; Rhoads and DeBusk, 1931), but none of these reports have been confirmed (Lovisolo, 2001). It is a major problem of sweet oranges in Brazil and Argentina (Rossetti, 1996). Some citrus species show mild reaction to leprosis infection but severe infection to sweet orange may result heavy loss to the crop. In Brazil, this disease was known as "Leprose" and "variola" (Bitancourt, 1955) which was found identical to leprosis disease in Florida (Knorr, 1968). In Argentina, this disease is known as "lepra explosiva" (Frezzi, 1940). Zonate chlorosis is a similar disease to leprosis but is not as harmful as leprosis. Zonate chlorosis is found in coastal regions of Brazil. Citrus leprosis disease is an emerging disease in Central America. It is currently known that two different virus types cause leprosis disease, the prevalent cytoplasmic type (CiLV-C) and the rare, nuclear type (CiLV-N).

SYMPTOMS

Affected trees show heavy defoliation, fruit drop and dieback followed by death of infected tree. Leprosis causes symptoms on leaves, twigs, and fruits. On leaves of infected trees, symptoms appear as chlorotic patches with necrotic centre and chlorotic transluscent halo. Older lesions become large and may show cocentric brown rings occasionally impregnated with gum (Plate 8). These lesions are either dark brown with raised surface or brilliant yellow areas. Such lesions are called "ciclosis", "Ombrosis" or "crisosis" due to gum exudation on old lesions (Bitancourt, 1955). Leaves of severely affected trees may fall. Foliar symptoms also develop on tangerines and mandarins.

Disease symptoms also appear on twigs and branches as yellowish or chlorotic flat patches or lesions. These patches later developed into raised large brown or reddish areas which subsequently turn greyish and may be confused with citrus canker (Plate 9). Lesions may fuse togather and kill the stem by girdling which will not allow the photosynthate to travel to roots via phloem resulting death of tree. Twigs may show dieback symptoms similar to psorosis (Rossetti, 1980). Zonate chlorosis produces chlorotic areas on leaves of infected plants. These chlorotic areas can cover the whole leaf area and develops concentric lines in a marble-like pattern. Symptoms in fruits first appear as flat yellow patches which become larger, black and brown with somewhat depressed areas when fruits mature. These lesions may be in large number on fruits but remain resticted only to external part of the fruit. Fruits in severely affected trees may fall prematurely (Rossetti, 1996). Lesions are more common on unripe fruits of sweet orange and mandarin which subsequently become larger on ripening of fruits. The degree of susceptibility for different citrus varieties plays an important role in the lesion size and consequently on lesion morphology (Rodrigues, 2000). Anatomy of leaf, fruit and stem lesions reveals the presence of hypertrophied spongy parenchyma cells and hyperplastic palisade parenchymal cells interspersed with necrotic cells in advanced lesions (Marques *et al.*, 2004).

TRANSMISSION

Leprosis disease is transmitted by grafting (Knorr, 1968; Chagas and Rossetti, 1984). Symptoms development in inoculated plants may take from 4–12 months. Both forms of leprosis are transmitted by mite vector(s). Vectors of leprosis disease are different species of mites. Mite, *Brevipalpus phoenicis* is a vector of leprosis in Brazil. Leprosis disease is more efficiently transmitted by larvae of *B. phoenicis* than by adults (Chagas *et al.*, 1984; Rossetti, 1996). *B. phoenicis* is also suspected to be the vector of citrus zonate chlorosis and coffee ring spot but recent evidences show that these diseases are different (Rossetti *et al.*, 1975). In Brazil, both leprosis and zonate chlorosis are transmitted by *B. phoenicis* vector. In Florida, leprosis is transmitted by another mite species, *B. californicus* and in Argentina and Venezuela, the vector is *B. obovatus*. This disease is also transmitted by mechanical inoculation from citrus to citrus and herbaceous hosts (Colariccio *et al.*, 1995; Lovisolo *et al.*, 1996).

HOST RANGE

Sweet oranges and mandarins are highly susceptible to leprosis. Kumquats, *Poncirus trifoliata* (trifoliate orange), and sour orange are susceptible but grapefruits, Tahiti lime, and rough lemon are comparatively less susceptible. Valencia late sweet orange in Florida is symptomless host of leprosis virus. Lemons and limes appear to be immune to leprosis (Rodrigues *et al.*, 2003; Bastianel *et al.*, 2006). Symptoms on some citrus and noncitrus hosts are given below:

* ***Pera Sweet orange***: Round or elongated lesions measuring 5–12 mm with chlorotic or necrotic centre develop on inoculated plants (Plate 8).
- ***Bahianinha Sweet orange***: Leaves show large chlorotic lesions without necrotic centre.
- ***Caipira sweet orange and Tahiti lime***: Show typical leaf symptoms of leprosis.
- ***Cleopatra mandarin***: Large chlorotic areas sometimes covering the whole leaf.

A non-citrus rutaceous plant (*Swinglea glutinosa* (Blanco) Merr) was found naturally infected by CiLV-C in Colombia (G.E. Leon, J. Freitas-Astua and E.W. Kitajima, *unpublished data*). CiLV-C type leprosis was transmitted by mechanical inoculation to herbaceous hosts, *Chenopodium quinoa, C. amaranticolor* and *Gomphrena globosa* (Colaricio *et al.*, 1995). CiLV-C was also experimentally mite-transmitted to, *Solanum violaefolium* (Rodrigues *et al.*, 2005), *Phaseolus vulgaris* (Groot *et al.*, 2006), *Malvaviscus arboreus, Hibiscus rosa-sinensis, Grevilea robusta, Bixa orellana* and *Commelina benghalensis* (Nunes *et al.*, 2006). *Arabidopsis thaliana* was infected by citrus leprosis virus C through mite (Kubo *et al.*, 2010). However, CiLV-N infects only sweet oranges and mandarins. Symptoms of CiLV-C on some herbaceous hosts are described below:

- ***Chenopodium amaranticolor, C. quinoa, C. album, C. murale***: Conspicuous chlorotic lesions with necrotic centre and transluscent halo developed on inoculated leaves.
- ***C. capitatum, C. foliosum***: Necrotic local lesions.
- ***C. polyspermum***: Brown spots with chlorotic halo
- ***Gomphrena globosa***: Local lesions of brown colour develop on inoculated leaves.

PHYSICAL PROPERTIES

Thermal inactivation point (TIP) of leprosis virus in crude extract is between 55–60°C, dilution end-point (DEP) 10^3 and longevity *in vitro* of frozen leaves (–80°C) is approximately 4 months and about one month at –15°C (Lovisolo *et al.*, 1996).

THE PATHOGEN

Bacilliform particles measuring 100–110 × 40–50 nm similar to rhabdoviruses as associated with coffee ring spot virus and with some orchid viruses (orchid fleck virus) were observed in nucleus and cytoplasm of leprosis infected cells (Kitajima *et al.*, 1972). There are two known genera of plant-infecting rhabdoviruses, *Cytorhabdovirus* and *Nucleorhabdovirus*. Cytorhabdoviruses replicate in the cytoplasm of infected cells in association with masses of thread-like viroplasm structure, and virus morphogenesis occurs in association with vesicles of the ER (van Regenmortel *et al.*, 2000). Nucleorhabdoviruses accumulate in the perinuclear space with some particles scattered in the cytoplasm (Hull and Matthews, 2002). Both types of particles (CiLV-C and CiLV-N) have been reported in citrus leprosies (Guerra-Moreno *et al.*, 2005). In cells infected with CiLV-N, a characteristic nuclear, electron lucent viroplasm is detected as well as short rod-like particles in the nucleus or cytoplasm were present (Kitajima *et al.*, 1972). In ultrathin sections of lesions caused by CiLV-C, virions measuring 120–130 × 50–70 were present within membrane-bound vesicles and/or electron dense, circular to elliptical, vacuolated viroplasms, usually single, were present in the cytoplasm (Colariccio *et al.*, 1995). CiLV-C genome has been completely sequenced and it has been proposed as the type member of a new genus of plant viruses, Cilevirus (Kubo *et al.*, 2010). Leprosis virus (CiLV-N) belongs to the genus *Nucleorhabdovirus* in the family 01.062. *Rhabdoviridae*; order 01. Mononegavirales.

GENOME

Citrus leprosis symptoms are associated with the presence of two different viruses: cytoplasmic and nuclear. Complete sequence of the cytoplasmic *Citrus leprosis* virus (CiLV-C) has been done and its genome organization determined (Guerra-Moreno *et al.*, 2007). Characterization of genome of CiLV-C including the subgenomic RNAs revealed the presence of defective RNAs in infected plants. Sensitive RT-PCR assay based on a highly expressed open reading frame present in infected tissue has been developed. The genome consists of two segments (5 and 9 kb) of positive sense ssRNA with poly-A tail (Pascon *et al.*, 2006; Locali-Fabris *et al.*, 2006). CiLV-C has been shown to have two RNAs (RNA 1 and RNA 2). Probes of clones from RNA 1 hybridized with two RNAs of approximately 1.0 kb and 1.5 kb, while probes of clones from RNA 2 hybridized with four RNAs of approximately 4.7, 2.5, 1.6, and 1.0 kb in Northern hybridizations tests. RNA 1 and RNA 2 have been sequenced (Guerra-Moreno *et al.*, 2007). Viral genome encodes structural and non structural proteins. Its unique genomic sequence, structure, and organization led Locali-Fabris *et al.* 2006 to propose CiLV-C as the prototype of the *Cilevirus,* a new virus genus. Little is known about the CiLV-N genome. It may belong to the same group as Orchid fleck virus (OFV). OFV is also transmitted by mite, *B. californicus* and shares cytopathology and virion morphology with CILV-N. OFV is a bipartite rhabdovirus (proposed genus-*Dichorhabdovirus*) and its particles are short, rod-like measuring 40–50 nm × 100–110 nm (Kondo *et al.*, 2006).

DIAGNOSIS AND MANAGEMENT

Symptoms of lesions in sweet oranges and mandarins are of satisfactory diagnosis of leprosies in field trees. Diagnosis of leprosis disease is done by sap inoculation to *Chenopodium* spp. or to *Gomphrena globosa*. These hosts develop local lesions upon inoculation as described above. Electron microscopy is routinely used for leprosis virus detection (Kitajima *et al.*, 2003). Leprosis can be detected in Western blotting (Manjunath *et al.*, 2006), PTA-ELISA, RT-PCR (Locali *et al.*, 2003) and by immunofluorescence (Calegario *et al.*, 2010). Control of leprosis is based on elimination of inoculum by pruning of affected trees and to reduce mite population by applying acaricides (Bastianel *et al.*, 2006). The disease is extremely destructive if mite control is not applied. Successful control of leprosis using extracts of *Datura metal* has been reported (Guirado *et al.*, 1995).

Satsuma Dwarf

Satsuma orange (*Citrus reticulata* var. Satsuma) is a variety of mandarin orange grown extensively in Japan. Many people call satsuma as tangerine type of orange, but most people prefer to call it as *Satsuma mandarin*, which is botanically identified as *Citrus unshiu* Marcovitch. Home gardeners generally prefer satsuma oranges because of their small size, sweet and juicy taste and easy peeling technique. Origin of satsuma oranges appears to be from China, but it was reported from Japan about 700 years ago and now Japan has a dominating satsuma orange market. Satsuma trees are preferably grown in the cool subtropical regions of Japan, Spain, central China, Turkey, Russia, South Africa, South America and in California and Florida in United States of America. Like all other citrus fruits, satsumas are also an excellent source of vitamin C and are enriched with fibre and folate. Satsuma's skin is a bit tighter than clementine but it is a seedless citrus.

Satsuma dwarf is a common disease of satsumas and was first shown to be a graft transmissible disease in 1952 in Japan (Yamada and Sawamura, 1952). Satsuma dwarf disease is increasing due to widespread use of topgrafting in addition to its natural transmission. Citrus mosaic, Natsudaidai dwarf, navel orange infectious mottling and hyuganatsu are usually placed in the same category as Satsuma dwarf, but their incidence is limited only to some isolated areas of Japan. A disease similar to Satsuma dwarf has been reported from Cheju Island in Korea, Mainland China and Turkey (Azeri, 1973; Cui *et al.*, 1991). It appears that this disease was probably introduced from Japan to other countries as infected budlines. Because of their soilborne nature, these diseases are being given considerable attention. In Japan, citrus mosaic virus, Natsudaidai dwarf virus and navel infectious mottling virus are caused by related viruses (Iwanami and Koizumi, 2000). Satsuma dwarf affected trees are dwarfed with poor growth and decrease yield.

SYMPTOMS

Most characteristic symptom of the disease are boat and spoon-shaped malformed leaves, leaf crinkling, short internodes and a witches'-broom appearance of the twigs, downward curling and leaf deformation (Plate 10a). Boat-shaped leaves develop under temperatures ranging from 13/8°C to 23/18°C (day/night). Affected trees of Satsuma mandarin are stunted. Flecking of young leaves between the veinlets develop at lower temperatures. Exposure to temperatures above 28°C for about 12 hours per day will mask the symptoms of Satsuma dwarf disease. Severely affected trees bear small and malformed fruits but mildly affected trees bear normal fruits. Green patterns such as rings, mosaic, and blotches are developed on fruits of satsuma once the fruit has colored in citrus mosaic, a disease related to satsuma dwarf (Plate 10b). When Iyo or navel oranges are topgrafted onto affected trees, there is poor fruit set and fruits become malformed. In Japan, citrus mosaic virus, Navel orange infectious mottling virus and natsudaidai dwarf virus are serologically related and possibly are the strains of Satsuma dwarf virus (SDV) (Tanaka and Imada, 1976). Major symptoms of these diseases are enations, multiple flushing, stunting or dwarfing, reduction in number and size of leaves and shoots, shortened internodes, and small-sized fruits with thick peel. Fruits of severely affected trees are usually small, with a thick rind and poor taste. Fruit yield is much reduced and depend on infection of severe or mild strain (Iwanami *et al.*, 1993). Navel orange infectious mottling caused by one of Satsuma dwarf group viruses is characterized by formation of yellow mottling on leaves. Several new varieties of tangelo and tangor become dwarfed if affected by Satsuma dwarf (Shimizu and Miyoshi, 2003).

TRANSMISSION

Satsuma dwarf virus (SDV) is transmitted by grafting (Ushiyama, 1981; Yamaguchi, 1984). Mechanical inoculation by rubbing or knife slashing is successful not only to various herbaceous but also citrus plants (Kishi and Tanaka, 1964; Tanaka, 1972; Tanaka and Kishi, 1963; Tanaka and Imada, 1974). Soil-borne transmission has been confirmed by many experiments and observations in Japan (Izawa, 1966; Ushiyama and Ohgaki, 1970) but vector remains unknown. Field observations suggest that the disease is slowly transmitted from tree to tree presumably through soil, although no vector of the disease has been identified (Isoda and Gyoutoku, 1990). SDV is transmitted through seeds (8.6% transmission in *Phaseolus vulgaris* seeds).

HOSTS

Principal host of SDV is satsumas (*Citrus unshiu*) but most species of citrus are infected by grafting and a few by mechanical inoculation (Tanaka, 1972). Typical symptoms of the disease develop in satsuma mandarin but severe symptoms were also developed on navel cultivars of oranges (*Citrus sinensis*), possibly due to a new strain of SDV (Iwanami *et al.*, 1991). SDV infects nearly all of citrus and citrus relatives (Iwanami *et al.*, 1993; Miyakawa, 1968). The most sensitive citrus species is mandarin including satsuma. SDV also infects orange, lemon, Pummelo. Among citrus relatives, SDV also infects *Aegle marmelos, Aeglopsis chevalieri, Atalantia monophylla, Clymenia polyandra, Fortunella polyandra, Poncirus trifoliata, Swinglea glutinosa* (Iwanami *et al.*, 1993).

Hosts of Satsuma dwarf are: Citranges (Carrizo, Morton, Troyer), *Citrus excelsa, C. junos*, Clementine, Grapefruits, Eureka lemon, Lime (Kagzi, Key, Mexican, West Indian), Mandarin (Dancy, Cleopatra, King), Natsudaidai, *Poncirus trifoliata*, Rough lemon, Satsuma, Sour orange, Sweet oranges and Tangelos (Orlando). Other hosts like Etrog citron, Hassaku, Tahiti lime, Orlando tangelo, *C. latifolia, C. medica, Chenopodium amaranticolor, Cucumis sativus, Nicotiana glutinosa, Viburnum odoratissimum* are symptomless hosts and Citron (Etrog), Hassaku, Tahiti lime, Tangelo (Orlando), *Chenopodium amaranticolor, Cucumis sativus, Nicotiana glutinosa, Viburnum odoratissimum* are tolerant hosts to Satsuma dwarf. China laurestine (*Viburnum odoratissimum)* is a plant commonly used as windbreak for citrus in Japan is also susceptible to SDV (Koizumi *et al.*, 1988). It is thought that *V. odoratissimum* may act as a reservoir for Satsuma dwarf infection.

A few non-rutaceous and rutaceous species are susceptible by sap inoculation. These are: cowpeas (*Vigna unguiculata*), *Nicotiana tabacum, Petunia hybrida, Phaseolus vulgaris, Phaseolus aureus*, sesame (*Sesamum indicum*), *Chenopodium quinoa, Chenopodium capitatum, Vigna sinensis, Citrus unshiu and Physalis floridana* (Tanaka and Nakanishi, 1972; Usugi and Saito, 1977). Diagnostic symptoms on some mechanically transmitted host species are given below.

- ***Chenopodium quinoa*:** Chlorotic and necrotic local lesions develop on inoculated leaves.
- ***Physalis floridana*:** Systemic mottling and malformation of leaves.
- ***Sesamum indicum*:** Local lesions; systemic vein clearing, necrosis and leaf deformation.
- ***Vigna unguiculata*:** Local lesions; vein clearing and mottling, necrotic streaking of petioles and stems.
- ***Citrus unshiu*:** Stunting, leaf malformation and rosetting. Also used as indicator/assay hosts
- ***Satsuma mandarin*:** Stunting, boat- or spoon-shaped leaves (Plate 10a) develop after a few months of inoculation at low temperature.
- ***Sesame*:** Local lesions develop on mechanically inoculated leaves within a few days.

White sesame, satisfaction kidney bean and black eye cowpea are also used as indicator hosts. Symptoms in these hosts appear as: necrotic spots and streaks on leaves, leaves bent downward and become boat or spoon shaped. *Chenopodium quinoa* is a good local lesion host for virus assay. Virus can be propagated on *Physalis floridana* (Tanaka and Imada, 1974).

PHYSICAL PROPERTIES

Thermal inactivation point (TIP) in crude extract of *P. floridana* is between 50–55°C. Longevity *in vitro* (LIV) is 8–12 days at room temperature. Although the virus titer depends on host species, the decimal exponent (DEX) of the dilution end point is usually around 3–4.

THE PATHOGEN

Virus preparation contains few virions and 3 sedimenting component(s) in purified preparations (Usugi and Saito, 1979). Top component (T) consists of empty capsids. Middle component (M) and bottom component (B) contain genomic RNAs of buoyant densities in CsCl of about 1.43 and 1.46 g/cm^3, respectively. Virus particles are round with icosahedral symmetry and have a diameter of 26–28 nm (Saito and Hibino, 1962). SDV is best isolated from mesophyll and epidermal tissues. Virions are found in the cytoplasm. Infection with SDV results in formation of characteristic tubules containing single rows of virions in the infected cells (Hibino *et al.*, 1977). Intercellular inclusions and tubules containing arrays of spherical virus particles are found in leaf cells of satsuma and some herbaceous plants systemically infected with Satsuma dwarf virus (Saito and Hibino, 1962). SDV has been classified by ICTV as the type member of the genus 00.112.0.01. *Sadwavirus* in the family 00.018. *Comoviridae.*

GENOME

SDV genome is segmented and bipartite. Genome segements are distributed among 2 particle types of linear, positive-sense, single-stranded RNA (Iwanami *et al.*, 1996). All virions have two species of coat proteins with molecular weights of about 42,000 and 22,000 (Tanaka and Imada, 1974). Virions encapsulate two species of RNAs of about 7.0 kb and 5.4 kb both of which have a poly (A) sequence at the 3' end (Usugi and Saito, 1977, 1979; Iwanami *et al.*, 2001). Virions are encapsidated with two single-stranded, positive-sense RNAs, namely RNA1 and RNA2. Both RNAs are polyadenylated at the 3’-termini and encode a single polyprotein that is processed to yield mature proteins. Complete genome is 10100 nucleotides long and is divided among different particles, each segment encapsidated separately and the segments are distributed between 2 different types of particles. Proteins constitute about 63% of the particle weight (B), or 67% of the particle weight (M), or 100% of the particle weight (T). Viral genome encodes both structural and non-structural proteins. Virions consist of 2 structural protein(s) (Usugi and Saito, 1979).

Sequence analysis of a virus serologically related to *Satsuma dwarf virus* (SDV) from Hyuganatsu (*Citrus tamurana*) was determined. Sequences at the 3`-terminal were 1306 and 2160 nucleotides of RNAs 1 and 2 respectively. The partial RNA-dependent RNA polymerase region in RNA1 and the coat proteins (CPs) region in RNA2 of the virus tentatively named Hyuganatsu virus (HV) have 78.3–84.0% and 76.9–80.7% amino acid sequence identities to those of known SDV-related viruses (SDV-RVs), *i.e.*, SDV, Citrus mosaic virus, and Navel orange infectious mottling virus. Sequence analyses show that HV is a new SDV-RV species (Iwanami *et al.*, 1999, 2001; Ito *et al.*, 2004). Citrus mosaic virus (CiMV), Natsudaidai dwarf virus (NDV), and Navel orange infectious mottling virus (NIMV) are now considered to be strains of SDV, because they induce similar symptoms on satsuma mandarin and herbaceous hosts and share over 75% amino acid sequence identity with SDV (Iwanami *et al.*, 1999). SDV is serologically related to CiMV (Usugi *et al.*, 1986), NDV, and NIMV, but not to any other viruses. NIMV is more distantly related to SDV than to CiMV or NDV (Iwanami *et al.*, 1999). Molecular studies revealed close relationships among these viruses (Ito *et al.*, 2004; Iwanami *et al.*, 1999, 2001). These diseases will be described in following chapters.

DIAGNOSIS AND MANAGEMENT

In field affected trees diagnosis is done by observing spoon shaped leaves. White sesame (*Sesamum indicum*) is a good indicator host for SDV diagnosis. Sesame plants at the age of two-four-leaf stage are sap inoculated using 1/15 M phosphste buffer at pH 7.0 and maintained at 20–25°C. Local lesions will develop on inoculated leaves followed by subsequent necrosis and malformation on the upper portion of the plants. Blackeye cowpea and Satisfaction kidney bean can be used as alternative indicators (Tanaka and Kishi, 1963) showing chlorotic spots on inoculated leaves and systemic symptoms of mottling, vein clearing and necrosis.Tissue-graft inoculation onto citrus indicators is also used for virus diagnosis. Antisera to SDV are available, and ELISA is widely used in Japan for SDV, CiMV, and NDV indexing. ELISA with monoclonal antibodies is used to distinguish citrus mosaic strain from other SDV (Nozu *et al.*, 1986). DAS-ELISA, SDS-PAGE, and electroblot immune assay diversity are also used to detect Satsuma dwarf and related viruses (Iwanami *et al.*, 1993). A simple and rapid detection kit using an immunochromatographic assay (ICA) has been developed in Japan (Nario Kusano, 2006). The subject on progress on diagnostics since the first IOCV in 1957 has been reviewed by Lee and Garnsey (1996).

Satsuma dwarf and related viruses can be eliminated from citrus by heat-treatment (Ieki and Yamada, 1984) or by shoot-tip grafting.

Certification of budstock is the primary counter measures to control the disease. Citrus planting material should be certified free from SDV by methods such as those of the EPPO certification scheme for citrus (OEPP/EPPO, 1995). Propagation of budwood by topgrafting should be avoided. If affected trees are found, they should be immediately removed from the orchard. Soil should be fumigated with fumigated pesticides like chloropicrin and others (Isoda *et al.*, 1991).

Citrus Mosaic

A mosaic disease of citrus has been reported from Japan (Ishigai and Jinno, 1958). Mosaic disease in Japan is commonly occurring in Satsuma mandarin in most citrus growing areas of Japan (Yamamoto and Yamaguchi, 1980, Usagi et al., 1986). Mosaic diseases have also been reported from various parts of India (Ahlawat *et al.*, 1985; Bhagabati *et al.*, 1989; Pandey *et al.*, 1986; Dakshinamurty and Reddy, 1975; Naidu and Reddy, 1985). However, limited information is available on mosaic diseases reported from India and it is not known whether these diseases are similar to mosaic disease of citrus as has been reported from Japan. In 1995, a mosaic disease of citrus has been studied in India and a badnavirus was found to be associated with this disease (Ahlawat *et al.*, 1996). This disease in India has different virus morphology and other characters than mosaic disease in Japan and has been considered to be entirely different disease from Japanese mosaic and mosaic diseases earlier reported from India. To avoid confusion with citrus mosaic of Japanese citrus, the disease in India was renamed as citrus yellow mosaic in 13th IOCV meeting in China in 1995 because this disease produced yellow mosaic symptoms in most of the citrus hosts. Citrus yellow mosaic will be described in one of the following chapters of this book.

SYMPTOMS

In Japan, infected Satsuma trees show large diffuse blotches and mottling on leaves, with brown spots on the under surface of the leaf (Tanaka *et al.*, 1971). Similar symptoms also appear on the rind of fruit at the time of developing colour, near maturity or matured fruits (Usagi *et al.*, 1986). However, Indian scientists have reported different symptoms of mosaic in different cultivars and locations. In Maharashtra, affected mosambi sweet orange trees show irregular green/yellow patches on leaves (Pandey *et al.*, 1986). In Darjeeling hills, field infected Darjeeling orange (*Citrus reticulata*) trees show mosaic symptoms as severe mottling of leaves which is sometimes confused with greening symptoms. Initial symptoms appear as mild chlorosis on graft inoculated mosambi seedlings followed by severe mosaic pattern (Plate 11). On mature leaves, chlorosis was severe and the leaves become leathery in texture with wavy margin (Fig. 4). Affected trees are comparatively stunted (Ahlawat *et al.*, 1985).

TRANSMISSION

Mosaic disease of citrus in Japan is transmitted by graft and mechanical inoculations (Tanaka and Yamada, 1972; Tanaka and Imada, 1974). Field

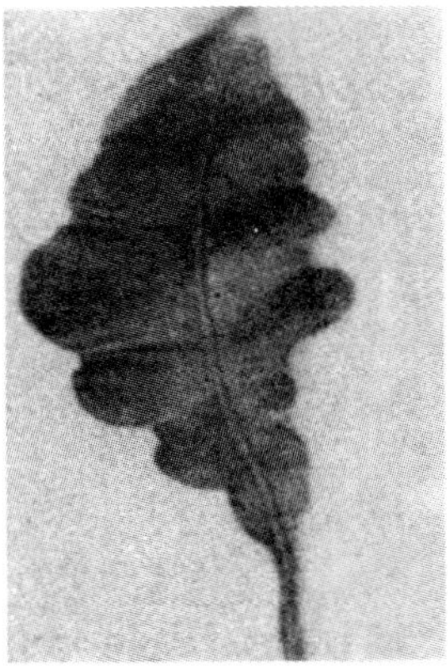

Fig. 4: Wavy margin of the leaf infected with citrus mosaic virus in India

observations in Japan suggested soil transmission of mosaic disease but no vector has been identified (Iwanami and Leki, 1996). Seed borne nature of the Indian mosaic has been suggested (Ahlawat *et al.*, 1985). In glasshouse, symptoms of mosaic disease on grafted/budded mosambi seedlings appear after 70–90 days. Comparatively higher transmission rate was observed when budwood collected from North-eastern part of India were scion grafted to sweet orange. Mosaic disease of citrus from South India has been transmitted by dodder laurel (Reddy *et al.*, 1985). In India, some isolates are transmitted by aphid vector, *Toxoptera citricidus* and also by sap inoculation (Pandey *et al.*, 1986; Dakshinamurty and Subbayya, 1981). The North-eastern region isolates of mosaic are transmitted by aphids, *Myzus persicae* and *Aphis craccivora* but not by *T. citricidus* (Ahlawat *et al.*, 1985).

HOSTRANGE

Japanese citrus mosaic has a very wide host range in citrus and in herbaceous plants as well (Usagi *et al.*, 1986). Hosts susceptible to Satsuma dwarf virus are also susceptible to mosaic virus in Japan. In India, citrus species/cultivars susceptible to mosaic infection are: Sweet orange (*Citrus sinensis*) cultivars mosambi, sathgudi, Washington navel; Indian acid lime (*C. aurantifolia*) cultivars kagzi nimbu; Rangpur lime (*C. limonia*), grapefruit (*C. paradisi*), mandarin (*C. reticulata*) cvs. Cleopatra mandarin, Nagpur santra, Sikkim or Darjeeling orange, Kinnow mandarin; lemon (*C. limon*) cv. Lisbon, pummelo (*C. grandis*) and bale (*Aegle marmelos*). Symptoms in these hosts appear as chlorosis of the leaves with clear mosaic pattern, extreme reduction of leaf lamina and wavy margin of the leaves (Fig. 4) (Ahlawat *et al.*, 1985). Host range of citrus mosaic virus in Japan is similar to Satsuma dwarf virus (Kishi and Tanaka, 1964).

THE PATHOGEN

Virus particles of Japanese citrus mosaic virus are polyhedrons measuring 28 nm in diameter. Purified preparation had three centrifugal components in sucrose density gradients (Tanaka and Imada, 1976; Usagi *et al.*, 1986). Although the virus is unclassified as yet but the properties of the virus indicate that the virus is similar to comoviruses and the nepoviruses (Bruening, 1977; Harrison and Murrent, 1977).

GENOME

Genome of Japanese citrus mosaic virus consists of two RNA species (RNA1 and RNA2) 7.0 kb and 5.4 kb (Iwanami *et al.*, 1993). RNA2 codes for coat proteins, which shows limited sequence similarity with other sequenced viruses at the 3′ region (Iwanami and Leki, 1996). Sequence of 3′terminal region of RNA1 has 2615 nucleotides and has a single ORF of 2401 nucleotides and a 3′untranslated region of 211 nucleotides upstream of the poly-A tail. Mosaic virus sequences showed maximum similarity (78%) with satsuma dwarf virus suggesting that citrus mosaic and Satsuma dwarf viruses are related to the viruses of the genera comoviruses and the nepoviruses (Iwanami and Leki, 1996). However, low similarity in sequences suggests that these two viruses are distinct.

Satsuma dwarf and citrus mosaic are known to be similar in biological, morphological, and serological properties (Tanaka and Yamada, 1972; Iwanami *et al.*, 1993; Usagi *et al.*, 1986). These viruses however, did not react serologically to several established comoviruses. The Darjeeling orange mosaic virus in India did not react with antiserum of Japanese citrus mosaic virus (Ahlawat and Chenulu, 1994) suggesting that mosaic diseases in India and Japan are caused by different viruses.

DIAGNOSIS AND MANAGEMENT

Mosaic disease is detected by grafting on to susceptible hosts or by sap inoculation to herbaceous hosts like sesamum. Japanese mosaic virus can be detected by serological tests like ELISA (Usugi and Tsuchizaki, 1982). Mosaic virus in Japan can be managed by same approaches as have been described for management of Satsuma dwarf. Details of detection of mosaic and its management have been described in "*Technical guidelines for the safe movement of citrus germplasm*" (Frison and Taher, 1991).

Navel Orange Infectious Mottling

Navel orange infectious mottling virus (NIMV) is known to infect citrus in Japan for a long time (Tanaka *et al.*, 1971). Symptoms of this disease consist of large diffused chlorotic blotches and mottling of leaves with brown spots on the undersurface. The virus is easily transmitted by graft inoculations from infected trees to citrus seedlings (Tanaka and Yamada, 1972). NIMV is also transmitted by mechanical inoculations to herbaceous hosts (Tanaka and Imada, 1974). Host range of citrus mosaic (Japan) and navel infectious mottling is similar to Satsuma dwarf virus described in earlier chapters. NIMV is considered as strain of Satsuma dwarf virus (Tanaka and Imada, 1976). The virus particles are isometric in shape measuring 23 nm in diameter which are smaller in size than that of citrus mosaic virus (28 nm). The genome of the virus is positive sense ssRNA. Citrus mosaic virus (CiMV), Natsudaidai dwarf virus (NDV) and Navel orange infectious mottling virus (NIMV) are considered to be distantly related strains of Satsuma dwarf virus (SDV) because:

1. Their host range is similar to that of "type" SDV.
2. They are serologically related although they form three different serogroups.
3. SDV protects Satsuma orange against infection of these viruses (except in case of NDV).
4. Sequence identities in the Large CP sequence is 81–85% between groups.
5. Sequence identity between NIMV and SDV is 82% in the proteinase-polymerase region.
6. Natural isolates of Satsuma mosaic virus have more than 90% sequence identity with each other in large CP as sequence and more than 94% in the RdRp region.

Vein Enation and Woody Galls

Vein enation was first observed in California on leaves of sour orange rough lemon and Mexican lime (Wallace and Drake, 1953). Woody galls were first described on rough lemon in Australia (Fraser, 1958). Vein enation and woody galls were later shown to be different symptoms of the same disease (Wallace and Drake, 1960, 1962). Now vein enation/woody gall disease is known to occur in many citrus growing countries, United States (California), South America (Peru, Brazil), Africa (South Africa, Kenya, Libya, Reunion, Tanzania), Oceania ((New South Wales, Australia), New Zealand, Kiribati) EPPO region (Spain, Libya and Turkey), Asia (China (Zhejiang)), India, Iran and Japan. This disease is also present in European Union countries wherever citrus is grown. Cooler coastal regions are more favorable to gall formation. In India, galls have been reported on Rangpur lime in Darjeeling and Sikkim hills, on Kagzi lime in Rajasthan (Chenulu and Ahlawat, 1994) and on rough lemon rootstock of nucellar Mosambi sweetorange in Maharashtra (Mali *et al.*, 1975, 1976; Mali and Chaudhuri, 1993) and in South India (Manjunath, 1987). Vein enation-woody galls disease is not of much economic importance for Indian citrus but reduced growth in young trees with extensive woody gall symptoms has been reported from Peru (Bazan de Segura and Ferrand, 1969).

SYMPTOMS

Symptoms that give the disease its name are the small enations (upto 1 mm) on the veins on under surface of leaves with corresponding depressions on the upper surface of leaves (Plate 12a) and galls on stem and branches of affected trees. Cauliflower-like swellings or galls of light grey colour develop on the bark of the stems, especially associated with thorns or wounds (Plate 12b) (Garnsey, 1988). Enations are not common on leaves of orchard trees. Enations are initiated as cytological abnormalities of phloem fibre primordial cells adjacent to protophloem sieve tubes (Hooper and Schneider, 1969). Woody galls develop from affected cells of procambial tissue between metaxylem and metaphloem of vascular bundles, and contain large amounts of abnormal xylem tissue.

Mesophyll and epidermal tissue on the abaxial side of leaf veins divide less prolifically and their growth ceases as leaves mature. Enations can also develop on thorns. In trees on rough lemon rootstocks, woody galls appear on the rootstock and can encircle the budunion. On other susceptible rootstocks, trees develop galls near graft/bud union. Cambial surface of gall tissue has a pitted appearance with numerous small holes.

Sprouts may develop on galls. Minor injuries to rough lemon and lime may initiate gall development.

TRANSMISSION

Vein enation-woody gall disease is bud and graft transmitted. It is also transmitted by dodder and by aphids, *Toxoptera citricida* (McLean, 1954), *Myzus persicae* (Wallace and Drake, 1959), *Aphis gossypii* (Laird and Weathers, 1961) and *Toxoptera aurantii* (Hermoso de Mendoza *et al*., 1993) in persistent manner. Latent period of the virus in vector, *T. citricida* is 2–3 days (Maharaj and da Graca, 1989). It is not seed transmitted.

HOSTRANGE

Host range of vein enation/woody gall virus is restricted to citrus species only. Vein enations on leaves of following inoculated host plants are observed: Citrons, Lemons, Mandarins, Sweet oranges, Rough lemon, Sour orange and West Indian lime but most severe vein enation symptoms are developed on rough lemon, West Indian lime and sour orange. Galls are formed on stems in hosts such as *Citrus volkameriana*, Rangpur lime, Rough Lemon, West Indian lime but severity of galls develops on *C. volkameriana*. Rough lemon (*Citrus jambhiri*), Sour orange and bitter sweet orange (*C. aurantium*) rootstocks. Other citrus species susceptible to woody galls include *C. junos*, citrons (*C. medica*), satsumas (*C. unshiu*) and various local Chinese varieties (Chen *et al*., 1992). In India, enations have been observed on grapefruit, *C. amblycarpa, C. macroptera, C. latipes* and *C. pennivesculata* but nearly all citrus species are susceptible and most of them remain symptomless (Mali *et al*., 1976). Other susceptible plant species in Rutaceae family are: Grapefruit (*C. paradisi*), Tangerine (*C. tangerine*), Mexican lime (*C. aurantifolia*), Rangpur lime, Palestine sweet lime, Rough lemon (*C. jambhiri*), Mandarins (*C. reticulata*), Fortunella and other citrus relatives and hybrids. More pronounced symptoms develop on sour orange and Mexican lime and less conspicuous on sweet orange and sour lemon (Wallace and Drake, 1953). In China, 11 citrus varieties and hybrids showed vein enation (CVEV) symptoms. CVEV was confirmed by topworking buds of virus-free Volkamer lemon onto field trees, and by indexing on seedlings of Rangpur lime, Bendizao mandarin, Volkamer lemon, rough lemon and Daidai sour orange (Chen Guoqing *et al*., 1996). Grapefruit appears to be a tolerant host.There is a report of synergistic effect between CVEV and the yellow vein pathogen resulting in a marked enhancement of the yellow vein symptom in Etrog citron, rough lemon and Mexican lime (Weathers, 1960). Vein enations are suppressed in Mexican lime and Pineapple sweet orange by CTV-T30 isolate, but not by other isolates, and are not suppressed in sweet orange by viroids (Vidalakis *et al.*, 2004).

INDICATOR HOSTS

- ***Indicator host for enation***: Kagzi lime or Small-fruited acid lime seedlings are used as indicator host for vein enaion. Vein enations are developed on underside of leaves within a few months after inoculation.
- ***Indicator host for woody gall***: Kagzi lime, *Citrus volkameriana* and rough lemon seedlings are good indicator hosts for woody gall symptoms. Galls on stems near thorns are formed in these hosts after a few months of inoculation. Injury near thorns can initiate early gall formation.

THE PATHOGEN

Isometric virus-like particle measuring 27 nm have been observed in electron microscope in the phloem cells of galls or vein enations (Iwanami *et al.*, 1992). Citrus vein enation virus is probably a member of Luteoviridae (Da Graca and Maharaj, 1991). Serological relationship of this virus to barley yellow dwarf (PAV and MAV strains) (Clark and de Graca, 2000) further suggested that this virus is a member of family Luteoviridae most closely related to cereal yellow dwarf-RMV and RPV strains now classified as Poleroviruses (Pringle, 1998). The associated virus has also been observed in salivary glands of viruliferous vector *T. citricidus* (Maharaj and da Graca, 1988).

DIAGNOSIS AND MANAGEMENT

Woody galls on stems and trunks and branches are diagnostic symptoms in field trees. Citrus vein enation-woody gall luteovirus is detected by grafting bark chips of suspected material on woody indicators. Enations are seen in 5–8 weeks after inoculation on underside of leaves of limes (*Citrus aurantiifolia*), rough lemons (*Citrus jambhiri*) or sour oranges (*Citrus aurantium*) at a temperature 24–27°C max. day/18–21°C min. night. Swellings or galls appear more slowly on stems of *C. jambhiri* or *C. volkameriana*. The degree of enation formation was higher in the spring shoots (20 to 88% of leaves) than in the autumn (0.5 to 19%). Number of enations varied from one to six per leaf. Bendizao and Zaoju varieties produce more enations than other varieties (Chen Guoqing *et al.*, 1996). A shock reaction in Mexican lime consisting of leaf droop is used in Spain as a diagnostic symptom. The virus can also be detected in electron microscopy. CVEV can also be detected by PAGE analysis of infected samples by observing double-stranded RNAs. CVEV has been detected in ELISA and ISEM using some commercial Barley yellow dwarf virus kits (Clark and da Graça, 2000) but attempts to amplify cDNA using luteovirus primers have not been successful, thus the luteovirus proposal remains unconfirmed.This disease can be managed by adopting practices such as to avoid the use of susceptible rootstocks and use clean planting material. Shoot-tip grafting and thermotherapy are effective to manage the disease (Roistacher and Kitto, 1977; Roistacher, 1979).

Tatter Leaf = Citrange Stunt

Tatter leaf disease of citrus was first discovered in 1962 on Meyer lemon trees (Wallace and Drake, 1962). Affected Meyer lemon trees were symptomless but when tissue grafts made from Meyer lemon to *citrus excelsa* plants, leaves of inoculated plants developed yellow blotches and edge malformation. Later it was reported that citrange and citremon also develop leaf and stem symptoms when inoculated with tatter leaf infected Meyer lemon (Wallace and Drake, 1963). Further investigations revealed that two viruses were involved in original Meyer lemon, tatter leaf causing symptom in *C. excelsa* and citrange stunt causing symptoms on citrange and citremon (Wallace and drake, 1968) but separation of two viruses could not be achieved by sequential passaging through a variety of citrus hosts (Roistacher *et al.*, 2000). Tatter leaf virus was therefore also known as Citrange stunt virus (Roistacher, 1988). Citrange stunt was first reported in 1968 as a new virus causing symptoms on citrange, but since it was eventually established that the two were one and the same, the original name 'citrus tatter leaf" was retained. Tatter leaf disease is wide spread in China (Zhang *et al.*, 1988), Japan, Korea Republic (Miyakava and matsui, 1976, Koizumi, 1989), Taiwan (Su and Chen, 1984), South Africa (Da Graca, 1977; Marais and Lee, 1986), and Australia (Fraser and Broadbent, 1979). It has also been reported from USA (California, Florida and Texas), Taiwan, Brazil, Thailand, and Philippines. Yield on infected mandarin (*Citrus reticulata*) trees on trifoliate orange (*P. trifoliata)* rootstock is 75% less than those of virus-free trees (Takahara *et al.*, 1988). This stock-scion combination also show symptoms of 'Bud-union crease', yellow ring groove, yellowing and decline of canopy (Miyakawa and Tsuji, 1988; Miyakawa, 1978; Zhang *et al.*, 1988). Trifoliate orange is very valuable rootstocks for citrus but presence of tatter leaf disease limits the use of trifoliate orange or its hybids wherever the disease is endemic (Roistacher *et al.*, 2000).

SYMPTOMS

Trees on susceptible rootstocks (Trifoliate orange and its hybrids) are stunted with chlorotic foliage. Seedlings of Mexican lime (*Citrus aurantifolia*) and Kalpi lime (*C. excelsa*) showed young leaf symptoms as chlorotic mottle and tearing-like or tattering of the margin of leaves. Based on these symptoms the name tatter leaf was assigned to the disease. Leaf symptoms may develop within 4–6 weeks after bud inoculation. When infected latent hosts are grafted on rootstocks of *P. trifoliata* or its hybrids, a bud-union crease, showing a yellow to brown line can be observed 1 year after grafting which is visible when the bark is removed. Affected plants bloom heavily and fruits on such trees mature early but often die. Distortion of fruit rind in *Cleopatra mandarin* has been associated with this disease (Lovisolo *et al.*, 2003). Citrus tatter leaf virus (CTLV) produces striking leaf symptoms and stunting in citremon (*Poncirus trifoliata* × *C. limon*) and Rusk and Troyer citranges (*P. trifoliata* × *C. sinensis*) (Wallace and Drake, 1963). Ringspot symptoms have also been observed on infected citrumelo leaves. Citrange leaves develop chlorotic spots, mottling and distortion (Plate 13c). Citrus tatter leaf virus (CTLV) infected Satsuma mandarin trees on troyer roots develope vertical corrugations on the stock and a groove at the bud union followed by decline of the tree (Calavan *et al.*, 1963). Bud-union crease develops one year after CTLV inoculation in sweet orange plants on trifoliate orange rootstock (*Poncirus trifoliata*). Stem of citrange plants may be deformed and have a zigzag growth pattern (Plate 13a) and develop chlorotic areas on stem (Plate 13b). Citremon, Citranges and citrange hybrids infected with CTLV often develop stem pitting symptoms like citrus tristeza virus although tatter leaf symptoms are apparently different from known citrus viruses (Garnsey, 1974).

TRANSMISSION

Tatter leaf virus is transmitted from citrus to citrus by grafting and mechanical inoculation. Mechanical transmission by knife slashes and leaf-abrasion is easily achieved from infected *Nicotiana clevelandii* to citron (Garnsey, 1974) and from citron to citron (Roistacher *et al.*, 1980). CTLV is seed transmitted in *Chenopodium quinoa,* cowpea, lily and soybeans seeds (Nishio *et al.*, 1982). It appears to be seed transmitted in Eureka lemon (Tanner *et al.*, 2010). No natural vector is known. Transmission by cutting or pruning tools may have occurred in Texas from a Meyer lemon tree to neighboring Eureka lemon and Cleopatra mandarin trees (da Graça and Skaria, 1996).

HOSTRANGE

Susceptible hosts of Tatterleaf-Citrange stunt virus are: *Citrus excelsa,* Citranges (Carrizo, Rusk, Troyer), Citremons (1448, 1449), Satsuma or sweet orange on *P. trifoliata* roots. Clearly visible and severe symptoms develop on citranges. *Aeglopsis chevalieri,* Lemons, Rough lemon, *Severinia buxifolia,* Sour orange, Sweet orange and West Indian lime are symptomless hosts of tatter leaf virus. CTLV also infects herbaceous hosts like *Chenopodium quinoa, Phaseolus vulgaris, Vigna sinensis (cowpea)* cv. Blackeye and Masterpiece bean. Tahiti lime is tolerant to this virus. Chlorotic leaf symptoms are produced in Mexican lime, *C. excelsa,* citranges and other trifoliate orange hybrids. *C. excelsa* leaves can initially be deformed (hence the name 'tatter leaf'), but infected plants often recover and remain symptomless. Citrange, citremon and citrumelo stems develop a 'zig-zag' growth pattern. The virus is irregularly distributed in citranges, and often cannot be recovered from symptomless areas.

Host range of CTLV is wide and susceptible hosts are both in citrus and herbaceous plant species (Calavan *et al.*, 1963; Garnsey and Drake, 1968; Miyakawa, 1980b; Miyakawa and Matsui, 1976; Wallace and Drake, 1968; Marais and Lee, 1986). These hosts are:

Citrus spp and hybrids

Citrus aurantifolia, C. excelsa, C. limon, C. medica, C. sinensis, P. trifoliate, C. sinensis × *Poncirus trifoliata, P. trifoliata* × *C. paradisi cv. Swingle, P. trifoliata* × *C. limon* are susceptible hosts. Reactions of CTLV to various susceptible hosts are given below:

- *Citrus limon* cv. Meyer - tatter leaf; blotchy and malformed leaves.
- *C. excelsa* (Kalpi lime)- tatter leaf or symptomless.
- *C. sinensis* - symptomless.
- *Poncirus trifoliata* - interveinal chlorosis.
- *Citrus sinensis* × *Poncirus trifoliata* cv. Rusk, *Poncirus trifoliata* × *Citrus paradisi* cv. Swingle - leaf malformation and twisted stems.

CTLV is also mechanically transmitted to herbaceous hosts (Nishio *et al.*, 1982; Tanaka and Imada, 1974). 19 species in nine families are susceptible to CTLV (Semancik and Weathers, 1965). These hosts are: *Amaranthus tricolor, Catharanthus roseus, Chenopodium amaranticolor, Cucurbita pepo, Dianthus barbatus, D. chinensis, Faba beans, Gomphrena globosa, N. debneyi, Petunia hybrida, Tetragonia tetragonioides, Vigna unguiculata* peas, tomatoes, and soybeans. Tolerant (symptomless carriers) hosts are: Sweet orange, sour orange, grapefruit, lemon and mandarin. Cleopatra mandarin may develop fruit symptoms. Other plants such as *Catharanthus roseus* (periwinkle), *Petunia axillaris, Zinnia elegans, Pisum sativum, Glycine max* and *Crotalaria spectabilis* are symptomless hosts. Symptoms produced on inoculated herbaceous hosts are given below:

- *Cowpea cv. Early Ramshorn*: shows diffuse necrotic lesions 3–4 days after inoculation.
- *Chenopodium quinoa* and *C. amaranticolor*: local lesions and systemic mottle
- *Ipomea purpurea* and *Phaseolus vulgaris*: necrotic local lesions
- *Nicotiana clevelandii*: etched ringspot, systemic mottle, vein clearing

- *N. glutinosa*: systemic mottle
- *Wild Lily*: yellowing symptoms

INDICATOR HOSTS

- ***Citrus excelsa seedlings***: Symptoms of deformed leaves with chlorotic zones develop within a few months after inoculation.
- ***Rusk citrange seedling***: Deformed leaves with chlorotic leaf blotch and stem with zigzag growth pattern develop within a few months after inoculation.
- ***Chenopodium quinoa***: Chlorotic local lesions and systemic chlorotic mottle develop in inoculated plants within a few days.

Virus culture can be multiplied and maintained on *Citrus excelsa, Nicotiana clevelandii, N. glutinosa, Vigna unguiculata*. This virus is assayed on either *Citrus excelsa* or on *Chenopodium quinoa,* a local lesion host.

THE PATHOGEN

Virus particles of tatter leaf are filamentous and flexuous with a clear modal length of 650 nm and 19 nm wide with cross banding pattern and with a helical construction of 3.4 nm pitch (Semancik and Weathers, 1965; Miyakawa and Matsui, 1976). Recent experiments suggest that this virus is probably a closely related strain of apple stem grooving *capillovirus* rather than a distinct virus (Ohira *et al.*, 1994). CTLV genomic RNA has been sequenced, and shown to be indistinguishable from apple stem grooving virus type member of genus 00.056.0.06. *Capillovirus* in family 00.056. *Flexiviridae* (Ohira *et al.*, 1995; Yoshikawa *et al.*, 1993).

CTLV has been isolated from wild lily with yellowing symptoms in Japan and virus particles have been observed in cells as banded aggregates (Inouye *et al.*, 1979). Crystallized virus bodies have been observed in phloem cells of citrange and mandarin, and mesophyll cells of *Chenopodium quinoa* and *Nicotiana clevelandii* (Chen *et al.*, 1995).

GENOME

Genome of CTLV is monopartite with only one particle size of linear, positive-sense, single-stranded RNA. Genomic RNA is partially sequenced and sequenced region is 2956 nucleotides long. RNA species has a molecular weight of 2.83 × 106 Da and produces a single protein band of molecular weight 27 × 103 Da in SDS-PAGE (Nishio *et al.*, 1989). Complete genome sequence of CTLV from Meyer lemon has also been determined to be 6,495 nucleotides (nts), with two overlapping open reading frames (ORFs) and a poly (A) tail at the 3' end. Genome organization is similar to other capilloviruses, with ORF1 (nts 37 to 6,354) encoding a putative 242-kDa polyprotein which contains replication-associated domains plus a coat protein (CP), and ORF2 (nts 4,788 to 5,750), which is located within ORF1 in a different reading frame and encodes a putative movement protein. Although the proteins encoded by CTLV of Meyer lemon possesses 84 to 96% amino acid sequence identity with strains of Apple stem grooving virus (ASGV), it is observed that two strikingly different regions in ORF1: variable region I (amino acids 532 to 570) and variable region II (amino acids 1,583 to 1,868) exist with only 15 to 18 and 56 to 62% identities, respectively, with the corresponding regions of ASGV strains (Satyanarayana *et al.*, 2009).

DIAGNOSIS AND MANAGEMENT

Bud union crease in trees on trifoliate orange and trifoliate hybrid rootstocks are the symptoms useful for field diagnosis of CTLV. Seedlings of Rusk citranges are used as indicators for citrus tatter leaf virus. Rusk citrange is budded on virus-free seedlings of rough lemon rootstock and the citrus tissues to be tested are also budded below the citrange bud. Optimum temperatures for symptom development should be 20–24°C (Miyakawa, 1978). Sap inoculation to *Chenopodium quinoa* is also used for detection and assay of virus which develops chlorotic or necrotic spots on the inoculated leaves, and its upper leaves show vein clearing, twisting and stunting. *Vigna unguiculata* is also used as diagnostic host, but symptoms vary markedly depending on virus isolate (Iwanami *et al.*, 1991). ELISA is used for detection of CTLV using polyclonal and monoclonal antibodies (Kawai and Nishio, 1990; Kawai *et al.*, 1996). By using PAGE, two typical dsRNA bands are observed in infected samples (Frison and Taher, 1991). PCR-based techniques have also been used for CTLV diagnosis (Hailstones *et al.*, 2000) and the virus can be detected in multiplex PCR with several viroids (Ito *et al.*, 2002). A real time RT-PCR assay using SYBR green has also been used to detect and quantify CTLV RNA-targets (Detection of citrus tatter leaf virus by real-time RT-PCR (Liu *et al.*, 2010).

Tatter leaf virus is managed by using virus-free budlines for propagation. *Citrus depressa or C. reshni* provide good results when used as rootstocks for CTLV-infected mandarins (Takahara *et al.*, 1988). CTLV cannot be eliminated by shoot-tip grafting alone (Roistacher and Kitto,

1977). Heat treatment for 30 days at 35–40°C/30°C (day/night) followed by shoot-tip grafting (STG) can be an effective therapy (Koizumi, 1984). Incubation of budsticks on synthetic medium *in vitro* for 10–14 days at 32°C, followed by shoot-tip grafting can produce CTLV-free plants with 30–50% success (Navarro *et al.*, 1989). Long-term heat treatment of affected plants for 90 or more days at 40°C/30°C (day/night) can also eliminate CTLV (Miyakawa, 1980a). In China, plants of citrus cultivars, Tanken, Xuegan, Gailian, Cheng orange and Taiwan Tanken are subjected to thermotherapy by alternating temperatures of 40–50°C for 8 h per day for 14–56 days followed by STG to eliminate CTLV infection. This process eliminates tatter leaf virus upto 82.6% (Wu *et al.*, 1996). Since it is mechanically transmissible, cutting and pruning tools need to be sterilized with sodium hypochlorite.

Indian Citrus Ring Spot

Indian citrus ringspot disease was originally recorded from kinnow mandarin trees at the Punjab Agricultural University Campus at Abohar in India by the author in 1985. Affected trees appeared yellow from a distance. Budwood from trees showing rings were graft inoculated to Mosambi sweet orange. Inoculated sweet orange plants showed symptoms like psorosis shock reaction, young leaf flecking symptoms and ringspot symptoms on matured leaves. Therefore, based on these symptoms, this disease was tentatively considered to be a strain of psorosis-A disease (Ahlawat, 1989). However, none of the affected trees in the field showed bark lesion symptoms –a characteristic symptom of psorosis-A. In inoculated plants, young leaf flecking symptoms persisted for longer period and did not disappear like psorosis young leaf symptoms within a few days suggesting that ringspot disease in India may be different from psorosis. Further, studies revealed that the ringspot disease in India is not related to psorosis (Byadgi *et al.*, 1993; Praveen *et al.*, 2003). During successive surveys, author came across to ringspot symptoms in over 90% Kinnow mandarin trees –a hybrid (Willow X king mandarins) in several orchards in or near Abohar in Punjab and on Mosambi sweet orange in Maharashtra and in satgudi sweet orange in Andhra Pradesh. Incidence of ringspot disease was up to 100% in most Kinnow mandarin orchards in Northern India, especially Punjab, in sweet orange 20–50% in Maharashtra, Andhra Pradesh and Karnataka, in rootstocks, Sohsarkar (*Citrus karna*) and Karna khatta, incidence of this disease was around 55%, but in trees on Troyer citrange roots, incidence was only 7% (Ahlawat and Pant, 2003). There is no report of this disease outside India so far but could be present. This disease is now known to be present in several states of India such as Delhi, Maharashtra, Punjab, Haryana, Karnataka, Uttar Pradesh and Andhra Pradesh. Interestingly, Indian citrus ringspot virus (ICRSV) infection was also recorded on lemon and Khasi mandarin in a place known as Digaru in Assam state in Northeast India (Bhagabati *et al.*, 1989). Flexuous filamentous virus-like particles were found in leaf extracts of graft-inoculated Mosambi sweet orange (Byadgi *et al.*, 1993) and the disease was further studied by Byadgi and Ahlawat (1995).

Field affected trees show typical yellow rings in most of the leaves and such trees look as ornamental trees (Plate 14a). The leaves on such trees had bright yellow rings. Indian ring spot disease was found in all the orchards which were surveyed from 1990 onwards in Punjab and other places in India. Although trees in several orchards show 100% infection but as an average the incidence of ringspot disease ranged from 5–83.8% and yield loss was estimated from 20.5 to 98.38 % (Byadgi and Ahlawat,

1995; Thind and Arora, 1997; Hoa and Ahlawat, 2004). Fruit quality on affected trees is reduced with less juice and TSS (Thind *et al.*, 1998). Based on detailed studies on ringspot disease and the virus associated with it, this disease was named as *Indian citrus ringspot disease* (Byadgi and Ahlawat, 1995) and the causal virus as *Indian citrus ringspot virus* (ICRSV) (Pant and Ahlawat, 1998). Although the origin of the disease is not known but it appears that a contaminated planting material of kinnow mandarin was introduced in the country. This disease has been published by the names of "citrus ringspot" and "psorosis-A" in the past but based on detailed studies, it is now popularly known as Indian citrus ringspot disease (ICRSD). It is also important to distinguish ICRSD from another ringspot diseases known to occur in North America and Europe which will be described in following pages.

SYMPTOMS

Natural infection of ICRSV has been observed in sweet oranges (Malta, Mosambi, Satgudi (Chini), mandarins (Nagpur, Sikkim, Coorg and Kinnow), Kagzi lime and Kagzi Kalan (*Citrus aurantifolia*). Leaves of affected trees showed typical symptoms as complete or broken rings with yellow margins and green centre (Plate 14b) ranging from mere dots to 4–18 mm in diameter (Byadgi *et al.*, 1993). Rings may be one to several on a leaf and their size may be variable. Fruits on affected trees remain smaller with less juice. In some affected trees, fruits also show streaks on green or ripe yellow fruits. Trees with severe infection show decline but they do not die. Affected trees become unprofitable. Experimentally inoculated sweet orange plants show vein clearing or flecking symptoms during the first spring flush, 2–4 months after graft inoculation (Plate 15b). However, inoculated sweet orange plants also showed shock reaction as rolling of young leaves and drooping of young shoot during the same spring (Plate 15a). Vein clearing symptoms persisted even after the leaves become mature. Matured leaves of inoculated plants show chlorotic spot or rings persist till the abscission of such leaves (Plate 15c) (Ahlawat, 1989; Byadgi *et al.*, 1993; Hoa and Ahlawat, 2004). Young leaf and mature leaf symptoms develop every year on inoculated plants during winter (from December to February). In field affected trees, symptoms develop only during winter season in north India.

TRANSMISSION

Indian citrus ringspot disease is transmitted by wedge, side and leaf patch grafting. Approach grafting is also used for transmission but not routinely.

Ring spot disease could not be transmitted by several aphid species (*Aphis gossypii, A. craccivora, A. citricola* and *Myzus persicae*) when tested by both persistent and non-persistent manners. However, natural transmission of the disease has been observed on kinnow mandarin seedlings in the vicinity of infected trees (Ahlawat, 2007) suggesting that a vector is involved. This disease is transmitted by dodder, *Cuscuta reflexa* from citrus to citrus (Byadgi and Ahlawat, 1995) and also by mechanical transmition from citrus to some herbaceous hosts (Pant *et al.*, 2000). It is not seed or soil transmitted (Byadgi and Ahlawat, 1995). No seed transmition was observed in seeds of citrus varieties tested and in *Phaseolus vulgaris* (Hoa *et al.*, 2004; Rutici *et al.*, 2000). However, virus particles of ICRSV were detected in pollens collected from ringspot infected trees in immunosorbent electron microscopy (ISEM) tests suggesting the possibility of pollen transmission of this virus (Pant and Ahlawat, 1998). The virus was detected in seed coat but not in embryo or endosperm (Prabha, 2010).The virus may have been present in the original Kinnow mandarin mother trees from where it might have spreaded widely through vegetative propagation.

HOSTRANGE

Citrus and several of its relative and hybrids are susceptible to Indian citrus ring spot virus (ICRSV) by graft inoculation. Typical symptoms of ringspot develop on Kinnow mandarin, Sweet oranges (Mosambi, Pineapple, Malta and Satgudi), (Plate 16a, b, c) Nagpur mandarin and Rough lemon (Plate 16d, e). Sour orange, khasi or Darjeeling mandarin, Rangpur lime, Kagzi lime and Kagzi kalan (*C. aurantifolia* (Christen) Swingle) and Karna khatta (*C. karna* Raf.) are also susceptible and are easily infected by graft-inoculation (Hoa and Ahlawat, 2004; Hoa *et al.*, 2004). Symptoms in these hosts varied from shock reaction to leaf flecking, vein banding, chlorotic spots and ring spots. Diagnostic symptoms developed on citrus and herbaceous hosts under experimental inoculation are described below (Byadgi *et al.*, 1993; Pant *et al.*, 1997, 2002; Hoa and Ahlawat, 2004).

Symptomatic Citrus species susceptible to ICRSV

- ***Kinnow mandarin*** (Willow X King mandarins): Young leaf flecking and chlorotic and yellow ring spots with green centre (Plate 16b)
- ***Malta, Satgudi and Mosambi sweet oranges*** (*C. sinensis*): Shock reaction, leaf flecking, chlorotic and ring spots (Plate 16a, c)
- ***Darjeeling orange, Nagpur orange*** (*C. reticulata*): Necrotic spots (Plate 16d)

- ***Calamondin*** (*Citrus limonia*): Chlorotic and ring spots on mature leaves.
- ***Etrog citron*** (*C. medica*): Chlorotic spots
- ***Indian wild citrus*** (*C. indica*): Chlorotic spot
- ***Kagzi kalan*** (a selection of acid lime): Chlorotic spots
- ***Kagzi lime*** (Key lime, Mexican lime) (*C. aurantifolia*): Leaf flecking, chlorotic spot
- ***Sour orange*** (*C. aurantium*): Vein banding
- ***Lemon*** (*C. limon*) *and* ***Grapefruit*** (*C. paradisi*), ***Pummelo (Shaddock)*** (*C. grandis*), ***Rough lemon*** (*C. jambhiri*) and ***Sweet lime*** (*C. limetta)*: chlorotic spots or Latent infection (Plate 16e)

Herbaceous hosts susceptible to ICRSV

- ***Chenopodium amaranticolor and C. quinoa***: Necrotic local lesions are developed on inoculated leaves.
- ***Phaseolus vulgaris cv Saxa, Singtami, gheusemi, alapatri*:** Local lesions with dark brown margins, followed by vein clearing, browning and necrosis of leaf veins and tissues, mottling of leaves (Plate 17a, b), smalling of leaves and blight.
- ***Glycine max (Soybean cv Hodgson)*:** Local necrotic lesions.
- ***Vigna unguiculata (Cowpea cv Black)*:** Local lesions
- ***Vicia faba major*:** Necrosis of leaf veins and blight.

Typical symptoms of ring spot disease in herbaceous hosts develop at an optimum condition of 28°C day temperature and 18°C night temperature with a relative humidity of 75%. Culture of the virus is best maintained in Kinnow mandarin and Mosambi sweet orange but can be multiplied in French bean Cv. Saxa for purification. Assay hosts are *C. amaranticolor, C. quinoa* and saxa bean.

Following plant species appear resistant: *Cucumis sativus* (cucumber, cv 'Marketer'), *Cucurbita pepo* (zucchini, cv 'Genovese'), *Datura stramonium, Gomphrena globosa, Nicotiana benthamiana, N. clevelandii, N. glutinosa, N. megalosiphon, N. tabacum* (cv 'White Burley'), *Petunia hybrida, Phaseolus coccineus* (runner bean, cv 'di Spagna') and *Pisum sativum* (garden pea, cv 'Nano') (Hoa and Ahlawat, 2004; Rustici *et al.*, 2000).

PHYSICAL PROPERTIES

Dilution end-point of ICRSV is between 10^{-4} and 10^{-5}, longevity *in vitro* is 8h at room temperature and thermal inactivation point is between 65°C and 70°C.

THE PATHOGEN

A filamentous virus associated with Indian citrus ringspot virus was first reported by Ahlawat and Chakraborty (1990). Purified preparations showed three types of particles in electron microscope (Fig. 5). The major filamentous and flexuous particles were with a modal length of 650 nm with clearly visible cross banding and the virus was tentatively identified as capillovirus (Byadgi *et al.*, 1993; Byadgi and Ahlawat, 1995; Pant and Ahlawat, 1998; Ahlawat *et al.*, 1997). Other components in purified preparation were highly flexuous particles measuring 690 × 9 nm and tubules measuring 2250 × 40 nm (Byadgi *et al.*, 1993). A tubular component is always co-purified with the virions when purification is done from citrus

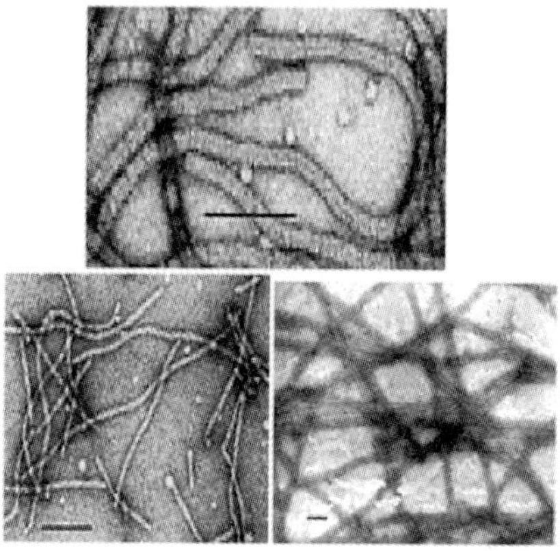

Fig. 5: Purified preparation showing three types of particles: capillo-type, thin particles and tubules

plants infected with ICRSV (Byadgi and Ahlawat, 1995; Byadgi *et al.*, 1993) but its role is not known. Both these components are eliminated if the virus is passaged through *Phaseolus vulgaris*. However, flexuous particles and tubular particles were later found to be the host components (Rutici *et al.*, 2000). Main virions associated with ICRSD are of 650 nm modal length and about 13 nm in diameter, are flexuous rods with a distinct zipper-like structure (Fig. 6). In thin sections of systemically infected 'Saxa' bean leaves, abundant virus particles are seen by EM in the cytoplasm (Rustici *et al.*, 2000). A260/280 ratio of purified preparion was 1.216. Indian citrus ringspot disease is now known to be caused by *Indian citrus ringspot virus*, the only current member of the genus *Mandarivirus* in the family *Flexiviridae* (Adams *et al.*, 2004, 2005). Further, based on serological and molecular studies, Indian cirus ringspot showed some affinities to potex, carla, fovea and allexiviruses but does not fall into any known genus and hence was proposed to be a new species. The virus particles are relatively robust, easily purified, and highly immunogenic. ICTV classified Indian citrus ringspot virus in family 00.056 *Flexiviridae,* genus 00.056.0.02 *Mandrivirus,* virus code 00.056.0.02.001 and type species *Indian citrus ringspot virus* (Fauquet *et al.*, 2005).

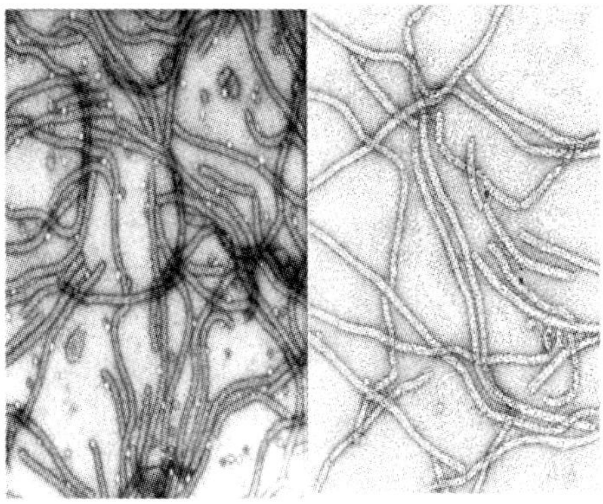

Fig. 6: Purified preparations of ICRSV particles. *Left*: from French bean cv saxa, *Right*: from glasshouse inoculated citrus

GENOME

ICRSV genome is positive sense single stranded RNA of about 7.5 kb without poly A tail and with a coat protein (CP) of 34 kDa. ICRSV particles have 6% nucleic acid (Byadgi and Ahlawat, 1995). The viral genome was fully sequenced which consists of 7560 nucleotides. It contains six open reading frames (ORFs) which encode putative proteins of 187.3, 25, 12, 6.4, 34 and 23 kDa respectively. ORF1 encodes a polypeptide that contains all the elements of a replicase; ORFs 2, 3 and 4 compose a triple-gene block; ORF5 encodes the capsid protein; function of ORF6 is unknown. Phylogenetic analysis of the complete genome and each ORF separately, and database searches indicate that ICRSV, though showing some similarities to potexviruses, is significantly different, as in the presence of ORF6, the genome and CP sizes, and particle morphology (Rutici *et al.*, 2000a). As a result a new species, *Indian citrus ringspot virus* (ICRSV) in the genus, *Mandarivirus,* within the family *Flexiviridae* was proposed and was finally approved by ICTV(Rustici *et al.*, 2002; Adams *et al.*, 2004; Hoa and Ahlawat, 2004; Adams *et al.*, 2005). Two ORFs encoding putative proteins of 34 kDa and 23 kDa were located in the 3' part of the RNA. The product of the 34 kDa ORF was confirmed as CP by expression in *E. coli.*

Hoa and Ahlawat (2004) have described several isolates of ICRSV. Analysis showed some variation near the N terminus of the coat protein but conservation in the core region. Isolates from Delhi (ICRSV-Dl), Abohar

(ICRSV-Ab), Ahmedabad (ICRSV-Ah) and Pune (ICRSV-Pu) were compared with the original isolate (ICRSV-D, Byadgi *et al.*, 1993; Rustici *et al.*, 2000, 2002). Isolates Ab and Dl were closely related to D (98% and 99.5% sequence identity, respectively). Isolates Ah and Pu were similar to each other (99.6% identity) but showed only 85% and 84.5% identity, respectively, to isolate D. All the isolates were serologically similar. When graft-inoculated to a panel of citrus cultivars (Kinnow, Mosambi, Malta, Rough lemon and King mandarin), the different isolates induced various combinations of vein flecking, vein clearing, chlorotic spots, ringspots and necrotic spots, sufficient for initial identification of the isolates (Hoa and Ahlawat, 2004). Coat protein genes of all the four isolates have been cloned and sequenced (Accession No. AY255006, AY255007, AY255008, AY255009). The analysis of both nucleotide and amino acid sequence levels when compared to earler report of Rutici *et al.* (2000), variation occurred at N terminus and conserved in the core region of the CP. The nucleotide and aminoacid sequence identity matrix and phylogenetic tree showed that the ICRSV-D and ICRSV-Ab shared close relation to the earlier reports. However, ICRSV-Ah and ICRSV-Pu were distantly related to earlier reports but the two are closely related to each other suggesting strainal variation in ICRSV (Hoa and Ahlawat, 2004). A graft transmissible yellow ringspot disease is known to occur on sweet orange and mandarin varieties in Spain, Mediterranean countries and Australia but no information is available on its etiology (Timmer *et al.*, 2000). Hence, no comparison can be made with ICRSV.

DIAGNOSIS AND MANAGEMENT

Diagnosis and management of ICRSV has been reviewed from time to time (Ahlawat *et al.*, 1995; Ahlawat and Pant, 2008). In Field trees, ICRSV is easily detected by yellow rings on leaves especially during winter period. A diagnostic panel of citrus species comprising of Kinnow mandarin, Mosambi sweet orange, Malta sweet orange, Rough lemon and King mandarin showed variable symptoms upon graft inoculation. The leaf symptoms on diagnostic panel were: vein flecking, vein clearing, chlorotic spots, ringspots and necrotic spots. This diagnostic panel can easily distinguish between strains of ICRSV (Hoa and Ahlawat, 2004). ICRSV is easily detected by sap inoculation on indicator plants such as *Chenopodium amaranticolor* and *C. quinoa* which react with local lesions. Systemic hosts like Kinnow mandarin and Mosambi sweetorange can also be used for disease diagnosis by grafting with test scions. Inoculated plants will show a shock reaction followed by young leaf vein flecking and ring spots on mature leaves. Virus particles can be seen in electron microscope. Polyclonal antibodies have been developed against ICRSV and used for virus detection in ELISA, ISEM (Fig. 7), and dot immunobinding assay (DIBA) (Fig. 8) (Pant and Ahlawat, 1998; Rutici *et al.*, 2000; Hoa and Ahlawat, 2004). Nucleic acid hybridization (Fig. 10) and western blotting are also used to detect ICRSV (Byadgi and Ahlawat, 1995; Hoa and Ahlawat, 2004; Hoa *et al.*, 2004; Ahlawat and Pant, 2002). Titre of ICRSV antiserum in tube precipitin test was 1:1024 and for EM decoration it was 1:3200. In DAS-ELISA, concentration of coating antibodies as 1 and 2 μg/ml and conjugate as 1:500 provided better results (Byadgi and Ahlawat, 1995). However, another lot of antiserum gave better results at a conjugate dilution of 1:2000 (Hoa and Ahlawat, 2004). Titre of antiserum in gel diffusion test was 1/64 and in EM decoration test (ISEM) the titre was 1/4096. Digoxigenin-labeled cDNA and RNA probes both complementary to the sequence of ICRSV detected it in infected plants (Rutici *et al.*, 2000). Virus specific primers have been designed and used for virus diagnosis including its strains (Dl, Pu, Ab, Ah) in RT-PCR (Figs. 9, 11)(Hoa and Ahlawat, 2004; Hoa *et al.*, 2004).

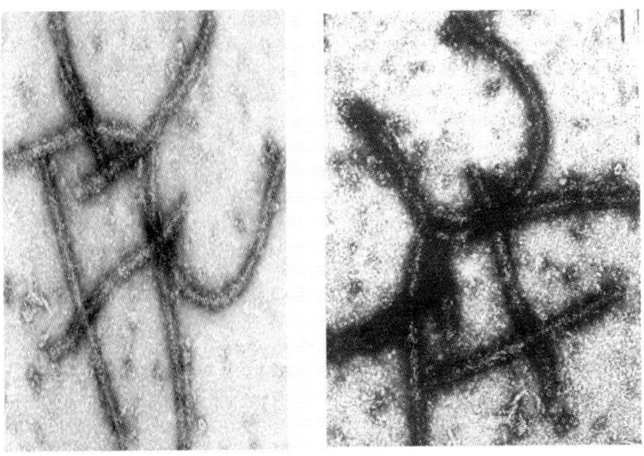

Fig. 7: ISEM decoration test: *Left*: antibodies developed from citrus preparations, *Right*: antibodies developed from ICRSV inoculated French bean

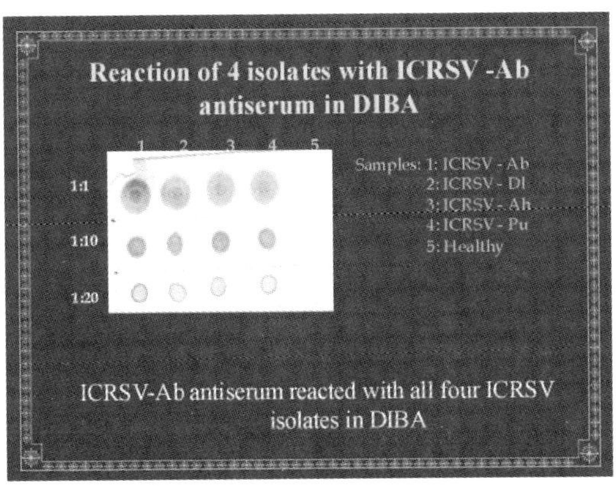

Fig. 8: All the four isolates of ICRSV detected by DIBA using antibodies from Ab isolate.

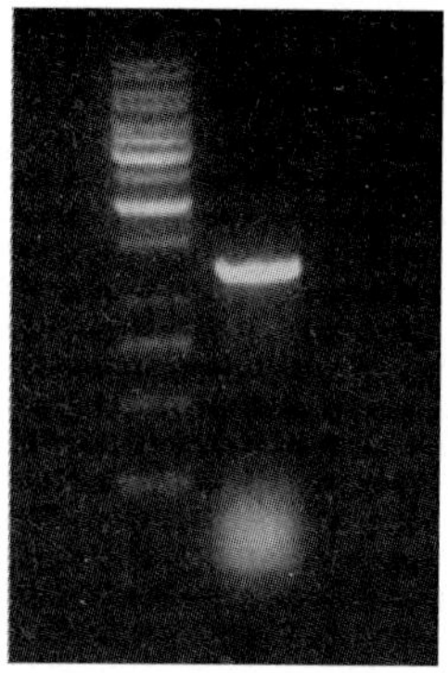

Fig. 9: PCR detection of ICRSV with virus specific primers

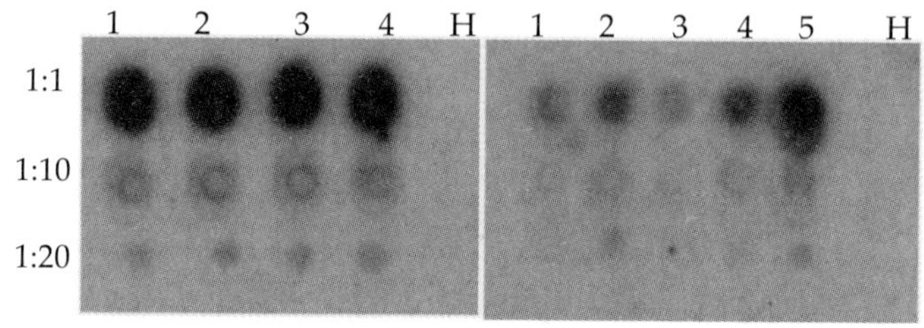

Fig. 10: Detection of ICRSV isolates by NASH using probes marked from 350bp (left) and 450 bp (right) amplicons of ICRSV-Ab

1: ICRSV – Pu 2: ICRSV – Dn
3: ICRSV – Ah 4: ICRSV – Ab
5: Bean inoculated of Abohar isolate H: Healthy control

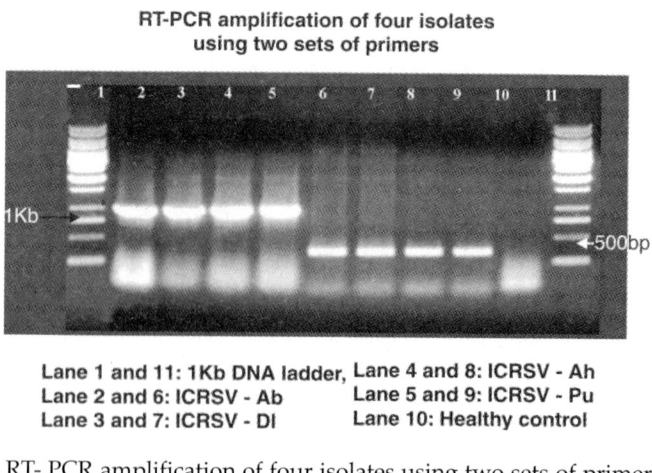

Fig. 11: RT- PCR amplification of four isolates using two sets of primers

Incidence of ICRSV is very high in Kinnow mandarin and sweet oranges in India. Masking of symptoms of the disease during hot climate is a common phenomenon and hence identification of ICRSV-free trees is difficult at the time of propagation. This is one of the reasons that the disease has not been identified earlier. It has been spreading for a long time inadvertently approaching high incidence of the disease and is believed to be one of the factors in citrus dieback in India (Ahlawat, 1997a). Use of virus-free planting material has been recommended to contain this disease (Ahlawat, 1997b; Ahlawat and Pant, 2003; Byadgi *et al.*, 1994). Virus-free propagation material in Kinnow mandarin and Mosambi sweet orange has been developed through shoot-tip grafting and regeneration of plants through somatic embryogenesis and by embryo culture (Hoa *et al.*, 2004; Praveen *et al.*, 2003). Another strategy developed for management of ICRSV was the use of CP gene construction. Gene constructed in pB1121 binary vector was mobilized to *Agrobacterium* strain LBA 4404 and co-cultivated in French bean var. saxa. Transformant was truncated version of CP gene of 900bp. The presence of gene in transformed plants was confirmed by PCR, southern hybridization and nucleic acid spot hybridization (Hoa and Ahlawat, 2004). This technology will be helpful to develop virus resistant plants in near future. Reliable means of detection have recently become available, and there appears to be no seed transmission. No natural vector is known and there are hopes that clean material planted in the field may remain virus-free.

Citrus Ring Spot

A ring spot disease in citrus was observed in California as early as 1949 (Reuther *et al.*, 1978; Wallace and Drake, 1968). Symptoms of psorosis-like leaf pattern on young leaves followed by rings on mature leaves were observed in some samples when grafted on sweet or sour orange, Eureka and Lisbon lemons and Brazilian Sweet Lime. However, in some cases no psorosis-like leaf pattern was observed and only ring spot symptoms developed on mature leaves. This isolate was maintained and studied at the University of California, Riverside. Citrus ring spot disease is now known to occur in Spain, Algeria, France, Greece, Italy (including Sicily), Turkey, Iran, Algeria, South Africa, Argentina, and Uruguay. This is an economically important disease in many areas of Argentina and other countries.

SYMPTOMS

Symptoms on tissue graft inoculated plants appear as faint chlorotic spots, clearing and chlorosis of some veins or part thereof. First symptom of the disease appears as vein flecking and chlorotic spotting similar to psorosis disease but later symptoms of bright yellow spots or distinct circular or irregular rings which may be gum impregnated differ from psorosis symptoms. In sour orange, symptoms of vein clearing similar to tristeza vein clearing develop upon tissue grafting. In sweet orange, the mature leaf symptoms are similar to that of psorosis-A except that the underside of the affected leves was not roughened or corky. Necrotic lesions may also develop on young stem which sometimes encircle the stem resulting death of the upper part of the tree. Although symptoms of ring spot disease were similar to psorosis-A virus infection but no bark lesions were developed in inoculated plants for several years after inoculation (Reuther *et al.*, 1978). Fruits may also show ringspot symptoms. Most *Citrus* spp. and their hybrids, especially grapefruits (*C. paradisi*) and oranges (*C. sinensis*) are susceptible but some ringspot isolates do not infect tangelos (*Citrus paradisi* × *C. reticulata*) or *Poncirus trifoliata* (Timmer *et al.*, 1978).

TRANSMISSION

Citrus ring spot disease is easily transmitted by tissue grafting. It is also transmitted from citrus to citrus, petunia and periwinkle and back through dodder, *Cuscuta subinclusa* (Desjardins *et al.*, 1969). No insect vecor is known but evidence of natural spread is there.

HOST RANGE

Citrus ringspot virus infects both citrus and herbaceous hosts. Citrus hosts susceptible to ringspot are: Citrange Troyer, Citremon (1448–1449), *Citrus excelsa*, Clementine, Grapefruit, Lemons, Lime (Kagzi, Key, Mexican, West Indian), Mandarin (Parson's Special, Szinkom), *Poncirus trifoliata* (Pomeroy), Rangpur lime, Rough lemon, *Severinia buxifolia*, Sour orange, Sweet oranges and Dweet tangor. Grapefruit is highly susceptible host to ringspot virus. Herbaceous hosts of the virus are: cowpea, bean, *Chenopodium quinoa, Gomphrena globosa, Crotalaria spectabilis, Petunia hybrida and Zinnia elegans.* An isolate of citrus ringspot (CRSV-6) from Florida showed typical symptoms of psorosis like shock reaction in young shoots, flecking and chlorotic blotches in young leaves, mature leaf symptoms and bark lesions were mechanically transmitted to herbaceous hosts and back to citrus (Garnsey and Timmer, 1988). Symptomatic hosts which are infected by mechanical inoculation are shown below:

- ***Chenopodium quinoa***: Necrotic local lesions develop after 7–10 days of inoculation on mature leaves.
- ***Gomphrena globosa***: Chlorotic blotches and spots.
- ***Etrog citron and Mexican lime***: Developed typical symptoms of ringspot, necrotic etch, leaf mottle and shock reaction when inoculated mechanically from inoculum obtained from *G. globosa.*

INDICATOR HOSTS

- ***Sweet orange seedlings***: Necrotic shoot reaction, ring pattern, mottled chlorotic areas on young leaves, chlorotic patterns on mature leaves and stem lesions develop a few weeks after inoculation.
- ***Chenopodium quinoa***: Necrotic local lesions on mechanically inoculated, fully expended leaves develop within few days after inoculation. No systemic infection develops.

Lemon, rough lemon, sour and sweet orange are also used as indicator hosts. Yellowish rings with green islands, vein clearing and stem lesions are developed on these hosts after 4–12 weeks of grafting.

THE PATHOGEN

Two components of infectious preparations of citrus ring spot virus were obtained by differential centrifugation. These two components depend on each other for infectivity (Derrick *et al.*, 1988). The virus particles associated with ring spot are of novel spiral filamentous type of two different length (Derrick *et al.*, 1988).They propose the name of the virus as 'Spirovirus". It was suggested that virus particles associated with ringspot disease resemble to Tenuiviruses similar to rice stripe virus (Navas-Castillo, 1993; Garcia *et al.*, 1994). Virus particles are in an open circular form and a linear form, plus intermediates. The linear form may appear as a sinuous structure about 10 nm diameter and 1500–2500 nm long; it is a two stranded structure with loops at the ends, formed from a collapsed circle. Open circular form is a highly kinked filament 3–4 nm in diameter and 300–500 nm long. Virion morphology suggests that ring spot virus represents a new genus of viruses similar to the Tenuiviruses in the bunyaviridae (Milne *et al.*, 1996; Garcia *et al.*, 1994). They proposed the name of the virus as "Ophiovirus"(from the Greek *ophis*, meaning 'serpent'). Ringspot virus has a 48kDa protein. Genome of Tenuiviruses is an ambisense divided ssRNA, each RNA is circularized by means of a panhandle structure of complementary base sequences at 5' and 3' termini.

DIAGNOSIS AND MANAGEMENT

Ringspot virus is difficult to detect by grafting on citrus hosts as similar symptoms of shock reaction, young and mature leaf symptoms are caused by psorosis-A virus. It is easily detected by sap inoculation to herbaceous host, *Chenopodium quinoa* which develops local lesions upon inoculation. Antisera to the 48 kDa protein which is presumed to be the capsid protein of the virus have been obtained (Derrick *et al.*, 1988; da Graca *et al.*, 1991) and used for virus detection in immunosorbent electron microscopy (ISEM). Normal methods for controlling graft-transmissible viruses of citrus (sanitation and certification) have partly failed in endemic areas (Frison and Taher, 1991). Budwood certification programme established against Citrus ringspot virus in some countries also failed to control the disease.

Citrus Yellow Mosaic

A mosaic diseases in india have been reported from Darjeeling hills on Khasi mandarin (Ahlawat *et al.*, 1985), from Andhra Pradesh on satgudi sweet orange (Dakshinamurti and Reddy, 1975) and from Maharashtra on Mosambi sweet orange (Pandey*et al.*, 1986). These mosaic diseases were either graft or aphid transmitted. However, no further information is available on these virus diseases. In 1992, a pummelo tree was observed with bright yellow mosaic and vein banding symptoms in an orchard near Indian Institute of Horticultural Research, Bangalore. Later such symptoms were observed in several Satgudi and Chini sweet orange trees in Karnataka and Andhra Pradesh. Budwood collected from these sources were grafted on Mosambi sweet orange seedlings which developed typical yellow mosaic symptoms. Bacilliform virions were observed in electron microscope from affected samples.

A mosaic disease of citrus has been reported from Japan (already described in earlier chapter of this book) and some from India as mentioned above. To avoid confusion, it was decided in 13th Conference of International Organization of citrus Virologists (IOCV) held in China in 1995 that a mosaic disease of citrus earler reported from India by the name of *'citrus mosaic disease'* (Ahlawat *et al.*, 1996) should be named as *'citrus yellow mosaic'* because yellow mosaic symptoms were observed on original host and other cultivars such as Satgudi and Mosambi sweet oranges. Symptoms of yellow mosaic have been observed in citrus species/cultivars in the states of Andhra Pradesh, Arunachal Pradesh, Assam, Delhi, Maharashtra, Manipur, Meghalaya, Rajasthan and Uttar Pradesh suggesting wider distribution of this disease in India. However, maximum incidence of yellow mosaic disease, from 10–70% was observed in Satgudi and Chini sweetorange trees in citrus orchards and nurseries in Andhra Pradesh (Ahlawat *et al.*, 1999). Yellow mosaic disease can cause yield reduction of fruits up to 77%. Juice on fruits from affected trees was 10% less and such fruits had 1.3% less ascorbic acid and were more acidic. Chlorophyll a, b, and IAA are reduced in mosaic affected leaves (Rao and Narasimhan, 1974). Citrus yellow mosaic disease is restricted in its distribution only to India so far.

SYMPTOMS

Satgudi sweet orange trees in field became completely yellow due to mosaic infection (Plate 18a). Symptoms of yellow mosaic disease in sweet orange and Pummelo trees consist of pale yellow to bright yellow blotches which are irregularly distributed on leaf surface (Plate 18b). Yellow flecking along the leaf veins is also seen quite often. Yellow mosaic virus is systemic in nature and almost all the leaves of an infected tree show symptoms. Because of bright yellow mosaic symptoms, affected trees sometimes appear like ornamental trees. Photosynthetic activities are reduced as most of the leaf surface becomes chlorotic resulting in reduced vigor and yield. Fruits on affected trees are comparatively smaller, have mosaic symptoms (Plate 18a) and often with bitter in taste.

TRANSMISSION

Yellow mosaic disease is transmitted by wedge grafting and T-budding from citrus to citrus. It is also transmitted by dodder, *Cuscuta reflexa*. Yellow mosaic virus is also transmitted by a mealybug vector, *Planococcus citri* (Plate 19) (Reddy, 1997; Ahlawat and Pant, 2003). Yellow mosaic virus is not transmitted by aphids, *Myzus persicae, Aphis gossypii, A. spiraecola, A. citricola, A. craccivora,* mealybugs, *Brevennia rehi* and *Saccharicoccus sacchari.* This disease is also transmitted by mechanical inoculation (stem slashing) to Pummelo, *Citrus decumana,* and Sathgudi sweetorange. Inoculated plants develop typical yellow mosaic symptoms of the disease after 90 days post-inoculation (Ahlawat *et al.*, 1996).

HOSTRANGE

Citrus yellow mosaic virus (CYMV) has been recorded mainly on sweetoranges (*Citrus sinensis*), but was experimentally transmitted on *Aegle marmelos, C. limonia, Fortunella* species, grapefruits (*C. paradisi*), lemons (*C. limon*), mandarins (*C. reticulata*), pummelos (*C. maxima*), *C. volkameriana, C. jambhiri, C. aurantium* and *C. mitis.* Milder symptoms develop in *C. limettioides, C. aurantiifolia* and *C. medica* (Ahlawat *et al.*, 1996). Symptoms of the disease equally develop during winter and summer suggesting that temperature variation does not affect development of symptoms. The reaction of yellow mosaic on some susceptible cultivars is given below:

- ***Sweet orange (Mosambi, Satgudi, Chini, Malta)***: Typical mosaic with scattered green and chlorotoc areas (Plate 18a).
- ***Rangpur lime***: Chlorotic spots, vein banding, mild mosaic, malformation.
- ***Volkamer lemon***: Yellow spot, yellow blotches, yellow mosaic (Plate 18c).
- ***Pummelo***: Yellow mosaic, smalling of leaves, golden yellow spot (Plate 18b).

- ***Grapefruit cv. Dunken***: Vein flecking, yellow mosaic.
- ***Sour orange***: Chlorotic and yellow spots, yellow mosaic.
- ***C. decumana***: Yellow mosaic, smalling of leaves, chlorotic spots, stunting.
- ***Citrus mitis***: Yellow rings and smalling of leaves.
- ***Rough lemon***: Yellow flecking, mosaic, yellow spot.
- ***Mandarin cv. Nagpur***: Vein banding.
- ***Sweet lime***: Mild mosaic.
- ***Citron***: Vein yellowing

THE PATHOGEN

A previously unreported badnavirus is associated with citrus yellow mosaic disease. Yellow mosaic virus was purified by differential centrifugation and a single band of virus particles was observed in cesium chloride density gradient which was removed and the preparation was evaluated for purity (Ahlawat *et al.*, 1996). UV absorption values of the purified preparation showed 0.694 OD at 260 nm and 0.542 at 280 nm. The 260/280 ratio was 1.28. Citrus yellow mosaic virus (CYMV) has non-enveloped bacilliform particles, 30 × 150 nm (Fig. 12), located in the cytoplasm, typical of badnaviruses. In ultrathin sections, large numbers of bacilliform particles were observed in the cytoplasm of tissue from CYMV infected plants (Fig. 13). Yellow mosaic bacilliform virus reacted specifically in ISEM test to antisera of seven badnaviruses, sugarcane mosaic badnavirus, commelina yellow mottle virus, banana streak virus, cacao swollen shoot virus, diascorea bacilliform virus, canna yellow mottle virus and kalanchoe top spotting virus but not to rice tungro bacilliform particles (Ahlawat *et al.*, 1996; Pant and Ahlawat, 1997). Its badnavirus

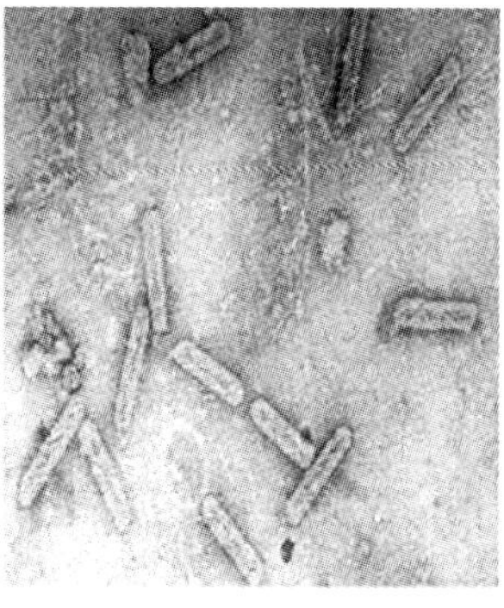

Fig. 12: Badnavirus particles associated with citrus yellow mosaic disease (CYMD)

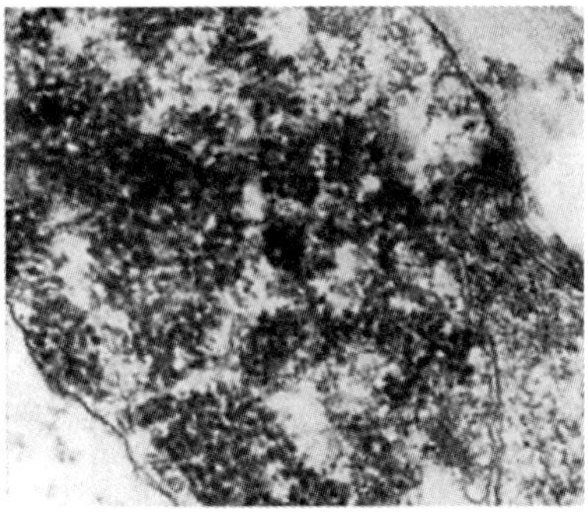

Fig. 13: Badnavirus particles in cytoplasm of CYMD infected cell

nature was further confirmed by amplifying its genomic nucleic acid in PCR using degenerated primers located on consenses sequences of badnaviruses (Fig. 16) (Ahlawat *et al.,* 1996). Based on its bacilliform morphology and serological reactivity, yellow mosaic virus was identified as a member of genus Badnavirus in family Caulimoviridae.

GENOME

Genome of yellow mosaic virus is double stranded DNA. A full length genome from Rangpur lime has been sequenced which consists of 7522 nucleotides with 6 ORFs. Intergenic region of 724 nucleotides between ORF6 and ORF1 consists of putative promoter elements (Singh *et al.,* 2009; Singh and Baranwal, 2009). SDS-PAGE and western blot analysis of purified virions showed one major band of 32 kDa and two minor bands of 22 and 19 kDa. 32 kDa is the major capsid protein and other two proteins are degraded products of 35 kDa.Nucleic acid from purified virions showed two bands of 21kb and 6.6 kb in non-denaturing agarose gel. The 6.6 kb is the linear form of the viral DNA and 21kb is the circular form of DNA (Ahlawat *et al.,* 1997).

Citrus yellow mosaic badnavirus (CYMBV) genome has been cloned and its complete nucleotide sequences determined. Viral DNA genome is 7559 bp in length and contains six putative open reading frames (ORFs), all on the plus-strand of the genome and each capable of encoding proteins with a molecular mass of greater than 10 kDa. ORF 3, the largest ORF, encodes a putative polyprotein for functions involved in virus movement, assembly and replication. The other ORFs encode proteins whose exact functions are not completely understood. The genome also contains a plant tRNA (met)-binding site, which may serve as a primer for minus-strand DNA synthesis in its intergenic region. Phylogenetic analysis of the badnaviruses revealed that Citrus yellow mosaic badnavirus (CYMBV) is most closely related to Cacao swollen shoot virus. It was demonstrated that a construct containing 1.4 copies of the cloned CYMBV genome could infect sweet orange via Agrobacterium-mediated inoculation (Huang and Hartung, 2001).

Two full-length genomes, one each from Sweet orange and Rangpur lime, are available in public accessible databases. Genetic variability in CYMBV has been studied from other hosts and based on in silico sequence analysis, possible function of protein encoded by one of the five ORFs was predicted. Nucleotide sequences obtained during variability studies have been submitted to the NCBI GenBank nucleotide sequence database and have been assigned Accession numbers, EU489744 and EU489745 (Borah *et al.*, 2009). In India, partial polyprotein gene of ORF III, partial ORF 6 and intergenic region, intergenic region of Tirupati isolates on Rangpur lime, acid lime and Sathgudi sweet orange have been sequenced (Accession No. AY950594, AY945935, AY944237, AY953373, AF347695).

DIAGNOSIS AND MANAGEMENT

Polyclonal antibodies against CYMBV have been prepared and used for virus trapping at a dilution of 1:500 of specific antiserum (Fig. 14) and decoration tests (1:50) in Immunosorbent electron microscopy (ISEM) (Fig. 15). CYMBV is also detected in DAS-ELISA (Ahlawat *et al.*, 1996). Viral DNA was cloned at Hind III site in P^{UC18} vector and transformed into competent *E. coli* strain NM 522 cells. Recombinant colonies were screened and finally two virus specific clones of 2.0 and 4.0 kb were identified. These clones were radiolabeled and used as probe to detect the virus in field trees and mealybug vector. 2.0 kb clone was more efficient probe and detected virus in DNA-DNA hybridization in even 25ng of the total DNA or 95pg of viral DNA or 25ng of tissue per spot (Reddy, 1997; Ahlawat *et al.*, 1997). Results of Table 7 clearly indicate that 2.0kb probe detected the virus even in asymptomatic trees suggesting that it is an important tool

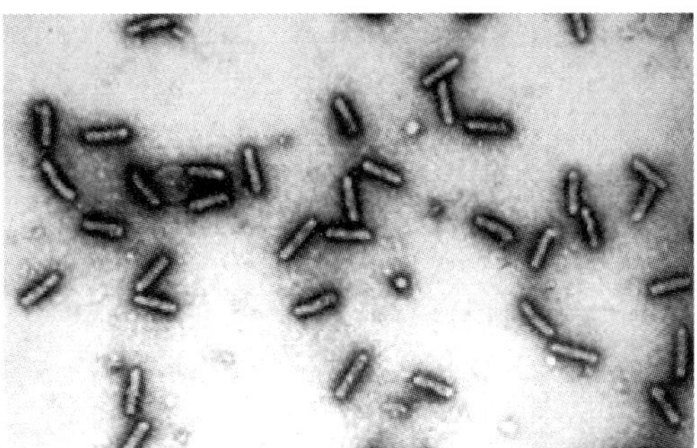

Fig. 14: Trapping of virus particles in ISEM

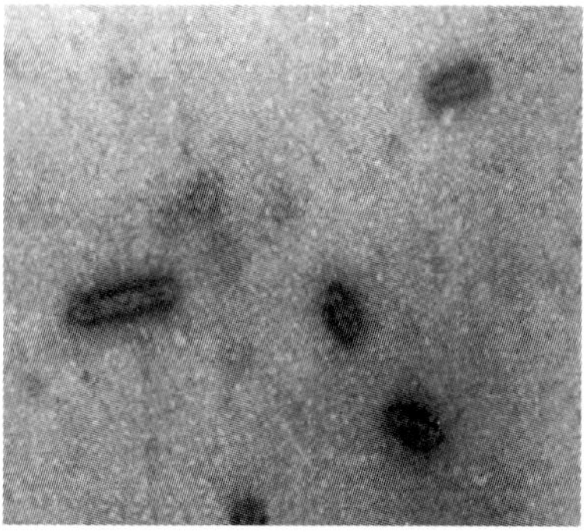

Fig. 15: Decoration of virus particles with specific antiserum

Table 7: Detection of CYMBV in dot-blot hybridization

Orchard No.	*Sample positive*	
	Visual observation	*Dot-blot hybridization*
1	0/11	10/10
2	4/11	8/11
3	3/23	9/23
4	4/11	8/11
5	2/10	5/10
6	2/9	2/9
7	0/7	7/7
8	0/5	1/5
9	0/2	½
Nursery	4/18	15/18
Total	**19/107**	**66/107**

for CYMBV detection. CYMBV is also detected in PCR using degenerate oligonucleotide primers (Fig. 16). A PCR product of approximately 1.5 kb and 1.0 kb in size was obtained (Ahlawat *et al.*, 1996; Baranwal *et al.*, 2003). By using two different sets of primer pairs specific for partial polyprotein gene of CYMBV genome, amplified product of 726bp and 245bp were obtained (Ghosh *et al.*, 2007). NCM (Nitrocellulose membrane) based DNA isolation method has been strandardized which does not require the use of liquid nitrogen for nucleic acid extraction and spotted membrane can be stored at room temperature for several months (Ghosh *et al.*, 2010). CYMBV was also detected in vector, *Planococcus citri* by PCR (Singh *et al.*, 2004). Citrus is a perennial crop and is subjected to infection by more than one virus in the field trees, hence mixed infections are not uncommon (Fig. 17). Simultaneous detection of CYMBV with other viruses or virus-like pathogens from mixed infection could be possible in multiplex PCR system (Fig. 18) which should be used preferably to save reagents and

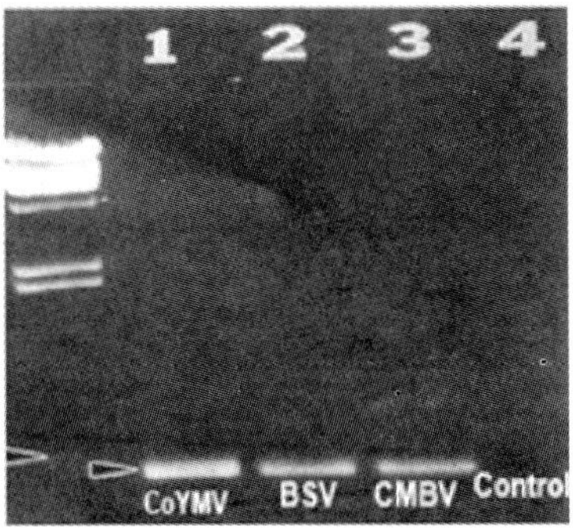

Fig. 16: Detection of CYMBV *CMBV) in PCR along with COYMV and BSV badnaviruses

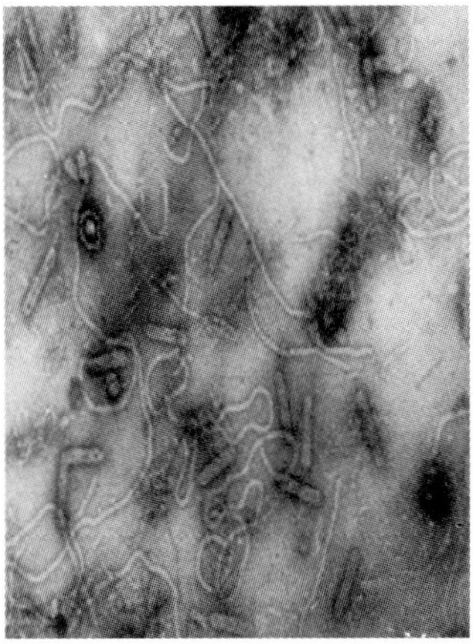

Fig. 17: Mixed infection of CTV, ICRSV and CYMBV isolated from naturally infected tree

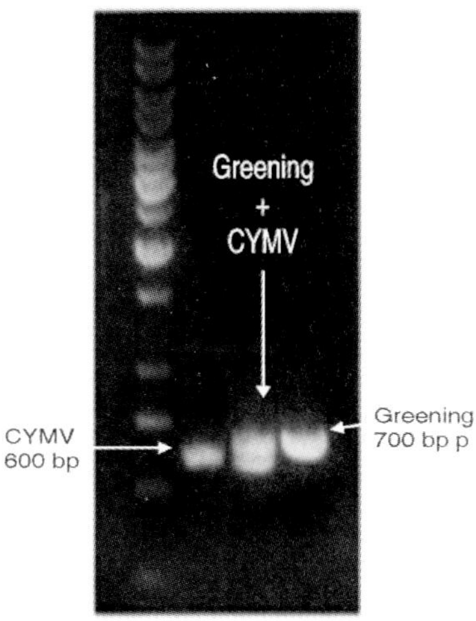

Fig. 18: Detection of CYMBV and Candidus *Liberibacter asiaticus* in duplex PCR

time for diagnosis (Baranwal *et al.*, 2005; Ghosh *et al.*, 2008). *Citrus decumana* and grapefruit are good indicator hosts of the virus which develop bright yellow mosaic symptoms and can be used for disease diagnosis by graft inoculation. Virus diagnosis is also possible by electron microscopy either by thin sectioning or by leaf dip preparations. Mealybug vector of CYMBV is not an efficient vector and hardly plays any role in natural spread. Therefore, this virus can be eliminated by using healthy planting material which may be obtained by shoot tip grafting and proper indexing.

Pummelo Yellow Vein Mosaic

Citrus yellow vein mosaic disease was initially observed on a Shaddock tree at Bangalore, India (Ahlawat *et al.*, 1993). Young leaves of naturally infected Pummelo trees showed yellow flecks or dots which later developed in to mottling and yellow flecking along the veins (Plate 20a). Some of the flecks fused together forming either rings or large chlorotic areas (Plate 20b). In advanced cases, leaves become smaller and clear raised spots are seen on leaves. This disease is transmitted by wedge and leaf patch grafting from Shaddock to Shaddock within 70 days of inoculation. It is not transmitted by sap inoculation to *Nicotiana glutinosa, N. benthamiana, Chenopodium amaranticolor, C. quinoa, Cathranthus roseus* and *Petunia hybrida.*

Natural and graft transmitted samples of Shaddock were screened under electron microscope in leaf dip preparations. Bacilliform particles with an outer membrane measuring 166 × 36 nm, typical of rhabdoviruses were observed both in natural and glasshouse inoculated plants (Fig. 19). In addition to this disease, rhabdovirus is known to be associated only with leprosis disease of grapefruit and sweet orange and found only in Brazil (Kitajima *et al.*, 1972). However, the symptoms of yellow vein mosaic are entirely different from leprosis. The name of present disease is tentatively proposed as 'pummelo yellow vein mosaic' and the associated virus as 'pummelo yellow vein mosaic rhabdovirus which is a tentative member of *Rhabdoviridae* (Ahlawat *et al.*, 1993). Further studies are required to establish the relationship of the two rhabdoviruses infecting citrus in India and Brazil. More information is required on the distribution of pummelo yellow vein mosaic in India and elsewhere.

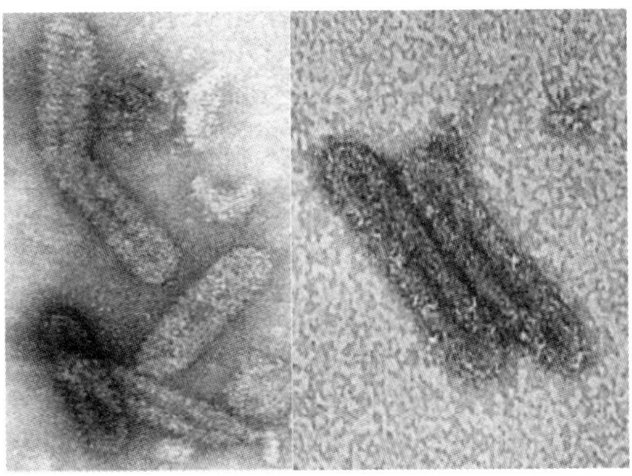

Fig.19: Rhabdovirus particles associated with Pummelo yellow mosaic virus

Citrus Yellow Vein Clearing

Citrus yellow vein clearing disease (CYVCD) appears to be widely distributed in Indian subcontinent including Pakistan. Yellow vein clearing disease has been reported from India (Ahlawat *et al.*, 2001; Alshami *et al.*, 2003), Pakistan (Bove, 1989; Catara *et al.*, 1993; Ashfaq *et al.*, 2005), and recently from Turkey (Onelge, 2003; Onelge *et al.*, 2010). In India, symptoms of yellow vein clearing were first observed by the author in 1996 on one tree of Etrog citron (*Citrus medica*) in the germplasm collection at Punjab Agricultural University (PAU) campus at Abohar (Plate 21a) Later, similar symptoms were observed on trees of Etrog citron, Sour orange, Lisbon lemon, *Citrus pectinifera* and Citranges in the germplasm block of PAU at Ludhiana. During surveys, this disease was also observed on Lison lemon and Rangpur lime at Anand in Gujarat suggesting wider distribution of yellow vein clearing disease in India. The diseases reported from India and Pakistan appears to be the same.

SYMPTOMS

Symptoms of yellow vein clearing disease appear as clearing of leaf veins which are clearly visible against transmitted light (Plate 21b). Water soaking along veins on underside of the leaf is the most characteristic symptom of the disease (Plate 21c). Clearing of lateral veins starts as flecks of varying length not involving the entire vein but subsequently most veins are involved (Plate 21d). The yellow flecks may extend laterally into the tissue adjacent to the veins. Midveins show little clearing even on leaves with severely affected lateral veins. Symptoms remain constant even after the leaf becomes mature and vein clearing symptoms on affected trees can be seen year around. Variable symptoms are observed on different *Citrus* species but clearing of veins is major symptom. Symptoms of yellow vein clearing are markedly different from those of citrus yellow vein disease occurring in California. This disease will also be described in this book. Disease severity is based on symptom expression and correlated with environmental factors. For example, in Pakistan, two lemon cultivars Eureka and Lisbon showed 100 and 64% infection respectively while Mesero and Verna were unaffected. Mean disease incidence at both locations in Pakistan ranged from 68.75 to 98.75 percent (Yasir Iftikhar *et al.*, 2010). Symptoms on major hosts are described below (Alshami *et al.*, 2003):

- ***Citrus medica (Etrog citron)*:** Leaf flecking, vein clearing, water soaking of veins (Plate 21a, b).

- ***Citrus aurantium (Sour orange)***: Vein clearing, water soaking of veins and leaf flecking.
- ***Citrus sinensis cv. Mosambi***: Vein clearing, chlorotic spots, leaf flecking and water soaking of leaves.
- ***Kinnow mandarin (King mandarin X Willow mandarin)***: Vein clearing, chlorotic spots, mottling, leaf flecking and water soaking.
- ***Citrus pectinifera***: Vein clearing, leaf flecking, oak-leaf pattern and water soaking of Veins.
- ***Citrus limon (galgal)***: Leaf flecking, water soaking of veins, vein yellowing and Vein clearing (Plate 21c).
- ***Citrus paradisi (grapefruit)***: Vein clearing, yellowing of veins and water soaking.
- ***Rangpur lime***: Vein clearing of full or partial veins (Plate 21d).

TRANSMISSION

Yellow vein clearing disease is easily transmitted by wedge and bud grafting from Etrog citron to various citrus species causing symptoms as described above. Indian isolates are mechanically transmitted from citrus to following herbaceous hosts:

- ***Phaseolus vulgaris cvs. Gheusami and Saxa***: Veinal necrosis, necrotic spots, blotching (Plate 22c) followed by blight symptoms.
- ***P. vulgaris cv. Singtamy***: Mottling of leaves (Plate 22b).
- ***Chenopodium quinoa***: Necrotic local lesions (Plate 22a).

Yellow vein clearing virus is not transmitted by dodder, *Cuscuta reflexa*, aphids, *Aphis gossypii, A. citricola, Myzus persicae*, whitefly, *Bemisia tabaci* and mealybug, *Planococcus citri*. It is also not seed transmitted. However, Turkey isolate of CYVCV has been reported to be transmitted by *A. craccivora* from infected lemon to bean plants upto 69% and with *A. spiraecola* upto 42% (Onelge *et al.*, 2010).

HOSTRANGE

In addition to herbaceous hosts described above, following citrus species/ variety are susceptible by experimental transmission causing typical symptoms of the disease: Etrog citron, Sour orange, Rough lemon, Rangpur lime, Mosambi and Malta sweet orange, Grapefruit, Kinnow mandarin, Nagpur orange, galgal (*C. limon*), Pummelo, Decumena, Calamondin, Karna Khatta and *C. pectinifera* (Alshami *et al.*, 2003). Yellow vein clearing seems to be widespread in Pakistan. In Pakistan, this disorder was observed on sour orange Eureka lemon; and on the following lemon varieties or selections: Allen, Cascade and Cook Eureka; Frost Eureka; Frost Lisbon; and Foothill and Limoneira 8A Lisbon, Corona Eureka; Caver's old line, Caver's new line, Monachello new line and Molla Mehmet (Catara *et al.*, 1993). In Turkey, the disease is transmissible to sour orange and lemon cultivars Interdonata, Kütdiken, Italian and Lamas but not to sweet oranges (Madam Vinous, Pineapple, Navelina, Washington navel, Valencia), mandarins (Clementine, Fremont, Nova, Parsons's Special, Kara), grapefruits (Star Ruby, Rio Red, Marsh seedless, Duncan), Tahiti lime and rough lemon (Onelge *et al.*, 2010).

PHYSICAL PROPERTIES

Thermal inactivation point (TIP) of the virus in crude extract of French bean is between 60°C and 70°C, dilution end-point (DEP) is 10^{-2} and longevity *in vitro* (LIV) is 2h at room temperature.

THE PATHOGEN

Initial results of immunosorbent electron microscopy (ISEM) tests with antiserum of Indian citrus ringspot virus (ICRSV) showed decorated and undecorated particles in samples of yellow vein clearing disease (Fig. 20). These results suggested that a filamentous virus with similar morphology to ICRSV was present in field affected samples. However, through experiments yellow vein clearing virus (CYVCV) was separated from mixed infection and brought to pure culture (Ahlawat *et al.*, 2001). Purified preparations of the virus showed filamentous virions measuring 685 nm in length and 14 nm in width in electron microscope (Fig. 21) (Alshami *et al.*, 2003). OD values of yellow vein clearing virus preparations at 260 and 280 nm were 0.6442 and 0.5735 respectively. A260/A280 ratio of purified virus prepation was 1:123. Filamentous virus particles with a modal length of 600–700 nm and diameter of 13–14 nm have also been observed in leaf dip preparation electron microscopy with Pakistani isolates (Grimaldi and Catara, 1996). They also reported aggregation of filamentous particles in phloem cells of thin sections from lateral symptomatic veins. Genome of the virus is ssRNA with a coat protein of 32kDa (Alshami *et al.*, 2003). Studies conducted on this virus so far indicate that this is an unreported virus of citrus species and needs more studies before its final classification by ICTV.

Fig. 20: Electron micrograph of yellow vein clearing virus (CYVCV) in ISEM with antiserum of ICRSV. Only ICRSV particles were decorated but not yellow vein clearing suggesting that two viruses are different

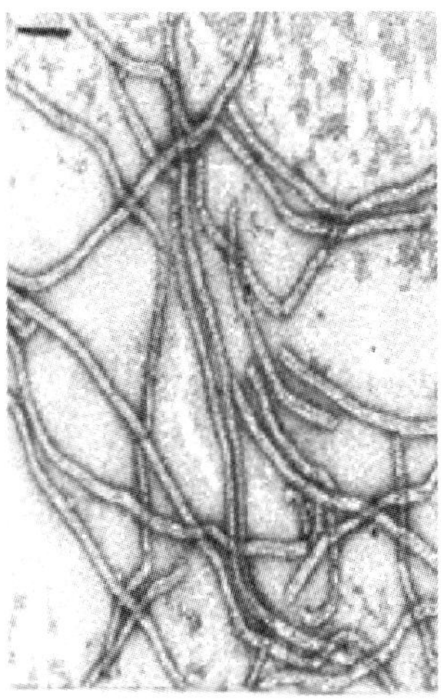

Fig. 21: Purified virus particles of yellow vein clearing virus

DIAGNOSIS AND MANAGEMENT

Polyclonal antibodies against the Indian isolates of yellow vein clearing virus have been prepared and used for serodiagnosis and serological relationship. This virus is not serologically related to nine filamentous viruses including Indian citrus ringspot mandarivirus in ISEM tests (Byadgi and Ahlawat, 1995). Yellow vein clearing virus specific antibodies decorated all particles in purified sample from glasshouse raised cultures

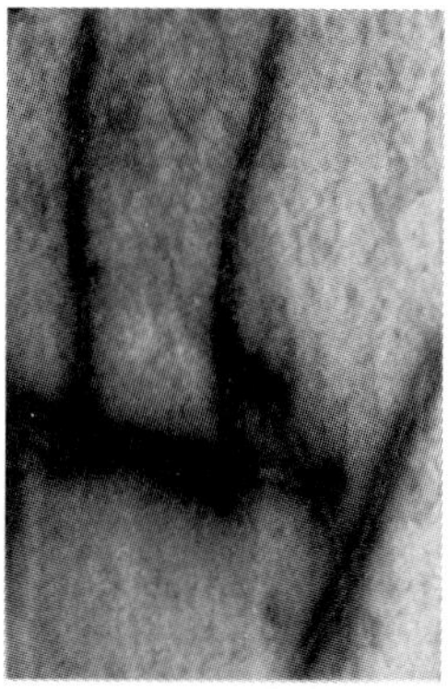

Fig. 22: ISEM of CYVCV with specific antibodies

(Fig. 22). Antibodies of yellow vein clearing virus (CYVCV) have been effectively used for virus detection from citrus and non citrus hosts both by DAS and DAC-ELISA. Homologous antibodies successfully detected the virus in decoration tests of ISEM. CYVCV could not be detected by antiserum of Indian citrus ring spot virus or vice versa further suggesting that two viruses are distinct. CYVCV could not be amplified with ICRSV primers in PCR. Since these two viruses have similar morphology and are always found as mixed infection, there could be a risk of CYVCV movement through planting material if both the viruses are not indexed. CYVCV can also be detected by sap inoculation to *C. quinoa*. Yellow vein clearing disease can be easily identified by the characteristic symptom of water soaking of veins on the lower side of leaves. This typical symptom is not found with ICRSV. Use of virus-free planting material is the only recommended method to contain yellow vein clearing virus.

Dweet Mottle = Leaf Blotch Virus

Dweet mottle symptoms were first observed in California in 1960s when graft inoculations were made from a symptomless but sick looking Cleopatra mandarin tree from Florida to dweet tangor. Inoculated plants of Dweet tangor showed chlorotic blotching and a mild exocortis reaction in Etrog citron (Roistacher and Blue, 1968). The pathogen associated with dweet mottle remains symptomless in several hosts but symptoms appear only on Dweet tangor. Therefore, the name of the disease was given "Dweet mottle". In 1984, Spanish Scientists discovered a new disease showing bud union incompatibilities in certain scion/rootstock combinations (Navarro *et al.*, 1984). Symptoms of chlorotic blotches similar to dweet mottle were observed in inoculated Dweet tangor plants but additionally vein clearing symptoms in pineapple sweetorange and bud union crease in Troyer citrange and citrumelo rootstocks were also observed (Navarro *et al.*, 1984; Galipienso *et al.*, 2000, 2001). This disease was given the name Citrus Leaf Blotch. Based upon reaction in indicator plants, citrus leaf blotchvirus (CLBV) was found to be closely related to Dweet mottle virus (DMV). CLBV reacts similarly to DMV in the Dweet indicator but also produces different symptoms in other indicators as well as causing bud union problems in some instances (Krueger *et al.*, 2005). The disease has been reported from different citrus varieties in Japan, Australia, Florida and Spain. Recently, CLBV has been found in Nagami kumquat (Guardo *et al.*, 2007a) and Calamondin (*Citrus mitis* or *madureusis*) (Guardo *et al.*, 2007b) in Italy, and in several citrus varieties in New Zealand (Harper *et al.*, 2008). Economic importance of CLBV for the citrus industry is presently unknown since its incidence in different citrus areas has not been evaluated and the actual involvement of CLBV with other pathogen in the bud union crease has not yet been proved.

Genome sequencing of the two viruses showed nucleotide identity between CLBV isolate SRA-153 and Dweet mottle sources DMV-931 and DMV-932 was 96.6 and 96.8% in replicase and 98.7 and 98.5% in the coat protein gene, respectively confirming that the two viruses are identical (Krueger *et al.*, 2005). Dweet mottle disease is probably caused by Citrus leaf blotch virus (Vives *et al.*, 2001, 2005, 2010). An infectious full-genome cDNA clone from CLBV isolate only induced chlorotic blotching in Dweet tangor plants and stem pitting in Etrog citron plants, but no vein clearing in Pineapple sweet orange or bud union crease on trifoliate rootstocks. These results suggest that bud union crease and vein clearing symptoms are likely caused by a different pathogen or by an interaction between CLBV and other biotic factors.

SYMPTOMS

Dweet tangor disease is found in trees showing bud union crease on trifoliate rootstocks and stem pitting in etrog citron. Stem pitting on etrog citron may be confused with citrus tristeza virus symptom and chlorotic blotches in dweet tangor are similar to be caused by psorosis, cristacortis, concave gum and impietratura and Indian citrus ring spot viruses. However, major symptom of the disease is chlorotic blotching on leaves of infected dweet tangor (Plate 23). Several trees grafted on citrange rootstock from Valencia (Spain) showed decline and bud union crease (Galipienso *et al.*, 2001, 2004; Vives *et al.*, 2002).

TRANSMISSION

CLBV is graft transmitted from citrus to citrus and shows a bud union disorder of Nagami kumquat and Calamondin scions grafted on trifoliate rootstocks. CLBV can also be transmitted by contamination of knife blade (Roistacher *et al.*, 1980). In Spain, CLBV is seed-transmitted causing concern in the nursery industry (Krueger *et al.*, 2005; Robert *et al.*, 2005). Low percentage (*ca.* 2.5%) of seed transmission has been reported through seeds of Carrizo citrange, kumquat and sour orange (Guerri *et al.*, 2004). It is experimentally transmitted from citrus to citrus by stem-slash inoculation and from citrus to *Nicotiana occidentalis* and *N. benthamiana* by mechanical inoculation with partially purified virus preparations.

HOST RANGE

Host range of the virus is within citrus species and some herbaceous hosts are also susceptible to CLBV. Susceptible hosts to CLBV are: Cleopatra mandarin, Dweet tangor, Nagami Kumquat, Etrog citron, Pineapple sweet orange, Nules Clementine and Troyer citrange. Dweet tangor and etrog citron are more sensitive and other hosts are tolerant. The virus remains symptomless in inoculated plants of *Nicotiana occidentalis, Nicotiana cavicola* and *Nicotiana benthamiana*. Infection in these hosts can be confirmed by positive response in RT-PCR using virus-specific primers and by electron microscopy (Guardo *et al.*, 2009). Dweet tangor seedlings are used as indicator hosts in which chlorotic blotches are developed upon inoculation.

THE PATHOGEN

Citrus leaf blotch virions are filamentous measuring 960 × 14 nm. Citrus leaf blotch virus (CLBV) is type species of the putative new genus *Citrivirus* in family *Flexiviridae* (Mayo and Haenni, 2006).

GENOME

Genome of CLBV is a single-stranded positive sense RNA (ssRNA) of about 9 kb (Mr 3 × 10^6) containing a 42 kDa protein. Infected tissue contained three species of double-stranded RNA (dsRNA) of Mr 6, 4.5 and 3.4 × 10^6 (Galipienso *et al.*, 2001). CLBV genomic RNA has been completely sequenced from kumquat, Nagami kumquat SRA-153 (Galipienso *et al.*, 2004; Galipienso *et al.*, 2001; Vives *et al.*, 2001). The CLBV genomic RNA contains three open reading frames (ORFs) encoding a polyprotein involved in replication, a potential movement protein and the capsid protein. Dweet mottle virus (DMV) genome has also been sequenced which has 8,747 nucleotides (nt) excluding the 32 poly-(A) tail. DMV genomic RNA contains three putative open reading frames (ORFs) and untranslated regions of 73 nt at the 52 and 541 nt at 32 termini. Nucleotide sequences of DMV were compared with members of the families *Alphaflexiviridae* and *Betaflexiviridae* (Hajeri *et al.*, 2010). Protein database searches showed 98–99% matches of DMV ORFs with CLBV sequences. Phylogenetic analysis based on the RdRP core domain revealed that DMV is closely related to CLBV as a member of the genus *Citrivirus*. DMV did not satisfy the molecular criteria for demarcation of an independent species within the genus *Citrivirus*, family *Betaflexiviridae*, and hence, DMV can be considered a CLBV isolate (Hajeri *et al.*, 2010).

DIAGNOSIS AND MANAGEMENT

CLBV induces chlorotic blotching leaf symptoms in Dweet tangor, stem pitting in Etrog citron, vein clearing in pineapple sweet orange and other cultivars and bud union crease when propagated in Troyer citrange. These species are appropriate diagnostic indicators. Chlorotic leaf blotching in Dweet tangor is faster to appear and easier to observe, therefore, inoculation of this indicator is recommended for routine indexing. Indicator plants should be decapitated after inoculation and preferably incubated in a temperature controlled glasshouse (26°C/18°C day/night). The symptoms of leaf blotching usually appear in young shoots located above or below the inoculation point (Galipienso *et al.*, 2000).

Citrus leaf blotch virus is detected by dot-blot hybridization, tissue print hybridization and by one-step RT–PCR in citrus plants growing both in the greenhouse and in the field. Dot blot with digoxigenin-labeled cDNA probes allowed CLBV detection in total RNA preparations equivalent to 5 and 0.1 mg of infected tissue (Galipienso *et al.*, 2004). These techniques of CLBV detection are being used in leaf samples of different citrus varieties in Spain, Australia, France, Japan and USA (Vives *et al.*, 2002; Galipienso *et al.*, 2001; Galipienso *et al.*, 2004). Recently, ELISA systems have been developed to detect DMV and CLBV using an antibody specific for DMV. Dweet mottle and CLBV are difficult to eliminate by shoot-tip grafting or thermotherapy (Vives *et al.*, 2005). It can also be detected by RT-PCR and quantitative real time PCR. Further molecular and sequencing data revealed that CLBV does not develop vein clearing in pineapple sweet orange or bud union crease on trifoliate rootstock, suggesting that the two latter symptoms are probably caused by a distinct pathogen or by an interaction between CLBV and other biotic factors (Vives *et al.*, 2010). CLBV is seed transmitted hence virus-free buds and rootstocks should be used.

REFERENCES

Adams, M.J., Accotto, G.P., Agranovsky, A.A., Bar-Joseph, M., Boscia, M. *et al.* 2005. Family Flexiviridae. pp. 1089–1124. *In*: Fauquet, C.M., Mayo, M.A., Maniloff, J., Desselberger, U. and Ball, L.A. (*Eds.*). Virus Taxonomy, Eighth Report of the International Committee on Taxonomy of Viruses, Elsevier Academic Press, London.

Adams, M.J., Antoniw, J.F., Bar-Joseph, M., Brunt, A.A., Candresse, T., Foster, G.D., Martelli, G.P., Milne, R.G. and Fauquet, C.M. 2004. The new plant virus family Flexiviridae and assessment of molecular criteria for species demarcation. *Arch. Virol.*, 149: 1045–1060.

Ahlawat, Y.S., Chakraborty, N.K., Varma, A. and Raychaudhuri, S.P. 1992. Diagnosis of diseases caused by viruses and fastidious Prokaryotes in citrus in India. pp. 749–751. *In*: Proc. Intematl. Soc. Citriculture, VII, Intl. Citrus Cong.

Ahlawat, Y.S. 1989. Psorosis 'A' disease of citrus in India. *Indian Phytopath.* 42: 21–25.

Ahlawat, Y.S. 1997a. Viruses, greening bacterium and viroids associated with citrus (*Citrus* spp.). *Indian J. Agric. Sci.*, 67: 51–57.

Ah1awat, Y.S. 1997b. Status of virus and virus-like pathogens infecting citrus in India and future strategies for citrus improvement. *Indian Phytopath.* 50: 192–193.

Ahlawat, Y.S. 2007. Citrus decline: Problems and Preventions. *Indian Phytopath.* 60: 1–12.

Ahlawat, Y.S. and Chakraborty, N.K. 1990. Etiology of psorosis 'A' disease of citrus. *Indian Phytopath* 42: 239.

Ahlawat, Y.S. and Chakraborty, N.K. 1990. Etiology of psorosis-A disease of citrus. p. 239. *In*: Proc. 42nd Ann. Meet. Indian Phytopath. Soc., Tirupati, Jan. 3–5, 1990 (Abs.).

Ahlawat, Y.S. and Chenulu, V.V. (*Eds.*) 1994. Virus and mycoplasma diseases of fruit crops in India. ICAR, New Delhi, p. 73.

Ahlawat, Y.S. and Pant, R.P. 2003. Major virus and virus-like diseases of citrus in India, their diagnosis and management. *Plant Pathol* 2: 447–474.

Ahlawat, Y.S. and Pant, R.P. 2008. Indian citrus ringspot virus. pp. 97–107. *In*: Characterization, diagnosis and management of plant viruses (*Eds.*) Rao, G.P., Arben, Myrta and Kai, Shu Ling, Studium press Houston, USA.

Ahlawat, Y.S. and Raychaudhary, S.P. 1988. Status of *Citrus tristeza* and dieback disease in India and their detection, pp. 871–879. *In*: Proc. 6th Internatl. Citrus Congress, (*Eds.*) Goren, R. and Mendel, K., Tel Aviv, Isarel.

Ahlawat, Y.S. and Sardar, K.K. 1976. Distribution of greening and tristeza diseases of citrus in various blocks of Darjeeling district. *Sci. and Cult.* 42: 175–277.

Ahlawat, Y.S. and Sardar, K.K. 1976. Lemon crinkly leaf virus disease in India. *Indian J. Hort.* 33: 168–171.

Ahlawat, Y.S., Al-Shami, A.A.A., Pant, R.P. and Prasad, Ramji 2001. Association of a filamentous virus causing yellow vein symptoms in citrus in India. p. 86. *In*: 15th Conf. IOCV, Paphos, Cyprus, Nov. 11–16, 2001 (Abs).

Ahlawat, Y.S., Byadgi, A.S., Pant, R.P. and Varma, A. 1997. Citrus ringspot capillovirus and Citrus mosaic badnavirus: two emerging pathogens infecting citrus in India. pp. 211–225. *In*: Management of threatening Plant diseases of National importance (*Eds.*). Agnihotri, V.P. , Sarbhoy, A.K. and Singh, D.V., Malhotra Pub. House, New Delhi.

Ahlawat, Y.S., Byadgi, A.S., Varma, A. and Chakraborty, N.K. 1995. Detection of ringspot virus and greening BLO in citrus and their management, pp. 27–44. *In*: Detection of plant pathogens and their management (*Eds.*) Verma, J.P., Varma, A. and Kumar, D., Angkor Publishers, New Delhi.

Ahlawat, Y.S., Chakraborty, N.K., Jagdishchandra, K., Srivastava, M. and Varma, A. 1993. Association of a Rhabdovirus with yellow vein mosaic, a new disease of citrus. pp. 455–457. *In*: Proc. 12th Conf. IOCV, IOCV, Riverside.

Ahlawat, Y.S., Chenulu, V.V., Bhagabati, K.N., Pandey, P.K., Viswanath, S.M. and Chakraborty, N.K. 1985. Studies on a mosaic disease of citrus. *Curr. Sci.* 54: 873–874.

Ahlawat, Y.S., Pant, R.P., Lockhart, B.E.L., Srivastava, M., Chakraborty, N.K. and Varma, A. 1996. Association of a badnavirus with citrus mosaic disease in India. *Plant Disease* 80: 590–592.

Ahlawat, Y.S., Reddy, B.V.B., Pant, R.P. and Varma, A. 1997. Diagnosis of virus and virus-like diseases of citrus in India. pp. 298–306. *In*: *Proc. Reg. Expert Consult. Applica. Biotech. Pl.* Pest Management. RAP publication: 1997/40, 25–28 Feb. 1997, IARI, New Delhi, India.

Ahlawat, Y.S., Reddy, D.R., Moses, J.E., Pant, R.P. and Singh, S.P. 1999. Distribution of yellow mosaic disease of citrus in South India. p. 7. Citrograph Magazine, July/August, 1999.

Ahlawat, Y.S., Varma, A., Pant, R.P., Shukla, A. and Lockhart, B.E.L. 1996. Partial characterization of a badnavirus associated with citrus yellow mosaic disease in India. pp. 208–217. *In*: Proc. 13th Conf Housenl. Org. Citrus Virologists, IOCV, Riverside, USA.

Alshami, A.A.A., Ahlawat, Y.S. and Pant, R.P. 2003. A hitherto unreported yellow vein clearing disease of citrus in India and its viral etiology. *Indian Phytopath* 56: 422–427.

Ananthakrishnan, G., Venkataprasanna, T., Roy, A. and Brlansky, R.H. 2010. Characterization of the mixture of genotypes of a *Citrus tristeza* virus isolate by reverse transcription quantitative real-time PCR. *J. Virol. Meth.* 164: 75–82.

Araya-González, J. 2000. Informe sobre la prospección de la lementin de los cítricos en la zona fronteriza sur (Costa Rica-Panamá). Ministerio de Agricultura Ganaderia. p. 5.

Ashfaq, M., Iftikhar, Y., Mughal, S.M. and Khan, M.A. 2005. Occurrence, distribution and transmission of yellow vein clearing virus (YVCV) of lemon in Faisalabad. *Pak. J. Phytopath* 17(1): 101–104.

Azeri, T. 1973. First report of lement dwarf virus disease on lement mandarins inTurkey. *Plant Disease Reporter* 57: 149–153.

Bahia Bennani, Celso Mendes, Mustapha Zemzami, Houssine Azeddoug and Gustavo Nolasco 2004. Citrus variegation virus: Molecular variability of a portion of the RNA 3 containing the coat protein gene and design of primers for RT-PCR detection. *European Journal of Plant Pathology* 108: 155–162.

Balaraman, K. and Ramakrishnan, K. 1977. Studies on strains and strain interaction of citrus tristeza virus. U.S.A. Tech. Series No. 19, p. 62, *Univ. Agric. Sci.*, Hebbal. Bangalore.

Balaraman, K. and Ramakrishnan, K. 1980. Strain variation and cross protection in citrus tristeza virus on acid lime. pp. 60–68. *In*: Proc. 8th Conf: IOCV, IOCV, Riverside, CA.

Baranwal, V. K., Mazumder, S., Ahlawat, Y.S. and Singh, R.P. 2005. A novel approach for simultaneous detection of citrus yellow mosaic virus and citrus greening bacterium by multiplex polymerase chain reaction. *Indian Journal of Biochemistry* 4: 528–533.

Baranwal, V.K., Majumder, S., Ahlawat, Y.S. and Singh, R.P. 2003. Sodium sulphite yields improved DNA of higher stability for PCR detection of Citrus Yellow Mosaic Virus from citrus leaves. *J. Virological Method.* 112: 153–156.

Barbarossa, L. and Vovlas, C. 2010. Molecular characterization of *Citrus tristeza* virus Isolates from Epirus (Greece). *In*: Proc. 17th Conf. IOCV, IOCV, Riverside, CA.

Bar Joseph, M., Garnsey, S.M. and Gonsalves, D. 1979. The Closterovirus: a distinct group of elongated plant viruses. *Adv. Virus Res.* 25: 93–168.

Bar Joseph, M., Loebenstein, G. and Cohen, J. 1970. Partial purification of virus-like particles associated with the citrus tristeza disease. *Phytopathology* 60: 75–78.

Bar Joseph, M., Marcus, R. and Lee, R.F. 1989. The continuous challenge of citrus tristeza virus control. *Ann. Rev. Phytopathol.* 27: 291–316.

Bar-Joseph, M. and Lee, R.F. 1989. Citrus tristeza virus; Description of Plant Viruses No. 353 (No. 33 revised). Common. Mycol. Inst./Assoc. Appl. Biol. Kew, Surrey, England.

Bar-Joseph, M., Roistacher, C.N. and Garnsey, S.M. 1983. The epidemiology and control of citrus tristeza disease. *In*: Plant virus epidemiology (*Ed.* Plumb, R.T. and Thresh, J.M.), pp. 61–72. Blackwell scientific publications, Oxford, U.K.

Barthe, G.A., Ceccardi, T.L., Manjunath, K.L. and Derrick, K.S. 1998. Citrus psorosis virus: Nucleotide sequencing of the coat protein gene by hybridization and RT-PCR. *Journal of General Virology* 79: 1531–1537.

Bastianel, M., Freitas-Astua, J., Kitajima, E.W. and Machado, M.A. 2006. The citrus lementin pathosystem. *Summa Phytopathologica,* 32: 211–220.

Baulcombe, D.C. 1996. Mechanisms of pathogen-derived resistance to viruses in transgenic plants. *Plant Cell* 8: 1833–1844.

Bazan de Segura, C. and Ferrand, A. 1969. Woody gall, its distribution and importance in new and old citrus plantings in Peru. *Proc. lst Int. Citrus Symp.* 3: 1449–1451.

Benatena, H.N. and Portello, M.M. 1984. Natural spread of psorosis in sweet orange seedlings. pp. 241–243. *In*: Proc. 9th Conf. IOCV, IOCV, Riverside, CA.

Bhagabati, K.N., Ahlawat, Y.S., Chakraborty, N.K. and Borthakur, B.C. 1989. The distribution of greening, tristeza and mosaic diseases of citrus in North-Eastern States of India. *Indian Phytopathol.* 42: 552–555.

Bioletti, F.T., Gowrie, W. and Cillie, P.J. 1904. Citrus culture in Cape colony. Report of Commission of inquiry into the causes of the failure of citrus trees in Cape Colony. *Agr. Jour., Cape of Good Hope* 21: 1–20.

Biswas, K.K. 2008. Molecular diagnosis of citrus tristeza virus in mandarin (*Citrus reticulata*) orchards of Darjeeling hills of West Bengal. *Indian J. Virol.* 19: 26–31.

Biswas, K.K., Manjunath, K.L., Lee, R.F. and Ahlawat, Y.S. 2004. Identification of diversified citrus tristeza virus population isolates from eastern Himalayan region of india by heteroduplex mobility assay and nucleotide sequence analysis. p. 9 *In*: Nat. symp. Molecular diag. for management of viral *Dis. Div. Pl. Path.* IARI, New Delhi.

Biswas, K.K., Sunita Kumari, Tarafdar, A. and Ramachandran, P. 2008. Biological and genetic diversity of citrus tristeza virus population from Northeastern Himalayan region of India. *Indian J. Virol.* 19: 59.

Biswas, K.K., Tarafdar, B,K,, Jayakumar, B.K. and Pun, K.B. 2009. Strategy for the production of citrus tristeza virus-free mandarin (*Citrus retuculata*) planting material in the Darjeeling hills of India. *Indian Phytopath.* 62: 376–380.

Bitancourt, A.A. 1937. Leprose e a proxima colheita. *Obiológico* 3: 37–40.

Bitancourt, A.A. 1955. Estudos sobre a leprose dos citrus. I-Distribuicao geografica e sintomatologia. Ii-transmissao natural as folhas. III. Transmissao natural as frutas. IV-Experiencias de tretamento. *Arq. Inst. Biol.* 22: 161–231.

Borah, B.K., Anthony Johnson, A.M., Sai Gopal, D.V.R. and Dasgupta, I. 2009. Sequencing and computational analysis of complete genome sequences of Citrus yellow mosaic badna virus from acid lime and lemen. *Virus Genes* 39: 137–140.

Bové, J.M. 1989. Virus and virus-like diseases of citrus in Pakistan. FAO Report, p. 56.

Bove, J.M. and Vogel, R. (*Eds.*). 1981. Description and illustration of virus and virus-like diseases of citrus. Setco-IRFA, Paris, France.

Bridges, G.D., Youtset, C.O. and Nixon, R.R. 1965. Observations indicating psorosis transmission by seed of Carrizo citrange. *Proc. Fla. Sta. Hort. Soc.* 78: 48–50.

Brlansky, R.H., Damsteegt, V.D., Howd, D.S. and Roy, A. 2003. Molecular analysis of citrus tristeza virus subisolates separated by aphid transmission. *Plant Dis.* 87: 397–401.

Bruening, G. 1977. Comovirus group. CMI AAB Desc. Plant Viruses, No. 199.

Brunt, A.A., Crabtree, K., Dallwitz, M.J., Gibbs, A.J., Watson, L. and Zurcher, E.J. (*Eds.*) 1996 (onwards). 'Plant Viruses Online: Descriptions and Lists from the VIDE Database. Version: 20th August 1996. URL http://biology.anu.edu.au/Groups/MES/vide/

Byadgi, A.S. and Ahlawat, Y.S. 1995. A new viral ringspot disease of citrus (*Citrus* spp.) in India. *Indian. J. Agric. Sci.* 65: 763–770.

Byadgi, A.S., Ahlawat, Y.S., Chakraborty, N.K. and Varma, A. 1994. Management of citrus ringspot disease. *Indian Phytopath.* 47: 326.

Byadgi, A.S., Ahlawat, Y.S., Chakraborty, N.K., Varma, A., Srivastava, M. and Milne, R.G. 1993. Characterisation of a filamentous virus associated with citrus ringspot in. India. pp. 155–162. *In*: Proc. 12th Conf. IOCV, IOCV, Riverside, CA.

CABI. 2010. Citrus tristeza virus. [Distribution map]. Distribution Maps of Plant Diseases, 2010, April, Map 289 (Edition 7).

Calavan, E.C. 1974. Registration and certification of citrus propagative material in California. *In*: Jackson, L.K., Krezdorn, A.H. and Soule, J. (*Eds*.). Proc. 1st inernl. Citrus short course – citrus rootstocks pp. 137–145 Fla *Coop. Exr. Serv. Inst.* Food and *Agric Sci., Univ.* Florida, Gainesville, Fla.

Calavan, E.C., Christiansen, D.W. and Roistacher, C.N. 1963. Symptoms associated with tatter-leaf virus infection of Troyer citrange rootstocks. *Plant Disease Reporter* 47: 971–975.

Calavan, E.C., Roistacher, C.N. and Nauer, E.M. 1972. Thermotherapy of citrus for inactivation of certain viruses. *Plant Dis. Reptr*. 56: 976–80.

Calegario, R.F., Kitajima, E.W., Salaroli, R.B., Rezende, J.A.M. and Locali-Fabris, E.C. 2010. Immunofluorescence studies on the distribution of citrus leprosies virus (CiLV-C) antigen in citrus leaf lesions. p. 55. *In*: Abs. 18th Conf. IOCV held in Caminas/SP, Brazil Nov. 7–12, 2010.

Calvert, L.A., Lee, R.F. and Hiebert, E. 1988. Characterization of the RNA species of citrus variegation virus and complementary DNA clones. pp. 327–333. *In*: Proc. 10th Conf. IOCV, IOCV, Riverside, CA.

Cambra, M., Camarasa, E., Gorris, M.T., Garnsey, S.M. and Carbonell, E. 1991. Comparison of different immunosorbent assays for citrus tristeza virus (CTV) using CTV specific monoclonal and Polyclonal antibodies. pp. 38–45 *In*: Proc. 11th Conf. IOCV, IOCV Riverside, CA.

Cambra, M., Camarasa, E., Gorris, M.T., Garnsey, S.M., Gumpf, D.J. and Tsai, M.C. 1993. Epitope diversity of citrus tristeza virus isolates in Spain. pp. 35–38. *In*: Proc. 13th Conf. IOCV, IOCV Riverside, CA.

Cambra, M., Gorris, M.T., Roman, M.P., Terrada, E., Garnsey, S.M., Camarasa, E., Olmos, A. and Colomer, M. 2000. Routine detection of citrus tristeza virus by direct immunoprinting ELISA method using specific monoclonal and recombinant antibodies. pp. 34–41. *In*: Proc. 14th Conf. IOCV, IOCV, Riverside, CA.

Cambra, M.,Olmos, A.,Gorris, M.T., Marroquin, C., Esteban, O., Garnsey, S.M., Llauger, R., Batista, L., Pena, I. and Hermosa de Mendoza, A. 2000. Detection of citrus tristeza virus by print capture and squash capture-PCR in plant tissue and single aphids. pp. 42–49. *In*: Proc. 14th Conf. IOCV, IOCV, Riverside, CA.

Capoor, S.P. 1961. Kazilime: An indicator plant of citrus decline virus in India. *Indian Phytopath.* 14: 109–112.

Capoor, S.P. 1963. Decline of citrus trees in India. *Bull. Nat. Inst. Sci. India,* 24: 48-64.

Capoor, S.P. 1965. Presence of seedling yellows complex in the citrus of South India. pp. 30–35. *In*: Proc. 3rd Conf. IOCV, IOCV, Riverside, California.

Capoor, S.P. 1975. Role of the tristeza virus in citrus dieback complex. *Indian J. Hort.* 32:1-6.

Capoor, S.P. and Rao,D.G. 1967. Tristeza virus infection of citrus in India. *In*: Proc. Intl. Symp. on Subtropical & Tropical Horticulture. Hort. Soc. of India, Bangalore, pp. 723–736.

Catara, A., Azzaro, A., Davino, M. and Polizzi, G. 1993. Yellow vein clearing of lemon in Pakistan. pp. 364–367. *In*: Proc. 12th Conf. IOCV, IOCV, Riverside, CA.

Cevik, B., Lee, R.F. and Niblett, C.L. 2006. Genetic transformation of *Citrus paradisi* with antisense and untranslatable RNA-dependent RNA polymerase genes of

Citrus tristezaclosterovirus. *Turkish Journal of Agriculture and Forestry* 30: 173–182.

Chagas, C.M., and Rossetti, V. 1984. Transmission of leprosis by grafting. pp. 215–217. *In*: Proc. 9th Conf. IOCV, IOCV, Riverside, CA.

Chagas, C.M., Rossetti, V. and Chiavegato, L.G. 1984. Effectiveness of the different life cycle stages of Brevipalus phoenicis Geijskes in lementin transmission. pp. 211–214. *In*: Proc. 9th Conf. IOCV, IOCV, Riverside, CA.

Chakraborty, N.K. and Ahlawat, Y.S. 2001. Indexing of cross protection and non-cross protected citrus cultivars/species for citrus tristerza virus by enzyme linked immunosorbent assays. *Plant Disease Research* 16: 10–16.

Chakraborty, N.K., Ahlawat, Y. S., Varma, A., Jagdish Chandra, K., Ramapandu, S. and Kapur, S.P. 1993. Serological reactivity in citrus tristeza virus strains in India. pp. 108–112. *In*: 12th Proc. IOCV, IOCV, Riverside, CA.

Chen Guoqing, Wong Hongxiang, Yan Senxiang and Roistacher, C.N. 1996. Occurrence and detection of citrus vein enation virus in Huangyan Zhejiang, China. pp. 218–219. *In*: Proc. 13th Conf. IOCV, IOCV, Riverside, CA.

Chen, G.Q., Yan, S.X. and Roistacher, C.N. 1992. First report of citrus vein enation disease in China. *Plant Disease* 76: 1077.

Chen, J., Zhang, T., Zhou, X. and Li, D. 1995. A study on citrus tatter leaf virus. *Acta Microbiol. Sin.* 35: 173–180.

Chenulu, V.V. and Ahlawat, Y.S. 1994. Virus and mycoplasma diseases of citrus in India. ICAR publication, New Delhi, India.

Childers, C.C., Rodrigues, J.C.V., Derrick, K.S., Achor, D.S., French, J.V., Welbourn, W.C., Ochoa, R. and Kitajima, E.W. 2003. Citrus leprosis and its status in Florida and Texas: past and present. *Exp. Appl. Acarol.* 30: 181–202.

Childers, C.C., Rodrigues, J.C.V., Kitajima, E.W., Derrick, K.S., Rivera, C. and Welbourn, W.C. 2001. A control strategy for breaking the virus-vector cycle of *Brevipalpus* spp. and the Rhabdovirus disease, citrus leprosis. *Manejo Integrado de Plagas* 60: 76–79.

Clark, C.C. and da Graca, J.V. 2000. Detection of citrus vein enation virus using cereal yellow dwarf virus ELISA kits. pp. 357–359. *In*: Proc. 14th Conf. IOCV, IOCV, Riverside,CA.

Colariccio, A., Lovisolo, O., Chagas, C.M., Galleti, S.R., Rosetti, V. and Kitajima, E.W. 1995. Mechanical transmission and ultrastructural aspects of citrus leprosis disease. *Fitopatol. Bras.* 20: 208–213.

Corbett, M.K. and Grant, T.J. 1967. Purification of citrus variegation virus. *Phytopathology* 57: 137.

Costa, A.S., Grant, T.J. and Moreira, S. 1949. Investigacoes sobre a tristeza dos citrus. *Bragantia* 9 : 59–80.

Cui, P.F., Gu, C.F. and Roistacher, C.N. 1991. Occurrence of lement dwarf virus in Zhejiang Province, China. *Plant Disease* 75: 242–244.

D'Onghia A.M., Djelouah, K., Frasheri, D. and Potere, O. 2001. Detection of *Citrus psorosis* virus by direct tissue blot immunoassay. *Journal of Plant Pathology* 83: 139–142.

Da Graca, J.V. 1977. Citrus tatter leaf virus in South African Meyer lemon. *Citrus and Subtrop. Fruit J.* 529: 18.

da Graça, J.V. and Skaria, M. 1996. Citrus tatter leaf virus in the Rio Grande Valley of South Texas. pp. 198–200. *In*: Proc. 13th Conf. IOCV, IOCV Riverside, CA.

da Graca, J.V., Bar-Joseph, M. and Derrick, K.S. 1993. Immunoblot detection of citrus psorosis in Isreal using citrus ringspot antiserum. pp. 432–434. *In*: Proc. 12th Conf. IOCV, IOCV, Riverside, USA

Da Graca, J.V., Lee, R.F., Moreno, P., Civerolo, E.L. and Derrick, K.S. 1991. Comparison of citrus ringspot, psorosis and other virus like agent of citrus. *Plant Dis.* 75: 613–616

da Graqa, J.V. and Maharaj, S.B. 1991. Citrus vein enation virus, a possible luteovirus, p. 391–394. *In*: Proc. 11th Conf. IOCV., IOCV, Riverside.

Dakshinamurti, V. and Reddy, G.S. 1975. Mosaic – a transmissible disorder of sweet oranges. *Indian Phytopathology* 28: 398–399.

Dakshinamurty, V. and Subbayya, J. 1981. Studies on transmission of citrus mosaic – a New disease of *Citrus sinensis* in Andhra Pradesh. (Abs) 3rd International Symp. Plant Pathology, New Delhi, December 14–18, 1981.

Dauthy-Yot and Bove, J.M. 1965. Isolement du virus de la frisolee des citrus (crinkly leaf virus). Note prkliminaire. *Fruits* 19: 187–93.

Davino, M. and Garnsey, S.M. 1984. Purification, characterization and serology of a mild strain of citrus infectious variegation virus from Florida. pp. 196–203. *In*: Proc. 9th Conf. IOCV, IOCV, Riverside, CA.

Derrick, K.S., Brlansky, R.H., Graca, J.V. da, Lee, R.F., Timmer, L.W. and Nguyen, T.K. 1988. Partial characterization of a virus associated with citrus ringspot. *Phytopathology* 78: 1298–1301.

Derrick, K.S., Brlansky, R.H., Lee, R.F., Timmer, L.W. and Nguyen, T.K. 1988. Two components associated with the citrus ringspot virus. pp. 340–342. *In*: Proc. 10th Conf. IOCV, IOCV Riverside, CA.

DeSimone, E.R., Laranjeira, F.F., Neri, F.M., Cunniffe, N.J. and Gilligan, C.A. 2010. Modelling spread and control of Bahia bark scaling of citrus. p. 49. *In*: Abs. 18th Conf. IOCV held in Campinas/SP, Brazil, Nov. 7–12, 2010.

Desjardins, P.R. and French, J.V. 1970. Purification of the citrus infectious variegation virus by density gradient electrophoresis. *Virology* 40: 746–151.

Desjardins, P.R. and Wallace, J.M. 1966. Host to citrus infectious variegation including wide diversity of virus strains. *Phytopath.* 56: 347–348.

Desjardins, P.R., Drake, R.J. and French, J.V. 1969. Transmission of citrus ringspot virus to citrus and noncitrus hosts by dodder (*Cuscuta subinclusa* Dur. and Hilg). *Plant Dis. Reptr.* 53: 947–948.

Desjardins, R. 1969. Purification, electron microscopy and serology of citrus viruses– A review? pp. 1481–1488. *In*: Proc. 1st Intern. Citrus Symp. 3. (Chapman, H.D. *Ed.*) held at the University of California, Riverside.

Dickson, R.C., Flock, R.A. and Johanson, M.M. 1951. Insect transmission of citrus quick decline. *J. Econo. Ent.* 44: 172–176.

Djelouah, K. and d'Onghia, A.M. 1986. Occurrence and spread of Citrustristeza in the Mediterranean area. pp. 14–16. *In*: Proc. 10th Conf. IOCV, IOCV, Riverside.

Djelouah, K., Potere, O., D'Onghia, A.M. and Savino, V. 2000. Production of monoclonal antibodies to citrus psorosis virus. pp. 152–158. *In*: Proc. 14th Conf. IOCV, IOCV, Riverside, California.

Dodds, J.A., Jordan, R.L., Roistacher, C.N. and Jarupat, T. 1987. Diversity of citrus tristeza virus isolates indicated by dsRNA analysis. *Intervirology* 27: 177–188.

Dodds, J.A., Morris, T.J. and Jordan, R.L. 1984. Plant viral double stranded RNA. *Ann. Rev. Phytopathol.* 22: 151–168.

Dominguez, A., Guerri, J., Cambra, M., Navarro, L., Moreno, P. and Pena, L. 2000. Efficient production of transgenic citrus plants expressing the coat protein gene of citrus tristeza virus. *Plant Cell Reports* 19: 427–433.

Dominguez, F.S., Bernal, A., Childers, C.C. and Kitajima, E.W. 2001. First report of *Citrus leprosis* virus in Panama. *Plant Dis.* 85: 228.

Fauquet, C.M., Mayo, M.A., Maniloff, J., Desselberger, U. and Ball, L.A. (*Eds.*) 2005. 8th Report of the International Committee on Taxonomy of viruses. Academic press, p. 1162, Elsevier publication 27th May 2005.

Fawcett, H.S. 1911. Scaly bark or nail-head rust of citrus. Florida Agricultura Experimental Station, Bulletin 106, p. 41.

Fawcett, H.S. 1932. New angles on treatment of bark lesions of citrus. *California Citrograph* 17: 106–108.

Fawcett, H.S. 1933. New symptoms of psorosis, indicating a virus disease of citrus. *Phytopathology* 23: 930.

Fawcett, H.S. 1936. Citrus Diseases and their Control. McGraw Hill, New York, London, p. 656.

Fawcett, H.S. and Bitancourt, A.A. 1943. Comparative symptomatology of psorosis varieties on citrus in California. *Phytopath.* 33: 837–864.

Fawcett, H.S. and Klotz, L.J. 1938. Types of symptoms of psorosis and psorosis-like diseases. *Phytopathology* 28: 670.

Fawcett, H.S. and Klotz, L.J. 1939. Infectious variegation of citrus. *Phytopathology* 29: 911–912.

Fawcett, H.S. and Wallace, J.M. 1946. Evidence of virus nature of citrus quick decline. *Calif. Citrog.* 32: 88–89.

Fawcett, H.S. and Lee, H.A. 1926. Citrus Diseases and their Control . New York, USA: McGraw-Hill.

Febres, V.J., Moore, G.A. and Lee, R.F. 2005. Development of pathogen-derived transgenic Citrus for resistance against *Citrus tristeza* virus. p. 490 *In*: Proc. 16th. Conf. IOCV. IOCV.

Febres, V.J., Niblett, C.L., Lee, R.F. and Moore, G.A. 2002. Characterization of grapefruit plants (*Citrus paradisi* Macf.) transformed with *Citrus tristeza* closterovirus genes. Springer Verlag.

Fraser, L.R. 1958. Virus diseases in Australia. *Proc. Linn. Soc. NSW* 73: 9–19.

Fraser, L.R. 1961. Lemon crinkly leaf virus. pp. 205–210. *In*: Proc. 2nd Conf. IOCV, IOCV, Riverside, CA.

Fraser, L.R. and Broadbent, P. 1979. Virus and related diseases of citrus in New South Wales. Dept. Agric. New South Wales, Australia 52 pp.

Fraser, Lilian. 1952. Seedling yellows – an unreported disease of citrus. *Agr. Gaz. N.S. Wales* 63: 125–131.

Frezzi, M.S. 1940. La lepra lementin del naranjo –Investigaciones realizadas per el laboratorio de patologia de bella vista (lementine. Bol fruitas y hortalizas. Min. Agric. De la Nacion (Buenos Aires) 5: 16.

Frison, E.A. and Taher, M.M. 1991. FAO/IBPGR technical guidelines for the safe movement of citrus germplasm. FAO/IBPGR, Rome, Italy.

Galipienso, L., Navarro, L., Ballester-Olmos, J.F., Pina, J., Moreno, P. and Guerri, J. 2000. Host range and symptomatology of graft-transmissible pathogen causing bud union crease of citrus on trifoliate rootstocks. *Plant Pathology* 49: 308–314.

Galipienso, L., Vives, M.C., Moreno, P., Milne, R.G., Navarro, L. and Guerri, J. 2001. Partial characterization of citrus leaf blotch virus, a new virus from Nagami kumquat. *Archives of Virology* 146: 357–368.

Galipienso, L., Vives, M.C., Navarro, L., Moreno, P., and Guerri, J. 2004. Detection of citrus leaf blotch virus using Digoxigenin-labeled cDNA probes and PCR. *European J. Plant Path.* 110: 175–181.

Garcia, M.L., Arrese, L.E., Grau, O. and Sarachu, A.N. 1989. Citrus psorosis disease agent behaves as a two component ssRNA Virus. pp. 380–384. *In*: Proc. 11th Conf. IOCV. IOCV. Riverside.

Garcia, M.L., Bo, E.D., Grau, O. and R.G. Milne 1994. The closely related citrus ringspot and citrus psorosis viruses have particles of novel filamentous morphology. *J. Gen. Virol.* 75: 3585–3590.

García, M.L., De La Torre, M.E., Dal Bo, E., Djelouah, K., Rouag, N., Luisoni, E., Milne, R.G., and Grau, O. 1997. Detection of citrus psorosis-ringspot virus using RT-PCR and DAS-ELISA. *Plant Pathol.* 46: 830–836.

García, M.L., Grau, O. and Sarachu, A.N. 1991. *Citrus psorosis* is probably caused by a bipartite ssRNA virus. *Research in Virology* 142: 303–311.

Garnsey, S.M. 1968. A citrus crinkly-leaf-type virus recently discovered in Florida. *Proc. Fla. State Hort. Soc.* 81: 79–84.

Garnsey, S.M. 1968. Additional non-citrus hosts for the Florida isolate of citrus variegation virus. *Phytopathology* 58: 1433–1434.

Garnsey, S.M. 1974. Mechanical transmission of a virus that produces tatter leaf symptoms in *Citrus excelsa*. pp. 137–140. *In*: Proceedings 6th Conference IOCV, IOCV, Riverside, USA.

Garnsey, S.M. 1974. Purification and serology of a Florida isolate of citrus variegation virus, pp. 169–75. *In*: Proc. 6th Conf. IOCV, IOCV, Riverside.

Garnsey, S.M. 1975. Purification and properties of citrus leaf rugose virus. *Phytopathology* 65: 50–57.

Garnsey, S.M. 1988. Vein enation (woody gall). *In*: Compendium of citrus diseases (*Eds.*Whiteside, J.O., Garnsey, S.M. and Timmer, L.W.), p. 50. APS, St. Paul, USA.

Garnsey, S.M. 1993. Indexing procedure for 15 viruses. pp. 5–15 in Agriculture Handbook No. 333.

Garnsey, S.M. and Drake, R.J. 1968. Indexing of tatter leaf with cow pea. pp. 80–82. *In*: Agric. Handbook 3332 USDA Washington D.C.

Garnsey, S.M. and Gonsalves, D. 1976. Citrus leaf rugose. CMI/AAB Description of plant viruses no. 164. Commonwealth Mycological Institute, Kew, Surrey, England

Garnsey, S.M. and Muller, G.W. 1988. Efficiency of mechanical transmission of citrus tristeza virus. pp. 46–54. *In*: Proc. 10th Conf. IOCV, IOCV, Riverside, CA.

Garnsey, S.M. and Purcifull, D.E. 1969. Serological detection of a citrus virus in leaf extracts from field trees. *Proc. Fl. State Hort. Soc.* 82: 56–60.

Garnsey, S.M. and Timmer, L.W. 1988. Local lesion isolate of citrus ringspot virus induces psorosis bark scaling. pp. 334–339. *In*: Proc. 10th Conf. IOCV, IOCV, Riverside.

Garnsey, S.M., Civerolo, E.L., Gumpf, D.J., Paul, C., Hilf, M.E., Lee, R.F., Brlansky, R.H., Yokomi, R.K. and Hartung, J.S. 2005. Biological Characterization of an International Collection of *Citrus tristeza* virus (CTV) Isolates. pp. 75–93. *In*: Proc. 16th Conf. IOCV, IOCV, Riverside, CA.

Garnsey, S.M., Kano, T., Permar, T.A., Cambra, M., Koizumi, M. and Vela, C. 1989. Epitope diversity among citrus tristeza isolates. *Phytopathology* 79: 1174 (Asbtr.).

Garnsey, S.M., Permar, T.A., Cambra, M. and Henderson, C.T. 1993. Direct tissue Blot immunoassay (DTBIA) for detection of citrus tristeza virus (CTV). pp. 39–50. *In*: Proc. 12th Conf. IOCV, IOCV Riverside, CA.

Ge, X. and Scott, S.W. 1994. The nucleotide sequence of citrus leaf rugose ilarvirus RNA-2. *J. Gen. Virol.* 75: 2841–2846.

Gerola, F.M., Lombardo, G. and Catara, A. 2005. Histological localization of citrus infectious variegation virus (CVV) in *Phaseolus vulgaris*. *Protoplasma* 67: 319–326.

Ghosh, D.K., Aglave, B., Bhanare, K. and Baranwal, V.K. 2007. PCR based detection of citrus yellow mosaic disease from Vidarbha region of Maharashtra. *Indian Phytopath.* 60: 520–526.

Ghosh, D.K., Aglave, K. and Baranwal, V.K. 2008. Simultaneous detection of one RNA and one DNA virus from naturally infected citrus plants using duplex PCR technique. *Curr. Sci.* 94: 1314–1318.

Ghosh, D.K., Balaji Aglave, Roy, A. and Ahlawat, Y.S. 2009. Molecular cloning, sequencing, and phylogenetic analysis of coat protein gene of a biological distinct citrus tristeza virus isolate occurring in central India. *J. Plant Biochemistry & Biotechnology* 18: 105–108.

Ghosh, D.K., Baranwal, V.K. and Ahlawat, Y.S. 2010. Improved diagnostics and molecular characterization of citrus mosaic badnavirus prevalent in India. p. 87. *In*: Abstract 18th Conf. IOCV hels in Campinas/SP, November 7–12, 2010.

Ghosh, D.K., das, A.K. and Singh, S. 2003. Individual and mixed infection of citrus tristeza virus and greening librobacter and their association with sweet orange decline in the state of Maharashtra. *J. Mycol. Pl. Pathol.* 33: 69–72.

Gonsalves and Garnsey, S.M. 1976. Association of particle size with sedimentation velocity of the nucleoprotein components of citrus variegation and leaf rugose viruses pp. 109–115. *In*: Proc. 7th Conf. IOCV, IOCV, Riverside, CA.

Gonsalves, D. and Garnsey, S.M. 1974. Infectivity of the multiple nucleoprotein and RNAcomponents of citrus leaf rugose virus. *Virology* 61: 343–53.

Gonsalves, D. and Garnsey, S.M. 1975. Nucleic acid components of citrus variegation virus and their activation by coat protein. *Virology* 67: 311–318.

Gonsalves, D. and Garnsey, S.M. 1975a. Functional equivalence of an RNA component and coat protein infectivity of citrus leaf rugose virus. *Virology* 64: 23–31.

Gonsalves, D. and Garnsey, S.M. 1975b. Infectivity of heterologous RNA-Protein mixture from alfalfa mosaic, citrus leaf rugose, citrus variegation and tobacco streak viruses.*Virology* 67: 319–326.

Grant, T.J. and Corbett, M.K. 1960 Mechanical Transmission of the Infectious Variegation Virus of Citrus. *Nature* 188: 519–520.

Grant, T.J. and Smith, P.F. 1960. Infectious variegation of citrus found in Florida, *Plant Dis. Rep.* 44: 426–429.

Grimaldi, V. and Catara, A. 1996. Association of filamentous virus with yellow vein clearing of lemon. pp. 343–345. *In*: Proc. 13th Conf. IOCV, IOCV, Riverside.

Groot, T.V.M., Freitas-Astua, J. and Kitajima, E.W. 2006. *Brevipalpus phoenicis* transmits Citrus leprosis virus, cytoplasmic type (CiLV-C) to common bean (*Phaseolus vulgaris*) znder experimental conditions. *Virus Reviews & Research*, 11(supl.): 67.

Guardo, M., Sorrentino, G., Marletta, T. and Caruso, A. 2007a. First report of Citrus leaf blotch virus on Kumquat in Italy. *Plant Disease* 91: 1054.

Guardo, M., Castellano, M.A., Savino, V. and Caruso, A. 2007b. First detection of Citrus leaf blotch virus in Italy. *Journal of Plant Pathology* 89: S13.

Guardo, M., Potere, O., Castellano, M.A., Savino, V. and Caruso, A. 2009. A new herbaceous host of citrus leaf blotch virus. *J. Plant Pathology* 91: 485–488

Guerra-Moreno, A.S., Keremane, M.L., Brlansky, R.H. and Lee, R.F. 2007. The complete sequence of the cytoplasmic Citrus lementin virus (CCLV) and its genome organization. *Phytopathology* 97: S174.

Guerra-Moreno, A.S., Manjunath, K.L., Brlansky, R.H. and Lee, R.F. 2005. Citrus Leprosis Symptoms can be Associated with the Presence of two Different viruses: Cytoplasmic and Nuclear, the Former having a Multipartite RNA Genome. pp. 230–239. *In*: Proc. 16th Conf. IOCV, IOCV, Riverside, CA.

Guerri, J., Moreno, P. and Lee, R.F. 1990. Identification of citrus tristeza strains by peptide maps of virion coat protein. *Phytopathology* 80: 692–698.

Guerri, J., Pina, I.A., Vives, M.C., Navarro, L. and Moreno, P. 2004. Seed transmission of Citrus leaf blotch virus: implication in quarantine and certification programs. *Plant Disease* 88: 906.

Guirado, N., Muller, G.W., Betti, J.A. and Costa, A.S. 1995. Particulas de controle da leprose dos citrus com extratos vegetais. *Fitopatol. Bras.* 20: 286.

Gutierrez, M.A., Luth, D. and Murre, G.A. 1997. Factors affecting Agrobacterium-mediated transformation in Citrus and production of sour orange (*Citrus aurantium* L.) plants expressing the coat protein gene of citrus tristeza virus. *Plant Cell Reports* 16: 745–753.

Hailstones, D.L., Bryant, K.L., Broadbent, P. and Zhou, C. 2000. Detection of Citrus tatter leaf virus with reverse transcription-polymerase chain reaction. *Aust. Plant Pathol.* 29: 240–248.

Hajeri, S., Ramadugu, C., Manjunath, K., Vidalakis, G. and Lee, R. 2010. Nucleotide sequence and genome organization of Dweet mottle virus and its relationship to members of the family betaflexiviridae. *Archives of Virology* 155: 1523–1527.

Harper S.J., Chooi, K.M. and Pearson, M.N. 2008. First report of Citrus leaf blotch virus in New Zealand. *Plant Disease* 92: 1470.

Harper, S.J., Dawson, T.E. and Pearson, M.N. 2010. Isolates of Citrus tristeza virus that overcome *Poncirus trifoliata* resistance comprise a novel strain. *Archives of Virology* 155: 471–480.

Harrison, B.D. and Murant, A.F. 1977. Nepovirus group. CMI AAB Descr. Plant Viruses, No. 185.

Hermoso de Mendoza, A., Pina, J.A., Ballester-Olmos, J.F. and Navarro, L. 1993. Persistent transmission of citrus vein enation virus by *Aphis gossypii* and *Myzus persicae*. pp. 361–362. *In*: Proc. 12th Conf. IOCV, IOCV, Riverside, CA.

Hibino, H., Tsuchizaki, T., Usugi, T. and Saito, Y. 1977. Fine structures and delopmental process of tubules lementi by mulberry ringspot virus and lement dwarf virus infections. *Ann. Phytopathol. Soc. Japan* 43: 255–264.

Hoa ,N.V., Ahlawat, Y.S. and Pant, R.P. 2004. Production of virus-free Kinnow mandarin and Mosambi sweet orange nucleus planting material through shoot tip grafting. *Indian Phytopath* 57: 482–487.

Hoa, N.V. and Ahlawat, Y.S. 2004. Characterisation of four isolates of Indian citrus ringspot virus. *Indian Phytopath* 57(3): 296–302

Hooper, G.R. and Schneider, H. 1969. Anatomical investigations of leaf enations on *Citrus aurantifolia* and *C. jambhiri* infected with citrus vein enation virus. *Phytopathology* 57: 1006.

Huang, Q. and Hartung, J.S. 2001. Cloning and sequence analysis of an infectious clone of citrus yellow mosaic virus that can infect sweet orange via Agrobacterium-mediated inoculation. *J. Gen. Virol.* 82: 2549–2558.

Hughes, W.A. and Lister, C.A. 1949. Lime disease in the Gold coast. *Nature* 164: 880.

Hull, R. and Matthews, R.E.F. 2002. Plant Virology. Academic Press, San Diego, CA, USA, p. 1001.

Hume, H.H. 1901. Some fungous diseases of citrus and others fruits. *Proc. Florida State Hort. Soc.* 14: 64–70.

ICTVdB Management 2006. 00.056.0.06.001.00.002.002. Citrus tatter leaf virus, isolate. *In*: ICTVdB – The Universal Virus Database, version 4. Büchen-Osmond, C. (*Ed*), Columbia University, New York, USA.

Inouye, N., Maeda, T. and Mitsuhata, K. 1979. Citrus tatter leaf virus isolated from lily. *Ann. Phytopathol. Soc. Jap*. 54: 712–720.

Iracheta-Cárdenas, M.M., Metheney, P., Polek, M.L., Manjunath, K.L., Lee, R.F. and Rocha-Peña, M.A. 2009. Serological detection of *Citrus tristeza* virus with antibodies developed to the recombinant coat protein. *Plant Disease* 93: 11–16.

Ishigai, T. and Jinno, M. 1958. On citrus mosaic. *Annals of the Phytopathological Society of Japan* 23: 29.

Isoda, T. and Gyoutoku, Y. 1990. [Studies on citrus virus diseases. VII. Propagation of satsuma dwarf virus to adjacent trees in a citrus orchard]. *Proceedings of the Association for Plant Protection of Kyushu* 36: 64–67.

Isoda, T., Gyoutoku, Y. and Yamada, K. 1991. [Studies on citrus virus diseases. VIII. Post-measure of orchard attacked by satsuma dwarf virus]. *Proceedings of the Association for Plant Protection of Kyushu* 37: 55–58.

Ito, T., Ieki, H. and Ozaki, K. 2002. Simultaneous detection of six citrus viroids and Apple stem grooving virus from citrus plants by multiplex reverse transcription polymerase chain reaction. *J. Virol. Methods* 106: 235–239.

Ito, T., Iwanami, T., Ieki, H., Shimomura, K., Shimizu, S.T. and Ito, T. 2004. A new virus related to satsuma dwarf virus: the nucleotide sequence of the 3′-terminal regions of Hyuganatsu virus RNAs 1 and 2. *Archives of Virology* 149: 1459–1465.

Iwanami, T. and Ieki, H. 1996. Nucleotide sequence of the coat protein genes of citrus mosaic virus. *Virus Res*. 42: 181–186.

Iwanami, T. and Koizumi, M. 2000. Satsuma dwarf virus group. *In*: Compendium of Citrus disease, 2nd edn. (*Eds.*) Timmer, L.W., Garnsey, S.M. and Graham, J.H., APS Press.

Iwanami, T., Kano, T. and Koizumi, M. 1991. Pathogenic diversity of citrus tatter leaf virus isolates. *Annals of the Plant Pathological Society of Japan* 57: 74.

Iwanami, T., Kano, T. and Koizumi, M. 1992. Spherical virus-like particles associated with vein enation on Yuzu (*Citrus junos*). Bull. Fruit Tree Research Station No. 23: 137–143.

Iwanami, T., Kano, T., Koizumi, M. and Watanabe, Y. 1991. Leaf blotch and stem mosaic symptoms in sweet orange caused by a strain of satsuma dwarf virus (SDV). Bulletin of the Fruit Tree Research Station No. 21: 75–83.

Iwanami, T., Koizumi, M. and Ieki, H. 1993. Diversity of properties among satsuma dwarf virus and related viruses. *Annals of the Phytopathological Society of Japan* 59: 642–650.

Iwanami, T., Kondo, Y. and Karasev, A.V. 1999. Nucleotide sequences and taxonomy of satsuma dwarf virus. *Journal of General Virology* 80: 793–797.

Iwanami, T., Kondo, Y., Kobayashi, M., Han, S.S, and Karasev, A.V. 2001. Sequence diversity and interrelationships among isolates of Satsuma dwarf-related viruses. *Archives of Virology* 146: 807–813.

Iwanami, T., Yamao, F., Seno, T. and Ieki, H. 1996. Nucleotide sequence of the 3′-terminal region of RNA1 of satsuma dwarf virus. *Annals of the Phytopathological Society of Japan* 62: 4–10.

Izawa, H. 1966. Investigation on withering disease of *Citrus unshiu* Marc. (in Gamagori district, Aichi Prefecture). *Bull. Aichi Hort. Expt. Stn.*, 5: 1–9.

Karasev, A.N., Boyko, V.P., Gowda, S., Nikolaev, O.N., Hilf, M.E., Koonin, E.N., Niblett, C.L., Cline, K., Gumpf, D., Lee, R.F., Garnsey, S.M., Lewandowski, D.J. and Dawson, W.O. 1995. Complete sequence of the citrus tristeza virus RNA genome. *Virology* 208: 511–520.

Kawai, A. and Nishio, T. 1990. Detection of citrus tatter leaf virus by enzyme-linked immunosorbent assay (ELISA). *Annals of the Phytopathological Society of Japan* 56: 342–345.

Kawai, A., Tsukamoto, T., Namba, S. and Nishio, T. 1996. Citrus tatter leaf virus : a review of its properties and the development of a serological detection system. pp. 339–342. *In*: Proc. 13th Conf. IOCV, IOCV, Riverside, CA.

Kishi, K. and Tanaka, S. 1964. Studies on the indicator plants for citrus viruses. 2. Mechanical transmission of the virus causing satsuma dwarf to sesame (*Sesamum indicum* L.). *Ann. Phytopathol. Soc. Jap*. 29: 142–148.

Kitajima, E.W., Chagas, C.M. and Rodrigues, J.C.V. 2003. Brevipalpus-Transmitted plant virus and virus-like diseases: cytopathology and some recent cases. *Experimental and Applied Acarology* 30: 135–160.

Kitajima, E.W., Muller, G.W., Costa, A.S. and Yuki, W. 1972. Short rod-like particles associated with citrus leprosis. *Virology* 50: 254–258.

Kitajima, E.W., Silva, D.M., Oliviera, A.R., Muller, G.W. and Costa, A.S. 1964. Thread-like particles associated with tristeza disease of citrus. *Nature* 201: 1011–1012.

Klotz, L.J. and Fawcett, H. S. 1941. Color handbook of citrus diseases.Univ. Calif. Press, Berkeley and Los Angeles. p. 90.

Knorr, L.C. 1968. Studies on the etiology of leprosis in citrus. pp. 332–341. *In*: Proc. 4th Conf. IOCV, IOCV, Riverside, CA.

Knorr, L.C. and DuCharme, E.P. 1951. The relationship between Argentina's lepra explosis and Florida s scaly bark with implications for the Florida citrus grower. *Plant Dis. Reptr* 35: 70–75.

Knorr, L.C. and Price, W.C. 1958. Leprosis. *In*: Florida Guide to Citrus Diseases, and Nutritional Disorders in Color. Pratt, R.M. (*Ed*.), 112–113. Agricultural Experiment Station, Florida, USA.

Knorr, L.C., Denmark, H.A. and Burnett, H.C. 1968. Occurrence of *Brevipalpus* mites, leprosis and false leprosis on citrus in Florida. *Fla. Entomol.* 51: 11–17.

Koizumi, M. 1984. Elimination of tatter leaf-citrange stunt virus from lement mandarin by shoot-tip grafting following pre-heat-treatment. pp. 229–233. *In*: Proceedings 9th Conference IOCV, IOCV, Riverside, USA.

Koizumi, M. 1989. Citrus diseases in Japan: their control, and possible outbreaks in the rest of Asia. Extension Bulletin of Food and Fertilizer Technology Center, ASPAC No. 284, pp. 1–6.

Koizumi, M., Kano, T., Ieki, H. and Mae, H. 1988. China laurestine: a symptomless carrier of satsuma dwarf virus which accelerates natural transmission in fields. pp. 348–352. *In*: Proc. 10th Conf. IOCV., IOCV, Riverside, CA.

Kondo, H., Maeda, T., Shirako, Y. and Tamada, T. 2006. Orchid fleck virus is a rhabdovirus with an unusual bipartite genome. *Journal of General Virology* 87: 2413–2421.

Krueger, R., Bash, J.A., Rangel, B. and Lee, R.F. 2005. Dweet mottle virus and citrus leaf blotch virus in California. International Organization of Citrus Virologists Newsletter.

Krueger, R., Lee, R.F. and Bash, J.A. 2005. Dweet mottle virus and citrus leaf blotch virus. Subtropics Newsletter V3 No. 1, Pages 7–9.

Kubaa, R.A., Djelouah, K. and D'Onghia, A.M. 2010. Study of the genetic variability of citrus infectious variegation virus isolates from different Mediterranean areas. p. 84. *In*: Abs. 18th Conf. IOCV held in Campinas/SP, Brazil, November 7–12, 2010.

Kubo, K.S., Arena, G.D., Kitajima, E.W., Machado, M.A. and Freitas-Astua, J. 2010. Transmission of citrus leprosies virus C (CiLV) to *Arabidopsis thaliana* by *Brevipalpus phoenicis* mite. p. 85. *In*: Abs. 18th Conf. IOCV held in Campinas/SP, Brazil, November 7–12, 2010.

Laird, E.F. and Weathers, L.G. 1961. *Aphis gossypii*, a vector of citrus vein enation virus. *Plant Dis. Reptr*. 45: 877.

Lee, R.F. and Bar-Joseph, M. 2000. Tristeza. pp. 61–63. *In*: Compendium of citrus diseases, 2nd ed. Timmers, L.W., S.M. Garnsey and J.H. Grahameds. APS, St.Paul, MN.

Lee, R.F. and Garnsey, S.M. 1996. Citrus virus and virus-like pathogens : A continuing evolution of progress and problems. pp. 1–7. *In*: Proc. 13th Conf. IOCV, IOCV, Riverside, CA.

Lee, R.F., Febres, V.J. and Moore, G.A. 2007. Resistance to Citrus tristeza virus via RNA interference. Plant Cell Reports (published on line 20 Sep 07).

Lee, R.F., Pappu, H.R., Pappu, S.S., Roch-Pena, M.A., Febris, E.J., Manjunath, K.L., Nicolaeva, O.V., Karasev, A.N., Cevik, B., Akbulut, M., Benscher, D., Anderson, E.J., Price, M., Ochaoa, F. and Niblett, C.L. 1997. Progress on strain differentiation of citrus tristeza virus. *Rev. Mex. Fitopatol.* 79: 79–87.

Lee, R.F., Roistacher, C.N., Niblett, C.L., Rocha-Pena, M., Garnsey, S.M., Yokomi, R.K., Gumpf, D.J. and Dodds, J.A. 1992. Presence of *Toxoptera citricidus* in central America, a threat to citrus in Florida and the United States. *The Citrus Industry* 73: 8.

Leki, H. and Yamada, S. 1984. Inactivation of tristeza virus, lement dwarf virus and citrus tatter leaf virus by heat treatment. *Bull. Fruit Tree Res. Stn., Japan* B, 11: 71–87.

Liu, K., Song, Z., Zhou, Y., Li, Z. and Zhou, C. 2010. Detection of citrus tatter leaf virus by real-time RT-PCR. p. 90. *In*: Abs 18th Conf. IOCV held in Caminas/CP, Brazil, 7–12 Nov. 2010.

Locali, E.C., Freitas-Astua, J., Souza, A.A., Takita, M.A., Astua-Monge, G., Antonioli, R., Kitajima, E.W. and Machado, M.A. 2003. Development of a molecular tool for the diagnosis of leprosis, a major threat to citrus production in the Americas. *Plant Disease* 87: 1317–1321.

Locali-Fabris, E.C., Freitas-Astua, J., Souza, A.A., Takita, M.A., Astua-Monge, G., Antonioli, R., Rodrigues, V., Targon, M.L.P.N. and Machado, M.A. 2006. Complete nucleotide sequence, genomic organization and phylogenetic analysis of *Citrus leprosis* virus, cytoplasmic type. *Journal of General Virology* 87: 2721–2729.

Lovisolo, O. 2001. *Citrus leprosis* virus: Properties, diagnosis, agro-ecology and phytosanitary importance. *Bull. OEPP* 31: 79–89.

Lovisolo, O., Accotto, G.P., Masenga, V. and Colariccio, A. 2003. An isolate of Apple stem grooving virus associated with *Cleopatra mandarin* fruit intumescence. *Fitopatol. Bras*. 28: 54–58.

Lovisolo, O., Colariccio, A., Chagas, C.M., Rosetti, V., Kitajima, E.W. and Harakava, R. 1996. Partial characterization of *Citrus leprosis* virus. pp. 178–188. *In*: Proc. 13th Conf. IOCV, IOCV, Riverside, CA.

Maharaj, S.B. and da Graca, J.V. 1988. Observation of isometric virus-like particles associated with citrus vein enation-infected citrus and the viruliferous aphid vector *Toxoptera citricidus*. *Phytophylactica* 20: 357–360.

Maharaj, S.B. and da Graca, J.V. 1989. Transmission of citrus vein enation virus by *Toxoptera citricidus*. *Phytophylactica* 21: 81–82.

Mali, V.R. and Chaudhuri, K.G. 1993. Occurrence of citrus virus and virus-like diseases in Maharashtra, India. pp. 466. *In*: Proc. 12th Conf. IOCV, IOCV, Riverside (Abs).

Mali, V.R., Chaudhuri, K.G. and Rane, S.D. 1975. Vein enation and woody gall diseases of citrus. *FAO Plant Prot. Bull.*, 23: 190–191.

Mali, V.R., Chaudhuri, K.G. and Rane, S.D. 1976. The vein-enation virus disease of citrus in India. *Indian Phytopathology* 29: 43–45.

Mandal, B., Biswas, K.K., Ahlawat, Y.S., Pradhan, G.S. and Debnath, S. 2001. *In vitro* culture of citrus tristeza free citrus plants and their survivability in nursery and field condition. Abs. Nat. Semi. On Approaches for increasing agricultural productivity in hill and mountain ecosystem. Org. by Indian ass. Hill farming UMIAM Meghalaya and ICAR Res. Complex for NEH Reg. October 18–20, 2001.

Manjunath, K.L. 1987. Studies on Vein enation virus disease of *Citrus* in South India. *Indian J. Plant Path*. 5: 121–125.

Manjunath, K.L., Chandrika, R., Warren, C., Niblett, C.L. and Lee, R.F. 2005. Complete nucleotide sequence of a new genotype of *Citrus tristeza* virus from an isolate having a mixed infection. pp. 487. *In*: Proc. 16th Conf. IOCV. IOCV, Riverside.

Manjunath, K.L., Pappu, H.R., Lee, R.F., Niblett, C.L. and Civerolo, E. 1993. Studies on the coat protein genes of four isolates of citrus tristeza closterovirus from India : cloning, sequencing and expression. pp. 20–27. *In*: Proc. 12th Conf., IOCV, IOCV, Riverside, CA.

Manjunath, K.L., Rangel, E., Guerra-Moreno, A.S., Brlansky, R.H. and Lee, R.F. 2006. Serological detection of the cytoplasmic *Citrus leprosis* virus (CiLV) from infected citrus plants. *Phytopathology*, 96: S 163.

Manjunath, K.L., Sawant, S.D. and Sawant, I.S. 1989. Stem pitting in coorg mandarin caused by citrus tristeza virus. (Abstr) Proc. Symp. Indian Phytopath. Soc. (Southern Chapter), p. 17.

Marais, L.J. and Lee, R.F. 1986. Citrange stunt virus associated with decline of shamouti on Swingle citrumelo rootstock in South Africa. *Plant Disease* 70: 892.

Maria Cecilia Zanek, Eduardo Peña, Carina Andrea Reyes, Julia Figueroa, Beatriz Stein, Oscar Grau and Maria Laura Garcia 2006. Detection of Citrus psorosis virus in the northwestern citrus production area of Argentina by using an improved TAS-ELISA. *Journal of Virological Methods* 137: 245–251.

Marques, J.P.R., Apezzatto-da-Glória, B. and Kitajima, E.W. 2004. Anatomia das lesões foliares e de frutos de *Citrus* spp. Causadas pela doença leprose. *In*: Congresso Brasileiro de Botânica, 55. Viçosa, Sociedade Botânica do Brasil. CD Rom.

Martin, S., C. Lopez, M. L. Garcia, G. Naum-Ongania, O. Grau, R. Flores, P. Moreno and J. Guerri 2005. The complete nucleotide sequence of a Spanish isolate of citrus psorosis virus: comparative analysis with other ophioviruses. *Archives of Virology* 150: 167–176.

Martin, S., Daniela, A., Milne R.G., Garnsey, S.M., Garcia, M.L., José, G. and Moreno, P. 2004. Detection of *Citrus psorosis* virus by ELISA, molecular hybridization, RT-PCR and immunosorbent electron microscopy and its association with *Citrus psorosis* disease. *European J. Plant Pathology* 110: 747–757.

Martin, S.,Garcia, M.L., Troisi, G., Rubio,L., Legarreta, G., Grau, O., Alioto, D., Moreno, P. and Guerri, J. 2006. Genetic variation of populations of *Citrus psorosis* virus. *J. Gen.Virol.* 87: 3097–3102.

Mayo M.A. and Haenni A.L. 2006. Report from the 36th and 37th Meetings of the Executive Committee of the International Committee on Taxonomy of Viruses. *Archives of Virology* 151: 1031–1037.

Mc Clean, A.P.D. 1954. Citrus vein enation virus. *S. Afr. J. Sci.* 50: 147–151

Meneghini, M. 1946. Sobre a natureza a transmissibilidale do doencia"tristeza" dos citrus. *O. Biologico* 12: 285–287.

Milne, R.G., Djelouah, K., Garcia, M.L., Dal Bo, E. and Grau, O. 1996. Structure of Citrus Ringspot-Psorosis-Associated Virus Particles: Implications for Diagnosis and Taxonomy pp. 189–186. *In*: Proc. 13th Conf. IOCV, IOCV, Riverside.

Milne, R.G., Garcia, M.L. and Grau, O. 2000. Genus Ophiovirus. Type species Citrus psorosis virus (CPsV). In: Van RegenmortelMHV, Fauquet CM, BishopDHL, CarstensEB, EstesMK, LemonSM, ManiloffJ, MayoMA, McGeochDJ, PringleCR, WicknerRB, eds. Virus Taxonomy. 7th Report of the International Committee on Taxonomy of Viruses. San Diego, CA: Academic Press, 627–31.

Miyakawa, T. 1968. Susceptibility of trifoliate orange and some other citrus varieties to satsuma dwarf virus. pp. 153–159. *In*: Proc. 4th Conf. IOCV., IOCV, Riverside, CA.

Miyakawa, T. 1978. A bud-union disorder of Japanese citrus on *Poncirus trifoliata* rootstock caused by tatter leaf virus. *Review of Plant Protection Research* 11: 1–10.

Miyakawa, T. 1980a. Thermo-therapy for some citrus cultivars infected by tatter leaf virus. *Bulletin of the Tokushima Horticultural Experiment Station* 9: 7–11.

Miyakawa, T. 1980b. Occurrence and varietal distribution of tatter leaf- citrange stunt virus and its effects on Japanese citrus. p. 220–224. *In*: Proceedings of the 8th Conference IOCV, Riverside, USA.

Miyakawa, T. and Matsui, C. 1976. A bud union abnormality of Satsuma mandarin on *Poncirus trifoliata* rootstock in Japan. pp. 125–131 *In*: Proc. 7th Conf. IOCV, IOCv, Riverside, CA.

Miyakawa, T. and Tsuji, M. 1988. The association of tatter leaf virus with bud-union crease of trees on trifoliate orange rootstock. pp. 360–364. *In*: Proceedings of the 10th Conference IOCV, IOCV, Riverside, USA.

Moore, G.A., Luth, D., Febres,V.J., Garnsey, S.M. and Niblett, C.L. 2000. Agrobacterium-Mediated transformation of grapefruit (*Citrus paradisi* Macf.) with genes from citrus tristeza virus. *Acta Hort*. 535: 237–243.

Moreira, S. 1942. Observacoes sobre a 'tristeza' dos citrus ou podridao das radicelas. *O. Biologico* 8: 269–272.

Nagpal, R.L. 1959. Tristeza found in Bombay state. *Calif: Citrog*., 44: 392, 402–405.

Naidu, P.H. and Reddy, M.R.S. 1985. Movement and distribution of *Citrus mosaic* in host tissue. *Indian Phytopathol*. 38: 782.

Nariani, T.K. and Raychaudhuri, S.P. 1968. Occurrence of tristeza and greening viruses in Bihar, West Bengal and Sikkim. *Indian Phytopathol*., 21: 343–344.

Nariani, T.K., Raychaudhuri, S.P. and Bhalla, R.B. 1966. Citrus tristeza virus in northern and central India. Indian Phytopathol., 19: 397–399.

Nariani, T.K., Sahambi, H.S. and Chona, B.L. 1965. Occurrence of tristeza virus in citrus in northern India. *Indian Phytopathol*. 18: 220.

Nariani, T.K., Viswanath, S.M. and Menon, M.R. 1970. Occurrence of tristeza and greening Diseases of citrus in Kerala. *Agric. Res. J. Kerala*, 8: 123–124.

Nario, Kusano 2006. Simple and rapid diagnosis of Satsuma dwarf by Immunochromatographic assay using monoclonal antibodies. *Plant Protection* 60: 491–495.

Navarro, L., Civerolo, E.L., Juarez, J. and Garnsey, S.M. 1989. Improvement therapy methods for citrus germplasm exchange. *In*: Abstracts of the 11th Conference of IOCV, IOCV, Riverside, USA.

Navarro, L., Civerolo, E.L., Juarez, J. and Garnsey, S.M. 1989. Improvement therapy methods for citrus germplasm exchange. *In*: 11th Conference IOCV, IOCV, Riverside, CA (Abs).

Navarro, L., Jfiarez, J., Ballester, J.F. and Pina, J.A. 1980. Elimination of Some Citrus Pathogens Producing Psorosis-like Leaf Symptbms, by Shoot-tip Grafting *in vitro*. pp. 162–166. Proc. 8th Conf. IOCV, IOCV, Riverside, CA.

Navarro, L., Pina, J.A., Ballester-Olmos, J.F., Moreno, P. and Cambra, M. 1984. A new graft transmissible disease found in Nagami kumquat. pp. 234–240 *In*: Proc. 9th Conf. IOCV, IOCV, Riverside, CA.

Navas-Castillo, J., Moreno, P., Cambra, M. and Derrick, K. 1993. Partial purification of a virus associated with a Spanish isolate of citrus ringspot. *Plant Pathology* 42: 339–346.

Nishio, T., Kawai, A., Kato, M. and Kobayashi, T. 1982. A sap-transmissible closterovirus in citrus imported from China and Formosa. *Research Bulletin of the Plant Protection Service Japan* 18: 11–18.

Nishio, T., Kawai, A., Takahashi, T., Namba, S. and Yamashita, S. 1989. Purification and properties of citrus tatter leaf virus. *Annals of the Phytopathological Society of Japan* 55: 254–258.

Norman, P.A. and Grant, T.J. 1956. Transmission of tristeza virus by aphids in Florida. *Proc. Fla. State Hort. Soc.* 69: 38–42.

Nozu, Y., Usugi, T. and Nishimori, K. 1986. Production of monoclonal antibodies to satsuma dwarf virus. *Annals of the Phytopathological Society of Japan* 52: 86–89.

Nunes, M.A., Kitajima, E.W., Freitas-Astua, J., Hill, M.E., Gottwald, T.R. and C.A.L. 2006. Infecção de hibisco e malvavisco pelo leme da leprose dos citros através de Brevipalpus phoenicis (Acari: Tenuipalpidae). *In*: Resumos Simpósio Brasileiro de Acarologia I. Viçosa, Universidade Federal de Viçosa. p. 218.

Oberholzer, P.C.J., Mattews, I. and Stiemie, S.F. 1949. The decline of grapefruit trees in South Africa. A preliminary report on so called "stem pitting". Science Bulletin of the Dept. Agric. S. Africa No. 297.

OEPP/EPPO 1995. Certification schemes No. 12. Pathogen-tested citrus trees and rootstocks. Bulletin OEPP/EPPO Bulletin 25: 4.

Ohira, K., Ito, T., Kawai, A., Namba, S., Kusumi, T. and Tsuchizaki, T. 1994. Nucleotide sequence of the 3'-terminal region of citrus tatter leaf virus RNA. *Virus Genes* 8: 169–172.

Ohira, K., Nimba, S., Rosanov, M., Kusumi, T. and Tsuchizaki, T. 1995. Complete sequence of an infectious full length cDNA clone of citrus tatter leaf capillovirus: comparative sequence analysis of capillovirus genomes. *J. Gen. Virol.* 76: 2305–2309.

Omoto, C. 1998. Acaricide resistance management of leprosis mite (*Brevipalpus phoenicis*) in Brazilian citrus. *Pesticide Sci.* 52: 189–198.

Önelge, N. 2003. First report of yellow vein clearing of lemons in Turkey. *J. Turkey Phytopathol.* 32: 53–55.

Onelge, N., Bozan, O., Gok, M. and Satar, S. 2010. Yellow vein clearing of lemons in Turkey. Proc. 17th Conf. IOCV, IOCV, Riverside, CA. (Short Commun).

Onelge, N., Satar, S., Gok, M., Bozan, O. and Kamberoglu, M. 2010. Transmission studies of citrus yellow vein clearing virus (CYVCV). p. 51. *In*: Abs. 18th Conf. IOCV held in Campinas/SP, Nov. 7–12, 2010.

Palle, S.R., Miao, H., Seyran, M., Louzada, E.S., da Graca, J.V. and Skaria, M. 2005. Evidence for Association of Citrus psorosis virus with Symptomatic Trees and an *Olpidium*-like Fungus in Texas. pp. 423–426. *In*: Proc. 16th Conf. IOCV, IOCV, Riverside, CA.

Palmieri, M., Blanco, S., Cruz, N., Lopez, L., Howd, D., Brlansky, R.H. and Lee, R.F. 2003. Situation of leprosis in Guatemala. Proc. of the Pan American Plant Dis. Conf., p. 181. South Padre Island, Texas, USA. (Abstr.).

Palmieri, M., Donis, I., Salazar, A.L., Blanco, S., Porres, M., Brlansky, R.H., Guerra-Moreno, A.S., Manjunath, K.L. and Lee, R.F. 2005. Leprosis in Guatemala. p. 510. *In*: Proc. 16th. Conf., IOCV, IOCV, Riverside.

Pandey, P.K., Chavan, V.M. and Ahlawat, Y.S. 1987. Nasnaran –an indicator host of severe strain of *Citrus tristiza* virus. p. 23. *In*: Proc. 5th Nat. Citrus Seminar, Jorhat.

Pandey, P.K., Nimbalkar, M.R. and Dhale, M.G. 1986. Observations on a mosaic disease of citrus in Maharashtra State. *Journal of Maharashtra Agricultural University* 11: 254 .

Pant, R.P. and Ahlawat, Y.S. 1997. Partial characterization of citrus mosaic virus. *Indian Phytopath* 50: 557–564.

Pant, R.P. and Ahlawat, Y.S. 1998. Partial characterization of a filamentous virus associated with ringspot disease of citrus. *Indian Phytopath.* 51: 225–232.

Pant, R.P., Ahlawat, Y.S. and Milne, R.G. 1997. Studies on *Citrus* ringspot virus in India. National Sympossium on Citriculture, Nov. 17–19. 1997 at NRCC, Nagpur, 82–83 (Abs.).

Pant, R.P., Baranwal, V.K., Singh, S.P. and Ahlawat, Y.S. 2000. Mechanical transmission of Indian citrus ringspot virus (ICRSV). p. 1023. *In*: Proc. International Symposium on Citriculture, Nagpur, India.

Pappu, H.R., Pappu, S.S., Niblett, C.L., Lee, R.F. and Civerolo, E. 1993. Comparative sequence analysis of the coat protein of biologically distinct citrus tristeza closterovirus isolates. *Virus Genes* 7: 255–264.

Pascon, R.C., Kitajima, J.P., Breton, M.C., Assumpção, L., Greggio, C., Zanca, A.S., Okura, V.K., Alegria, M.C., Camargo, M.E., Silva, G.G.C., Cardozo, J.C., Vallim, M.A.I., Franco, S.F., Silva, V.H., Jordão Jr., H., Oliveira, F., Giachetto, P.F., Ferrari, F., Aguilar-Vildoso, C.L., Franchiscini, F.J.B., Silva, J.M.F., Arruda, P., Ferro, J.A., Reinach, F. and Silva, A.C.R. 2006. The complete nucleotide sequence and genomic organization of *Citrus leprosis* associated virus, cytoplasmic type (CiLV-C). *Virus Genes* 32: 289–298.

Passos, O.S. 1963. Absence of Young-Leaf Symptoms of Psorosis in the State of Bahia, Brazil. pp. 167–169. *In*: Proc. 3rd Conf. IOCV, IOCV, Riverside, CA.

Peña, L. and Navarro, L. 1999. Transgenic citrus. *In*: Biotechnology in Agriculture and Forestry, 39–54. Y.P.S. Bajaj, Springer, Berlin.

Pérez-M., B.E. and Ochoa, A. 1998. Regeneration of transgenic plants of Mexican lime from Agrobacterium rhizogenes–transformed tissues. *Plant Cell Rep*. 17: 591–596.

Permar, T., Garnsey, S.M., Gumpf, D.J. and Lee, R. 1990. A monoclonal antibody that discriminates strains of citrus tristeza virus. *Phytopathology* 80: 224–228.

Petri, L. 1931 Variegatura infecttiva delle foglie di "citrus vulgaris". *Risso. Boll. R. Stazione Pat. Veg.*, N.S. 11: 105–114.

Prabha, K. 2010. Assessment of genetic diversity and seed transmission of mandrivirus infecting citrus. Ph.D. thesis, Post-graduate School, Indian Agricultural Research Institute, New Delhi.

Praveen Shelly, Gary, Barthe, A., Derrick, K.S. and Ahlawat, Y.S. 2003. Comparison of isolates of Indian citrus ring spot virus and Citrus posorosis virus. *Indian Phytopath.* 56: 230–232.

Praveen Shelly, Pawar, V. and Ahlawat, Y.S. 2003. Somatic Embryogenesis and Plant Regeneration in Kinnow Mandarin. *J. Plant Biochemistry and Biotechnology* 12: 163–165.

Price, W.C. 1963. Transmission of Psorosis virus by Dodder. pp. 162–166. *In*: Proc. 3rd Conf. IOCV, IOCV, Riverside, CA.

Price, W.C. 1970. En. C.M.I./A.A.B. Description of plant viruses no. 33.

Pringle, C.R. 1998. Virus taxonomy- san Diego 1998. *Arch. Virol*. 143: 1449–1459.

Rana, G.L., Quacquarelli, A. and Martelli, G.P. 1974. Some properties of citrus variegation virus. P.165-168. In: Proc. 6th Conf. IOCV, IOCV, Riverside, CA.

Rao, M.R.K. and Narasimhan, B. 1974. The effect of different citrus viruses on photosynthetic pigments. *Curr. Sci*. 43: 86–87.

Rao,D.G and Capoor,S.P.1965.Transmission of tristeza virus by *Aphis craccivora* Koch. and *Dactynotus jaceae* L. *Ind. J. Agric. Sci.* 35: 85–89.

Raychaudhuri, S.P., Nariani, T.K. and Ahlawat, Y.S. 1977. Dieback of citrus in India. *Proc. Intl. Soc. Citriculture* 3: 914–918.

Reddy, B.V.B. 1997. Characterization of citrus mosaic virus and to develop methods for its quick detection. Ph.D thesis, Division of Plant Pathology, IARI, New Delhi, India.

Reddy, G.S. 1965. Citrus decline in south India. p. 225. *In*: Proc. 3rd Conf: IOCV, IOCV, Riverside, CA.

Reddy, G.S. and Govinda Rao, P. 1961. Is there tristeza in Andhra Pradesh, India? pp. 132–135. *In*: Proc. 2nd Conf: IOCV, IOCV, Riverside, CA.

Reddy, M.R.S., Naidu, P.H. and Raju, D.G. 1985. Dodder laurel transmission of citrus mosaic from sweet orange to acid lime. *Indian Phytopathol* 38: 172.

Reuther, W., Calavan, E.C. and Carman, G.E. 1978. (*Eds.*) The citrus Industry, Vol.IV, Univ. California, Div. Agr. Sci. pp. 362.

Rhoads, A.S. and DeBusk, E.F. 1931. Diseases of citrus in Florida. *Agricultural Experiment Station Bulletin* 229: 84–91 and 208–211.

Robert, K., Bash, J., Benjamin, R. and Lee, R.F. 2005.Dweet mottle virus and citrus leaf blotch virus in California. IOCV Newsletter.

Rocha-Pena, M.A. and Lee, R.F. 1991. Serological techniques for detection of citrus tristeza virus. *J. Virol. Methods* 34: 311–331.

Rodrigues, J.C.V. 2000. Relações patogeno-vector-planta no sitema leprose dos citrus. Ph.D. Thesis. p. 168. da Universidade de São Paulo, Piracicaba, Brasil.

Rodrigues, J.C.V., Kitajima, E.W., Childers, C.C. and Chagas, C.M. 2003. *Citrus leprosis* virus vectored by *Brevipalpus phoenicis* (Acari: Tenuipalpidae) on citrus in Brazil. *Exp. Appl. Acar.* 30: 161–179.

Rodrigues, J.C.V., Locali, E., Freitas-Astúa, J. and Kitajima, E.W. 2005. Transmissibility of *Citrus leprosis* virus by *Brevipalpus phoenicis* in *Solanum violaefolium*. *Plant Disease*. 89: 911–911.

Roistacher C.N. and Blue R.L. 1968. A psorosis-like virus causing symptoms only on Dweet tangor. pp. 13–18 *In*: Proc. 4th Conf. IOCV, IOCV, Riverside, CA.

Roistacher, C.N. 1979. Elimination of citrus pathogens in propagative budwood I. Budwood selection, indexing and thermotherapy. pp. 965–972. *In*: Proc. Interl. Soc. Citriculture Vol. 3 (*Ed.* by Grierson, W.). University of California, Riverside, USA.

Roistacher, C.N. 1988. Citrus tatter leaf virus: further evidence for single virus complex. pp. 353–369. *In*: Proc. 10th Conf. IOCV, IOCV, Riverside, CA.

Roistacher, C.N. 1993. Psorosis – A review. pp. 139–154. *In*: Proc. 12th Conf. IOCV, IOCV, Riverside,CA.

Roistacher, C.N. and Dodds, J.A. 1993. Failure of 100 mild *Citrus tristeza* virus isolates from California to cross protect against a challenge by severe sweet orange stem pitting isolates. pp. 100–107. *In*: Proc. 12th Conf. IOCV, IOCV, Riverside, CA.

Roistacher, C.N. and Kitto, S.L. 1977. Elimination of additional citrus viruses by shoot-tip grafting *in vitro*. *Plant Dis. Reptr*. 61: 594–596.

Roistacher, C.N., Bash, J. and Gumpf, D.J. 2000. Continued attempts over a 22-year period to separate components of the citrus tatter leaf-citrange stunt virus complex. pp. 179–184. *In*: Proc. 13th Conf. IOCV, IOCV, Riverside, CA.

Roistacher, C.N., D'Onghia and Djelouah 2000. Defining psorosis by biological indexing and ELISA. *In*: pp. 144–151. Proc. 14th Conf. IOCV, IOCV, Riverside, California.

Roistacher, C.N., da Graça, J.V. and Müller, G.W. 2010. Cross Protection against *Citrus tristeza* virus – a Review. Proc. 17th Conf. IOCV, IOCV, Riverside, CA.

Roistacher, C.N., Nauer, E.M. and Wagner, R.L. 1980. Transmissibility of cachexia, dweet mottle, psorosis, tatterleaf and infectious variegation viruses on knife blades and its prevention. pp. 225–229. *In*: Proc. 8th Conf. IOCV, IOCV, Riverside, USA.

Román, R.C.F., Gutiérrez, E.M.A., Pérez-M.E., Gutiérrez, C.R., Gómez, L.M.A., Palacios, T., E.C. and Robles, G.M.M. 2005. Expression of different genes to confer resistance to pathogens in transgenic citrus plants. pp. 392–398. *In*: Proc. 16th Conf. IOCV, IOCV, Riverside, CA.

Rossetti, V. 1980. Differenciacao entre cancro citrico e outras molestias. *Citrus* 1: 23–26.

Rossetti, V. 1996. Transmission of *Citrus leprosis* disease (CL) – A review. pp. 331–335. *In*: Proc. 13th Conf. IOCV, IOCV, Riverside, CA.

Rossetti, V., Chagas, C.M. and Muntaner, A.C. 1975. Recent studies on transmission of leprosis and zonate chlorosis. p. 31. *In*: Proc. 7th Conf. IOCV, IOCV, Riverside, CA.

Roy, A. and Brlansky, R.H. 2009. Population dynamics of a Florida *Citrus tristeza* virus isolate and aphid-transmitted subisolates: Identification of three genotypic groups and recombinants after aphid transmission. *Phytopathology* 99: 1297–1306.

Roy, A. and Brlansky, R.H. 2010. Genome analysis of an orange stem pitting *Citrus tristeza* virus isolate reveals a novel recombinant genotype. *Virus Res.* 151: 118–130.

Roy, A. and Ramachandran, P. 2002. Bi-directional PCR – a tool for identifying strains for *Citrus tristeza* virus. *Indian Phytopath.* 55: 185–186.

Roy, A., Ananthakrishnan, G., Hartung, J.S. and Brlansky, R.H. 2010. Development and application of a multiplex reverse transcription polymerase chain reaction assay for screening a global collection of *Citrus tristeza* virus isolates. *Phytopathology* 100: 1077–1088.

Roy, A., Fayad, A. Barthe, G. and Brlansky, R.H. 2005. A multiplex polymerase chain reaction methos for reliable, sensitive and simultaneous detection of multiple viruses in citrus trees. *J. Virol. Methods* 129: 47–55.

Roy, A., Ramachandran, P. and Brlansky, R.H. 2003. Grouping and comparison of Indian *Citrus tristeza* virus isolates based on coat protein gene sequences and restriction analysis patterns. *Arch. Virol.* 148: 707–722.

Rustici, G., Accotto, G.P., Noris, E., Mesenga, V., Luisoni, E. and Milne, R.G. 2000. Indian citrus ringspot virus: a proposed new species with some affinities to potex, carla, fovea and allexiviruses. *Arch. Virol.* 145: 1895–1908.

Rustici, G., Milne, R.G. and Accotto, G.P. 2000. Nucleotide sequence, genome organization and phylogenetic analysis of Indian citrus ringspot virus. *Archives of Virology* 147: 2215–2224.

Rustici, G., Noris, E., Accotto, G.P., Luisoni, E., Milne, R.G., Pant, R.P. and Ahlawat, Y.S. 2000. Further characterization and detection of Indian citrus ringspot virus. pp. 360–362. *In*: Proc. 14th conf. IOCV, IOCV, Riverside, CA.

Saito, Y. and Hibino, H. 1962. Electron microscopy of satsuma dwarf virus in host cells. pp. 217–222. *In*: Proc. 5th Conf. IOCV., IOCV, riverside, CA.

Sánchez de la Torre, E., Riva, O., Zandomeni, R., Grau, O. and García, M.L. 1998. The top component of citrus ringspot virus contains two ssRNAs, the smaller encodes the coat protein. Molecular Plant Pathology on Line http://194.247.68.33/mppol/1998/1019sanchez.

Sánchez de la Torre, M.E., García, M.L., Riva, O., Dal Bó, E., Jones, L., Zandomeni, R. and Grau, O. 2000. Genome organization of the top componentof citrus psorosis virus and identification of the coat protein gene. pp. 345–346. *In*: Proc. 14th Conf. IOCV, IOCV, Riverside.

Saponari M and Yokomi RK 2007. Characterization of citrus tristeza virus isolates by single-stranded conformation polymorphism analysis of the coat protein gene. p. 139. In: Proc. 17th Conf. IOCV, IOCV, Riverside, CA (Abs).

Saponari, M. and Yokomi, R.K. 2010. Use of the Coat Protein (CP) and Minor CP Intergene Sequence to Discriminate Severe Strains of *Citrus tristeza* virus (CTV) in three U.S. CTV isolate Collections. Proc. 17th conf. IOCV, IOCV, Riverside

Saponari, M., Manjunath, K.L. and Yokomi, R.K. 2008. Quantitative detection of *Citrus tristeza* virus (CTV) in citrus and aphids by real-time reverse transcription-PCR (TaqMan®). *J.Virological Methods* 147: 43–53.

Satyanarayana, T., Mohammad, A.R., Hilf, M.E., Gowda, S., Dawson, W and Garnsey, S.M. 2009. Molecular characterization of citrus tatter leaf virus historically associated with Meyer lemon trees.: complete genome sequence and development of biologically active *in vitro* transcripts. *Phytopathology* 99: 423–431.

Schneider, H. 1959. The anatomy of tristeza virus infected citrus. *In*: Wallace, J.M. (*Ed.*) Citrus virus diseases. pp. 73–84. Univ. Cal. Div. Agr. Sci. Berkeley, Ca.

Scott, S.W. and Ge, X. 1995. The complete nucleotide sequence of the RNA-3 of Lilac ring mottle ilarvirus. *J. Gen. Virol.* 76: 1801–1806.

Scott, S.W. and Ge, X. 1995a. The complete nucleotide sequence of the RNA-3 of citrus leaf rugose and citrus variegation ilarviruses. *J. Gen. Virol.* 76: 957–963.

Scott, S.W. and Ge, X. 1995b. The nucleotide sequence of citrus leaf rugose virus RNA-1. *J. Gen. Virol.* 76: 3233–3238.

Sekiya, M.E., Lawrence, M., McCaffery, M. and Cline, K. 1991. Molecular cloning and nucleotide sequencing of the coat protein gene of *Citrus tristeza* virus. *J. Gen. Virol.* 72: 1013–1020.

Semancik, J.S. and Weathers, L.G. 1965. Partial purification of a mechanically transmissible virus associated with tatter leaf of citrus. *Phytopathology* 55: 1354–1358.

Shimizu, S. and Miyoshi, T. 2003. Association of satsuma dwarf virus with dwarf-like symptoms on novel citrus varieties in Ehime Prefecture. Bull. Ehime Prefec. *Fruit Tree Exp. Stn.* 16: 41–46.

Singh, J. and Baranwal, V.K. 2009. Molecular characterization and full length genome sequencing of citrus yellow mosaic virus associated with Rangpur lime cultivar. p. 65. *In*: 21st internl. Conf. on virus and other graft transmissible diseases of fruit crops. Berichte aus dem JKI 148. (Abs).

Singh, J., Baranwal, V.K. and Ahlawat, Y.S. 2004. Detection of citrus yellow mosaic virus in its vector planococcus citri (mealy bug) by polymerase chain reaction. *Indian Phytopath.* 57: 335.

Singh, J., Baranwal, V.K. and Ahlawat, Y.S. 2009. Molecular characterization and full length genome sequencing of citrus yellow mosaic virus associated with Rangpur lime cultivar. pp. 296. *In*: 5th Intl. Conf. Pl. Path. *In*: Globalized era. Nov. 10–13, New Delhi, India (Abst.).

Storm, L.W. and Streets, R.B. 1963. Transmission of stubborn and psorosis-A viruses of citrus to an herbaceous host. *Phytopath.* 53: 1142 (Abs).

Su, H.J. and Chen, J.U. 1984. Occurrence and distribution of tatterleaf citrange stunt complex on Taiwanese citrus. pp. 42–47. *In*: Phytopathologist and Entomologist, National Taiwan University, Tapei, Tawan.

Sumiya, K.V., Baranwal, V.K. and Ahlawat, Y.S. 2004. isolation and characterization of antiviral principles from seed of *Celosia cristata* inhibiting tobacco streak ailarvirus. *Indian Phytopath.* 57: 528.

Swingle, W.T. 1909. The limitation of satsuma orange to trifoliate orange stock. U.S. Dept. *Agr. Bur. Plant Indus Cire* 46: 10.

Swingle, W.T. and Webber, H.J. 1896. The principal diseases of citrus fruits in Florida. *USDA Div. Veg. Physiol. Pathol. Bull.* 8: 42 .

Takahara, T., Kawase, K., Ono, S., Iwagaki, I., Hirose, K. and Yoshinaga, K. 1988. Rootstocks for Ponkan (*Citrus lementine* Blanco) in relation to tatter leaf virus. *Bulletin of Fruit Tree Research Station* D 10: 35–45.

Tanaka, H. 1972. Mechanical transmission of viruses of satsuma dwarf and natsudaidai dwarf from citrus to citrus. *Annals of the Phytopathological Society of Japan* 38: 156–160.

Tanaka, H. and Imada, J. 1974. Mechanical transmission of viruses of Satsuma dwarf, Citrus mosaic, Navel infectious mottling, and Natsudaidai dwarf to herbaceous plants, pp. 141–146. *In*: Proc. 6th Conf. IOCV, IOCV, Riverside.

Tanaka, H. and Imada, J. 1974. Physalis floridana, a good production host for Satsuma dwaf virus. *Plant Dis. Reptr.* 58: 603–605.

Tanaka, H. and Imada, J. 1976. Purification of viruses of citrus mosaic and navel orange infectious mottling. pp. 116–118. *In*: Proc. 7th Conf. IOCV, IOCV, Riverside, CA.

Tanaka, H. and Nakanishi 1972. *Ann. Phytopath. Soc. Japan* 38: 212.

Tanaka, H. and Yamada, S. 1972. Evidence for a relationship among the viruses of Satsuma dwarf, Citrus mosaic, Navel-infectious-mottling, Natsudaidai dwarf, Citrus variegation, and *Citrus crinkly* leaf, p. 71–76. *In*: Proc. 5th Conf. IOCV., IOCV, Riverside, CA.

Tanaka, H., Yamada, S. and Kishi, K. 1971. Symptoms and occurrence of navel orange infectious mottling and Natsudaidai dwarf. *Bull. Hort. Res. Stn.*, series B (Hiratsuka) 11: 141–147.

Tanaka, S. and Kishi, K. 1963. Studies on indicator plants for citrus viruses. I Mechanical inoculation on leguminous plants with sap from Satsuma dwarf tree. *Annl. Phytopath. Soc. Japan* 28: 262–269.

Tanaka, S., Shikata, E. and Sasaki, A. 1969. Studies on Hassaku dwarf virus. *In*: Proc. First Internl. *Citrus Symp.* 3: 1445–1448.

Tanner, J.D., Kunta, M., da Graca, J.V., Skaria, M. and Nelson, S.D. 2010. Evidence for citrus leaf tatter virus seed transmission in citrus. p. 50. *In*: Abs. 18th Conf. IOCV held in Campinas/SP, Brazil, Nov. 7–12, 2010.

Tarafdar, A., Kumari, S. and Biswas, K.K. 2009. Agrobacterium mediated transformation of *Citrus aurantifolia* (Kagzi lime) using coat protein gene of *Citrus tristeza* virus. p. 163. *In*: 5th Intl. Conf. Pl. Path. Globalized era. November 10–13, 2009, new Delhi, India (Abst.).

Thind, S.K. and Arora, P.K. 1997. Incidence and behaviour of ring spot virus in citrus. *Indian J. Virol.* 13: 159–160.

Thind, S.K., Sharma, J.N., Arora, P.K. and Cheema, S.S. 1998. Effect of citrus ringspot virus on quality of *Kinnow mandarin* fruits. *Indian J. Virol.* 14: 91–92.

Timmer, L.W. 1974. A necrotic ring spot of citrus ringspot virus and its relation to citrus psorosis virus. *Phytopathology* 64: 389–394.

Timmer, L.W. and Benatena, H.N. 1977. Comparison of psorosis and other viruses causing leaf flecking in citrus. *Proc. Intl. Soc. Citriculture* 3: 930–935.

Timmer, L.W., Garnsey, S.M. and Graham, J.H. (*Eds.*) 2000. p. 65. *In*: Compendium of citrus diseases, 2nd Edn, APS Press, St. Paul, MN.

Timmer, L.W., Garnsey, S.M. and McKitchie, J.J. 1978. Comparative symptomatology of Florida and Texas isolates of citrus ringspot virus on citrus and herbaceous plants. *Plant Dis. Reptr.* 62: 1054–1058.

Ushiyama, K. and Ohgaki, T. 1970. Studies on the satsuma dwarf virus disease. 1. Survey in Kanagawa Prefecture. *Bull. Kanagawa Hort. Exp. Stn.*, 18: 57–65.

Ushiyama, K. 1981. Studies on satsuma dwarf disease. II. Spread of the disease by Topworking satsuma trees with infected scions]. *Bulletin of the Kanagawa Horticultural Experiment Station* 28: 24–30.

Usugi, T. and Saito, Y. 1977. Some properties of satsuma dwarf virus. *Annals of the Phytopathological Society of Japan* 43: 137–144.

Usugi, T. and Saito, Y. 1979. *Citrus mosaic* virus. CMI/AAB Descriptions of Plant Viruses No. 208. Association of Applied Biologists, Wellesbourne, UK.

Usugi, T. and Tsuchizaki, T. 1982. Detection of citrus mosaic virus by enzyme-linked immunosorbent assay. *Annals of the Phytopathological Society of Japan* 48: 330–332.

Usugi, T., Yamamoto, S. and Tsuchizaki, T. 1986. Morphology, host range and serological properties of citrus mosaic virus causing mosaic diseases in satsuma mandarins. *Ann. Phytopathol. Soc. Jap.* 52: 349–354.

Van Regenmortel, M.H.V., Fauquet, C.M., Bishop, D.H.L., Carstens, E.B., Stes, M.K., Lemon, S.M., Maniloff, J., Mayo, M.A., McGeoch, D.J., Pringle, C.R. and Wickner, R.B. 2000. Virus taxonomy: The classification and nomenclature of viruses. The seventh report of the international committee on taxonomy of viruses. Academic Press, San Diego, CA, USA. p. 1167.

Varma, P.M., Rao, D.G and Capoor, S.P. 1965. Transmission of tristeza virus by *Aphis craccivora* Koch. and *Dactynotus jaceae* L. *Ind. J. Agric. Sci.* 35: 85–89.

Varma, P.M., Rao, D.G. and Capoor S.P. 1960. Additional vectors of tristeza disease of citrus in India. *Curr. Sci.* 29: 359.

Vasudeva, R.S and Capoor, S.P. 1958. Citrus decline in Bombay State. *FAO Plant Prot. Bull.* 6: 91.

Vasudeva, R.S., Varma, P.M., Capoor, S.P. and Rao, D.G. 1959. Transmission of Citrus decline virus by *Toxoptera citricidus* Kirk in India. *Curr. Sci.* 28: 418–419.

Vidalakis, G., Garnsey, S.M., Bash, J.A., Greer, G.D. and Gumpf, D.J. 2004. Efficacy of bioindexing for graft-transmissible citrus pathogens in mixed infections. *Plant Dis.* 88: 1328–1334.

Vijayakumari, N., Ghosh, D.K., Das, A.K. and Singh, S. 2006a. Elimination of CTV and greening pathogens through *in vitro* shoot grafting from citrus exotic germplasm. *Indian J. Agric. Sci.* 76: 209–210.

Vijayakumari, N., Ghosh, D.K., Das, A.K. and Singh, S. 2006b. Production of disease-free planting material of Nagpur mandarin (*Citrus reticulata* Blanco) by shoot tip grafting. *Indian Phytopath.* 59: 48–51.

Vives M.C., Galipienso, L., Navarro, L., Moreno, P. and Guerri, J. 2001. The nucleotide sequence and genomic organization of citrus leaf blotch virus: candidate type species for a new virus genus. *Virology* 287: 225–233.

Vives, M.C., Aguero, J., Pina, J.A., Navarro, L., Moreno, P. and Guerri, J. 2010. *Citrus* leaf blotch virus: A citrus pathogen transformed in a genetic tool. p. 63. *In*: Abs. 18th Conf. IOCV, held in Campinas/SP, Brazil, Nov. 7–12, 2010.

Vives, M.C., Pina, J.A., Juarez, J., Navarro, L., Moreno, P. and Guerri, J. 2005. Dweet mottle disease is probably caused by Citrus leaf blotch virus. pp. 251–256. *In*: Proc. 16th Conf. IOCV, IOCV, Riverside, CA.

Vives, M.C., Rubio, L., Galipienso, L., Navarro, L., Moreno, P. and Guerri, J. 2002. Low genetic variation between isolates of *Citrus* leaf blotch virus from different host species and different geographical origins. *J. Gen. Virol.* 83: 2587–2591.

Wallace, J.M. 1957. Virus strain interference in relation of symptoms of psorosis disease of citrus. *Hilgardia* 27: 223–246.

Wallace, J.M. 1959. Citrus virus diseases. Univ. Calif. Div. Agric. Sci., Berkeley, California.

Wallace, J.M. 1968. Recent developments in *Citrus psorosis* diseases. pp. 1–9 *In*: Proc. 4th Conf. IOCV, IOCV, Riverside, CA.

Wallace, J.M. and Drake, R.J. 1951. Recent developments in studies of quick decline and related diseases. *Phytopathology* 41: 785–793.

Wallace, J.M. and Drake, R.J. 1953. A virus-induced vein enation in citrus. *Citrus Leaves* 33: 22, 24.

Wallace, J.M. and Drake, R.J. 1959. Citrus vein enation. *In*: Wallace, J.M. (*Ed.*) citrus virus diseases pp. 163–165.

Wallace, J.M. and Drake, R.J. 1960. Woody galls on citrus associated with vein-enation virus infection. *Plant Disease Reporter* 44: 580–584.

Wallace, J.M. and Drake, R.J. 1962. Further studies on the woody gall disease of citrus. *Phytopathology* 52: 756.

Wallace, J.M. and Drake, R.J. 1962. Tatter leaf, a previously undescribed virus effect on citrus. *Plant Dis. Reptr.* 46: 211–212.

Wallace, J.M. and Drake, R.J. 1963. New information on symptom effects and host range of the citrus tatter leaf virus. *Plant Dis. Reptr.* 47: 352–53.

Wallace, J.M. and Drake, R.J. 1972. Use of seedling yellows recovery and protection phenomena in producing tristeza-tolerant, susceptible scion-root stock combinations. pp. 137–143. *In*: Proc. 5th Conf. IOCV, IOCV, Riverside, CA.

Wallace, J.M. and Drake,R.J. 1968. Citrange stunt and ringspot –two previously undescribed virus diseases of citrus. pp. 177–183. *In*: Proc. 4th Conf. IOCV, IOCV, Riverside, CA.

Wallace, J.M. 1945. Techniques for hastenning foliage symptomps of psorosis of citrus. *In*: Proc. 6th Conf. ICOV, ICOV, Riverside, CA.

Weathers, L.G. 1960. Yellow-vein disease of citrus and studies of interactions between yellow-vein and other viruses of citrus. *Virology* 11: 753–764.

Webber, H.J. 1925. A comparative study of the citrus industry of South Africa. S. Africa *Dep. Agr. Bull.* 6: 106.

Weng, Z., Liu, X., Gowda, S., Barthelson, R.A., Galbraith, D.W., Dawson, W.O. and Xiong, Z. 2010. Extreme genome stability of citrus tristeza virus. P. 25. *In*: Abstract 18th Conf. IOCV, Campinas/SP, Nov. 7–12, 2010.

Whiteside, J.O., Garnsey, S.M. and Timmer, L.W. (*Eds.*) 1988. Compendium of citrus diseases. APS Press St. Paul, p. 80.

Wu, T., Du, Y., Li, L., Deng, S., Wu, S. and Luo, Z. 1996. Detection and elimination of citrus tatter leaf virus. p. 411 *In*: Proc. 13th IOCV, IOCV, Riverside, CA., USA.

Yamada, S. and Sawamura, K. 1952. Studies on the dwarf disease of Satsuma orange, *Citrus unshiu* Marcovitch (Preliminary report). *Hort. Div., Tokai-Kinki Agric. Exp. Stn., Bull.* 1: 61–71.

Yamaguchi, A. 1984. Time requirement for graft transmission of Satsuma dwarf virus. *Bull. Fruit Tree Res. Stn.*, Japan B, 11: 63–69.

Yamamoto, S. and Yamaguchi, A. 1980. Spread of citrus mosaic through distribution of a new clone of satsuma mandarin, pp. 230–231. *In*: Proc. 8th Conf. IOCV., IOCV, Riverside. CA.

Yasir Iftikhar, Zafar Iqbal, Salman Ahmed, Awan, A.R.,Usman Saleem, Sarwar, G. and Moeen-ud-Din 2010. Effect of environmental factors on yellow vein clearing virus incidence in lemon. *J. Agric. Res.* 48: 87–92.

Yokomi, R.K., Metheney, P., Grafton-Cardwell, E.E. and Stewart-Leslie, J. 2010. Current status of citrus tristeza virus in central California. p. 23. *In*: Abstract 18th Conf. IOCV held in Campinas/SP, Brazil, Nov. 7–12, 2010.

Yokomi, R.K., Polek, M., Grafton-Cardwell, E.E., Vidalakis, G., O'Connell, N. and Saponari, M. 2010. Assessment of the *Citrus tristeza* virus isolates detected in Spring 2007 at the Lindcove Research and Extension Center, Exeter, California. *In*: Proc. 17th Conf. IOCV, IOCV, Riverside, CA.

Yokomi, R.K., Saponari, M. and Sieburth, P.J. 2010. Rapid differentiation and identification of potential severe strains of *Citrus tristeza* virus by real-time reverse transcription-polymerase chain reaction assays. *Phytopathology* 100: 319–327.

Yora, K., Raychaudhuri, S.P. and Ahlawat, Y.S. 1977. Infectious variegation of citrus in India. California *Citrogl. Aph.* 62: 342.

Yoshikawa, N.M., Imaizumi, T., Takahashi, N. and Inouye 1993. Striking similarities between the nucleotide sequence and genome organization of citrus tatter leaf virus and apple stem grooving capilloviruses. *J. Gen Virol.* 74: 2743–2747.

Yot Dauthy, D. and Bov', J.M. 1968. Purification of citrus crinkly leaf virus. pp. 255–263 *In*: Proc. 4th Conf. IOCV, IOCV, Riverside, CA.

Zhang, T.M. Liang, X.Y. and Roistacher, C.N. 1988. Occurrence and detection of citrus tatter leaf virus (CTLV) in Huangyan, Zhejiang Province, China. *Plant Disease* 72: 543–545.

Colour Plates

1a. Citron cv Buddha's hand.

1b. *Citrus australiasica*

1c. *Citrus medica*

2. Bark scaling due to psorosis on sweet orange.

3. Death of lime trees due to tristeza virus.

4. Pitting in stems and branches due to CTV.

5. Key lime leaf with vein flecking symptoms.

6. Eureka lemon leaves showing crinkling.

7. Variegation symptoms on lemon leaves.

8. Rings on leaves due to leprosis virus (CiLV).

9*. Branches showing lesions caused by CiLV.

10a*. Spoon shaped leaves of Satsuma tree

10b*. Satsuma fruit showing green rings

11. Mosaic symptoms on Mosambi sweet orange

* www. ivia.es.iocv

Note: Detailed captions are given in the beginning of the book

Colour Plates

12a*. Depressions corresponding to enations

12b*. Woody galls on stem of field tree

13a*. Zig-zag stem of Troyer due to due to CTLV

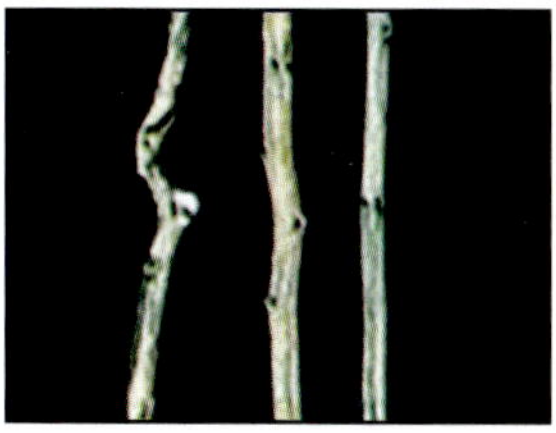

13b*. Bark blotches on CTLV-infected citrange

13c*. CTLV-infected distorted leaves of Troyer

14a. ICRSV affected Kinnow mandarin tree

14b. ICRSV affected leaves showing rings.

15a. Shock reaction due to ICRSV inoculation.

15b. Vein flecking due to ICRSV in young leaf

15c. Rings on mature leaf inoculated with ICRSV

16a. ICRSV Symptoms on young leaf of mosambi

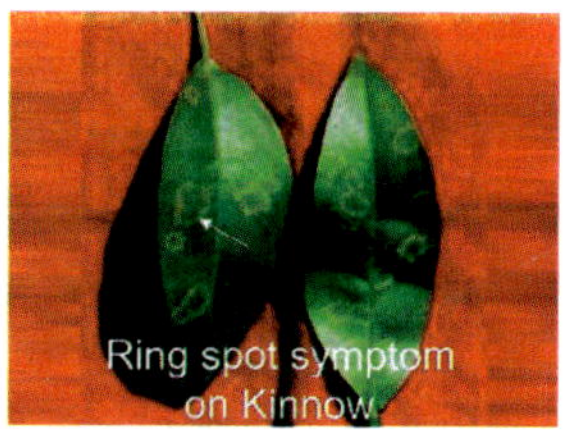

16b. Rings on kinnow mandarin due to ICRSV

16c. Rings on Mosambi due to ICRSV

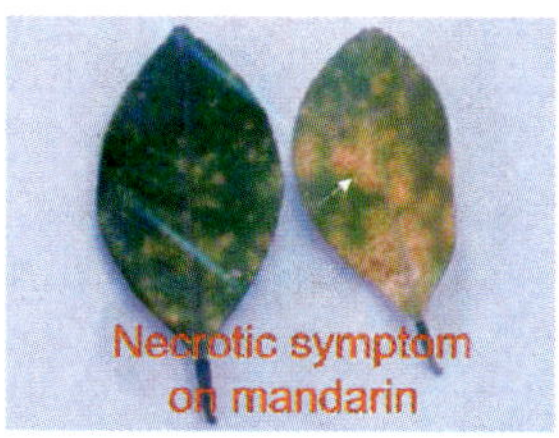

16d. Necrotic spots on Nagpur mandarin

---

* www. ivia.es.iocv

Note: Detailed captions are given in the beginning of the book

Colour Plates

16e. Spots on rough lemon leaf due to ICRSV.

17a. Lesions on ICRSV inoculated French bean

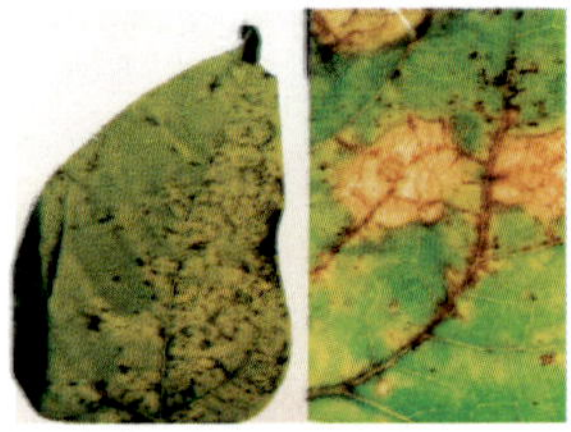

17b. Black veins and necrosis in French bean

18a. CYMBV infected Satgudi leaves and fruits

18b. Yellow mosaic on pummelo

18c. CYMBV symptoms on Volkamer lemon

19. *Planococcus citri*-vector of CYMBV

20a. Mosaic on pumello leaf infected with PYVM

20b. Necrotic spots on leaf due to PYMV infection

21a. CYVC symptoms on Etrog citron (E.C.) in field

21b. Yellow vein clearing symptoms on E.C. leaves

21c. Water soaking areas along veins of E.C. leaves

21d. Progress in CYVC symptoms

22a. Local lesions in C. amaranticolor leaf

* www. ivia.es.iocv

Note: Detailed captions are given in the beginning of the book

Colour Plates

22b. Mottling in CYVCV inoculated French bean

22c. Necrotic areas in a French bean cultivar.

23*. Chlorotic blotches on Dweet tangor leaf

24a. A blight affected tree in Florida

24b. Declinio affected trees in Brazil.

25a. A man showing water uptake in blighted tree

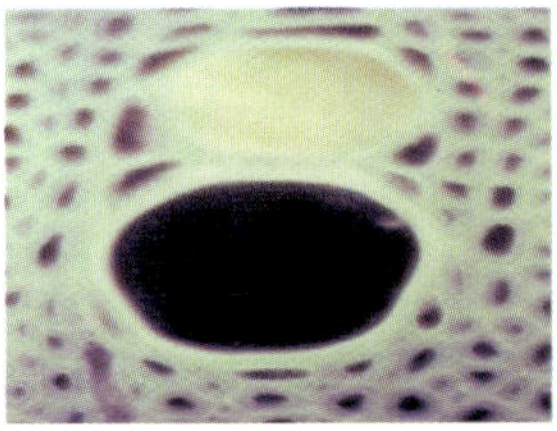

25b. Amorphous plug blocking xylem vessel

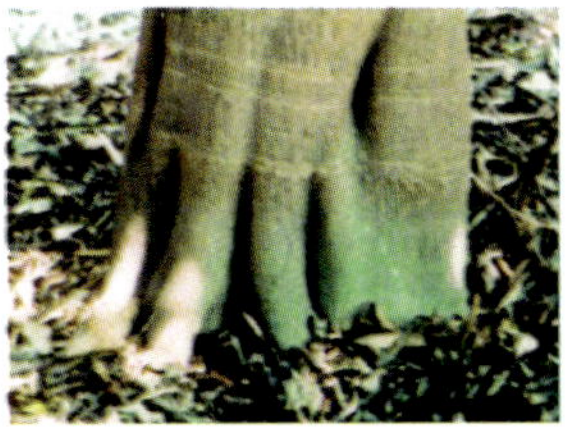

26a*. Deeper concavities due to blind pocket.

26b. Oak leaf pattern on concave gum affected leaf

27*. Citrus chlorotic dwarf on Aydin lemon

28a. Healthy and impietratura affected fruits

28b. Albedo gumming in infected fruit (right).

29. Symptoms of Citrus yellow vein disease

30a. Twigs of citrus showing curling of leaves

* www. ivia.es.iocv

Note: Detailed captions are given in the beginning of the book

Colour Plates

30b. Leaves showing curling. Healthy leaf on left.

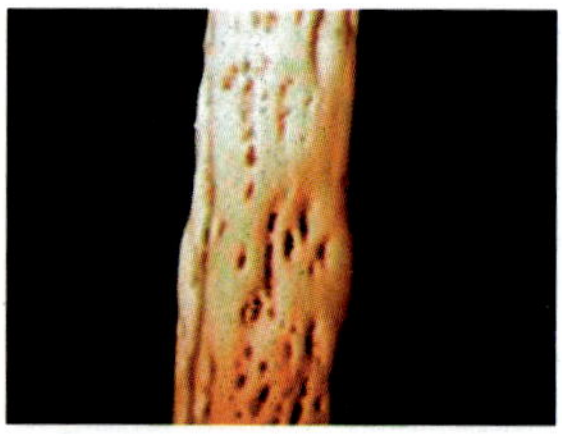

31a*. Pits and gum in trunk of a Webber tangelo.

31b*. Cristacortis induced stem pitting

32. Dieback on Nagpur mandarin in India

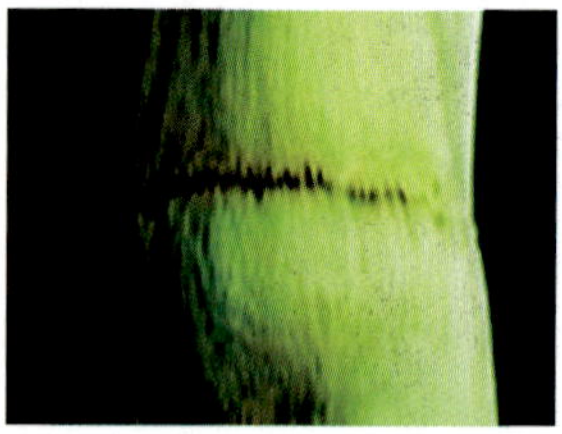

33a*. Bud union crease in Cadenera punxosa

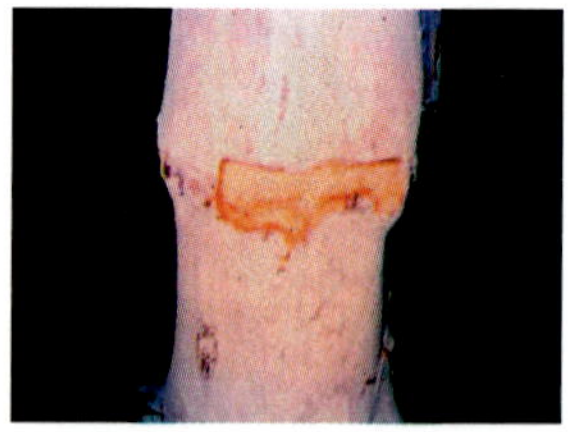

33b. Bud union crease in Nagpur mandarin

34*. Kassala induced gum streaks in trunk of grapefruit

35*. Rio Grande gummosis symptoms in wood

36a. HLB symptoms on khasi mandarin (right)

36b. Sweet orange tree showing mottled leaves.

36c. Leaves showing typical mottling symptoms

37a. Corking of veins due to HLB infection

37b. HLB affected fruits with unusual colouring

37c. Fruit sections showing aborted seed

---

* www. ivia.es.iocv

Note: Detailed captions are given in the beginning of the book

Colour Plates

38a. *Diaphorina citri*-vector of Asian HLB

38b. HLB transmitted to periwinkle by dodder

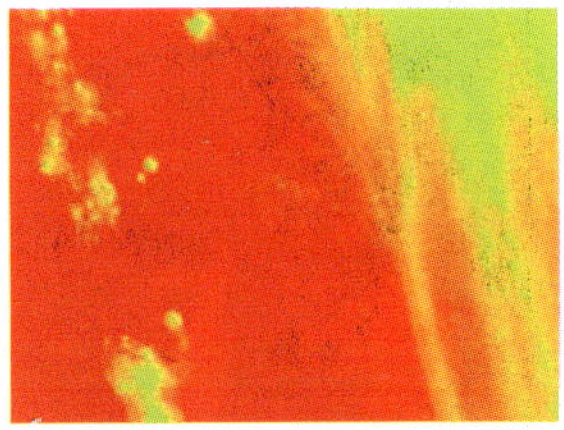

39a. Detection of HLB bacterium by IF test.

39b. Healthy section

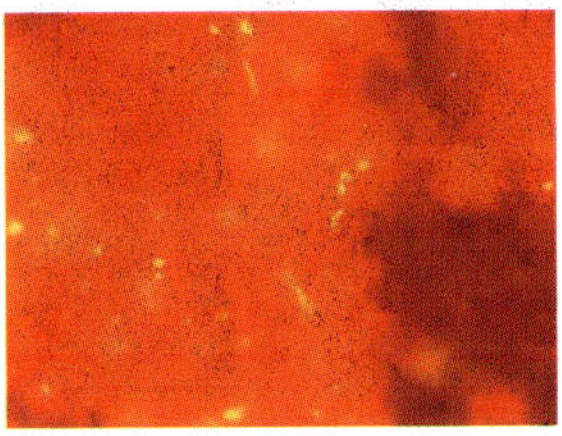

39c. IF detection in extract of infected plant.

40a*. CVC symptoms on sweet orange

40b*. CVC affected leaves with chlorosis

40c*. Sharpshooter leaf-hopper that transmit CVC.

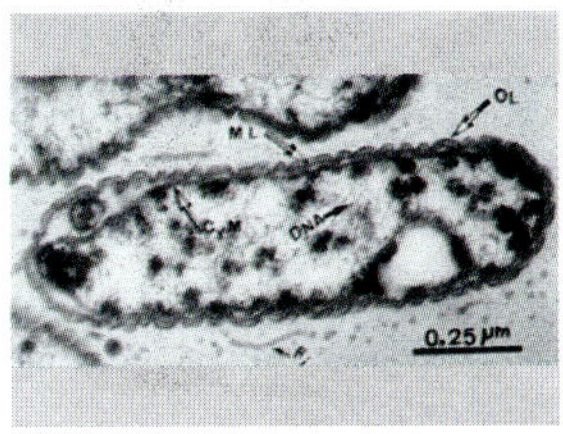

40d*. CVC bacterium with rippled cell wall

41a. Stubborn affected sweet orange tree in front

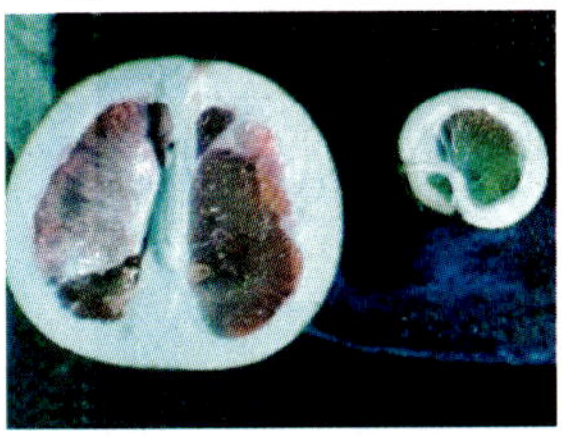

41b. Curved columella of grapefruit (Healthy Left)

41c-1. *Scaphytopius nitrides*-stubborn vector

41c-2. *Circulifer tenellus*-stubborn vector

42a. Healthy (left) and *S. citri* infected periwinkle

* www. ivia.es.iocv

Note: Detailed captions are given in the beginning of the book

Colour Plates

42b. *S.citri* infected marigold (left) Healthy-right

43. Lime trees infected with witches' broom (WB).

44a. Lime WB on Key lime seedlings.

44b. WB transmitted to Periwinkle plants by dodder

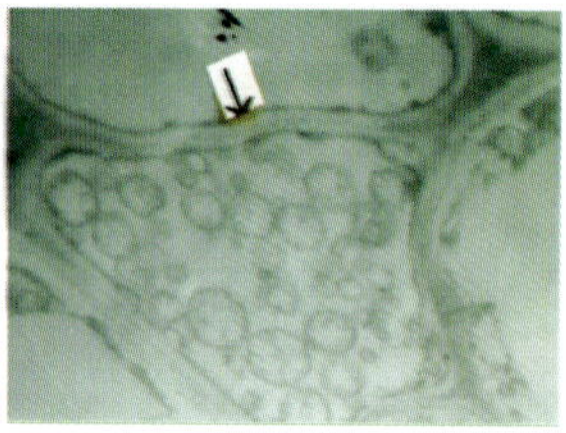

44c. Phytoplasmas in sieve tube of WB affected tissue

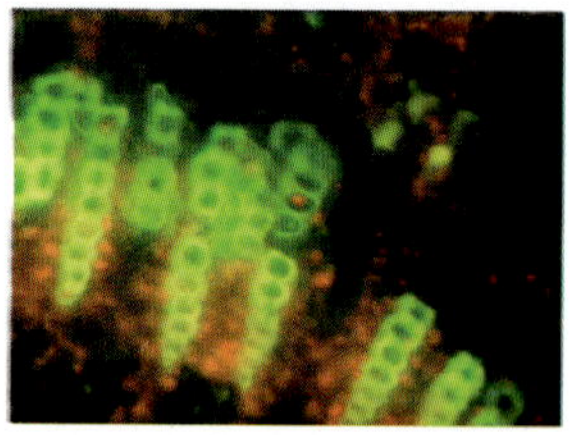

44d. DAPI test for phytoplasma detection.

45a. Mandarin tree with rubberywood (RW) symptoms

45b. Lemon plant graft inoculated with RW.

46a. Rangpur lime stock showing bark scaling

46b. Exocortis infected *Etrog citron*. Healthy-right.

47a*. P.S. mandarin with cachexia symptoms.

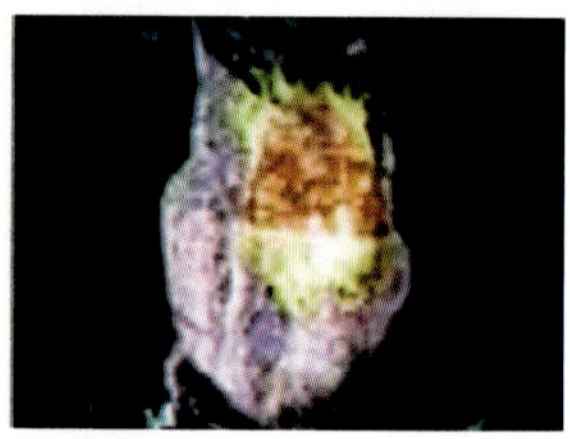

47b. Symptoms of cachexia in mandarin on sour orange

48a. Yellow corky vein symptoms on kagzi lime

48b. Yellow corky vein symptoms on Etrog citron

---

* www. ivia.es.iocv

Note: Detailed captions are given in the beginning of the book

Colour Plates

48c. Yellow corky vein symptoms on kinnow mandarin

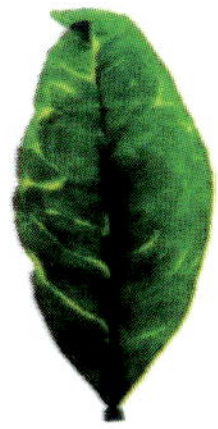

48d. Yellow corky vein symptoms on Rangpur lime

49a-1. Yellow corky vein (CYCV) on Nagpur mandarin

49a-2. Malformed leaves of Nagpur mandarin

49b-1. Corking on Kagzi lime leaf and twig due to CYCV

49b-2. Corking on trifoliate orange stem due to CYCV

49c. Corking on trifoliate orange leaf due to CYCV

49d. CYCV showing veinal necrosis in etrog citron

50*. Gummy bark in sweet orange on sour orange roots

51*. Gummy disease in radial L.S. of *P. trifoliata*

52. Eradication of ICRSV Shoot tip grafting.

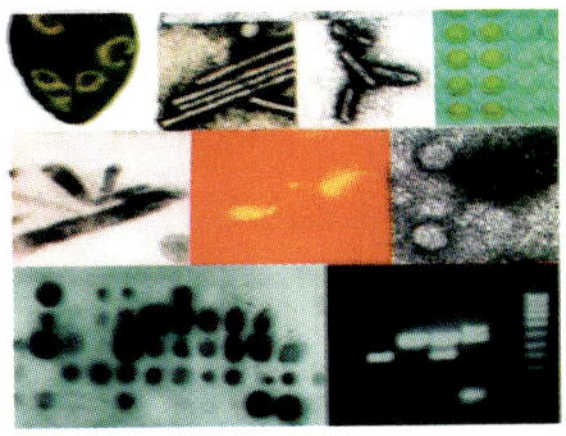

53. Method of detection of virus and virus-like pathogens

---

* www. ivia.es.iocv

Abbreviation used in captions of plates

CTV = Citrus tristeza virus
CTLV = Citrus tatter leaf virus
CYMBV = Citrus yellow mosaic badnavirus
CYVCV = Citrus yellow vein clearing virus
CVC = Citrus variegated chlorosis
WB = Lime witches' broom
CYCV = Citrus yellow corky vein
P. trifoliate = Poncirus trifoliata
CiLV = Citrus leprosies virus
ICRSV = Indian citrus ringspot virus
PYVM = Pumello yellow vein mosaic
HLB = Huanglongbing
S. citri = *Spiroplasma citri*
P.S. mandarin = Parsons' special mandarin
L.S. = Longitudinal section
DAPI = 4′,6-diamidino-2-phenylindole

Note: Detailed captions are given in the beginning of the book

3

VIRUS-LIKE CITRUS DISEASES OF UNKNOWN AETIOLOGY

Following 15 diseases of citrus have been considered to be of virus nature although the associated phathogen(s) have not been characterized as yet. These diseases are:

1. **Citrus Blight**
2. **Concave Gum – Blind Pocket**
3. **Citrus Chlorotic Dwarf**
4. **Impietratura Disease**
5. **Citrus Yellow Vein Disease**
6. **Leaf Curl Disease**
7. **Cristacortis**
8. **Citrus Leathery Leaf**
9. **Multiple Sprouting**
10. **Citrus Dieback Disease in India**
11. **Bud-Union Crease**
12. **Citrus Measles**
13. **Kassala Disease**
14. **Rio Grande Gummosis (RGG)**
15. **Citrus Sudden Death Disease**

Citrus Blight

Blight of citrus is most important among decline diseases. Due to blight disease, about five lakh trees in Florida and 10 million in Brazil are lost every year. In Florida, blight is present over 100 years (Smith and Reitz, 1977) especially in coastal citrus growing areas and common in trees on most of the commercial rootstocks. Blight disease was believed to be restricted to Florida (Rhoads, 1936) but it is now known to occur in several citrus growing states of United States of America, Brazil, Caribbean Basin, Central America, Colombia, Argentina, Venezuela, Uruguay, South Africa, Costa Rica and Australia (Timmer *et al.*, 1984, 2000; Childs, 1979; Lima, 1982; Villalobos *et al.*, 2005). Earlier in Florida, this disease was reported as Young tree decline, Sandhill decline and Rough lemon decline. In Brazil, blight is known as declinio, in Argentina as declinamiento and in Uruguay as marchtamiento repentino. It has been commonly observed that trees near the infected tree may get affected more quickly than trees at a distance (Yokomi *et al.*, 1984). Incidence of blight is more in warmer areas of the Transvaal of South Africa and Brazil (Lee *et al.*, 1984; Salibe and Jahen, 1984). Etiology of blight disease has not been determined although several attempts have been made and are being made to determine the etiology of blight disease (R.H. Brlansky-persnal communication, 2010).

SYMPTOMS

Blight disease is commonly found more in groves receiving above-average care in cultural practices. The symptoms of the disease are normally observed on bearing trees (8–10 yr old) but may also appear on adult tree with 4–6th year of age. Once a tree is affected, it does not recover. Blight causes a general decline of tree canopy with wilt, leaf loss, twig dieback, and poor growth flushes. Symptoms may be confined to one part of a tree and can be confused with those of other diseases like huanglongbing, tristeza or by burrowing nematodes (*Radophilus citrophilus*). Root and trunk sprouts are common on affected trees.

Blight affected trees may show zinc deficiency like symptoms in leaves. Foliage on affected trees become dull green as observed by the author in Florida and Brazil. Affected trees produce numerous water sprouts on the trunk and the lower branches followed by leaf drop and subsequent dieback (Plate 24a). Wilting of tree may be of a few branches or the entire canopy of a tree show wilting (Plate 24b). Delayed flush, thin foliage and delayed flowering are also symptoms of blight. Fewer but normal size fruits develop on affected trees but they have high acidity and soluble solids. Fibrous roots usually die in affected trees. Affected trees usually do not die but become nonproductive. Blight symptoms can appear at any stage after the bearing age. Stress symptoms of citrus trees are seen in successive stages of decline due to blight (Albrigo *et al.*, 1986).

TRANSMISSION

Blight disease is transmitted by natural root grafting from affected trees to healthy trees in USA, Brazil and South Africa suggesting the involvement of a graft transmitted pathogen (Tucker *et al.*, 1984; Albrigo *et al.*, 1992; Timmer *et al.*, 1991; Rossetti *et al.*, 1991; Beretta *et al.*, 1991; Marais and Lee, 1991). It is believed that a soil borne pathogen or a pathogen of very limited aerial movement is involved with blight disease (Yokomi *et al.*, 1984).

HOST RANGE

Citrus blight is common in all citrus cultivars on all rootstocks and seedling trees. Sweetorange and grapefruit varieties on rough lemon, Rangpur lime, trifoliate orange and Carrizo citrange rootstocks are more susceptible that on other rootstocks. Sweet oranges are the most affected but mandarins are generally not affected. Tahiti lime is most affected among lemon and lime cultivars (Donadio and Banzato, 1988). Blight mainly affects grapefruits (*Citrus paradisi*) and sweet oranges (*C. sinensis*) grafted on Rough lemon, *Citrus volkameriana* and Rangpur lime. Incidence of the disease is lower in lemons (*C. limon*) and mandarins (*C. reticulata*). Trees on *Citrus jambhiri, C. nobilis* and *Poncirus trifoliata* rootstocks are highly susceptible. Trees on *Citrus macrophylla, C. volkameriana* and *Citron* rootstocks are comparatively less susceptible but trees on *Citrus aurantium* and *C. reshni* rootstocks are tolerant, and those on *C. sinensis* are highly resistant (Young *et al.*, 1980). Rough lemon, Troyer citrange, Trifoliate orange, Empress and Cleopatra mandarin are the susceptible rootstocks to blight in South Africa (Marais and Lee, 1991). Once trees begin to show disease symptoms, they decline rapidly.

INDICATOR HOST

Although it is difficult to establish an indicator host for such diseases but Valencia sweet orange on rough lemon may be used as indicator host. Indicator plants when graft inoculated with root pieces from candidate tree, blight syndrome may develop in inoculated plants after several years of inoculation.

THE PATHOGEN

Attempts have been made to establish etiology of blight disease. Blight disease has been established as a transmissible disease through natural root grafts. Hence a soil pathogen was suspected as causal agent of the disease. *Fusarium solani* was suspected as a causal pathogen and was studied in detail but without any successful result. Gram negative bacterium, *Xylella fastidiosa* has also been shown to be constantly associated with blight disease (Hopkins *et al.*, 1991) but its role as etiological agent of blight was ruled out. Infectious nature of the disease has been established (Timmer *et al.*, 1992). Rod shaped virus particles were observed in extracts from blighted trees using a combination of sucrose density and cesium sulfate gradient centrifugation techniques (Brlansky and Hawd, 2001). Recently, an idaeovirus, similar to raspberry bushy dwarf virus has been shown to be associated with citrus blight (Derrick *et al.*, 2006). Raspberry bushy dwarf disease is caused by a mixed infection of an Idaeovirus and a second virus. Resistance to raspberry bushy dwarf disease has been obtained by transgenic silencing of raspberry Idaeovirus. Transgenic silencing of citrus Idaeovirus could lead to rootstocks that are resistant to citrus blight. Several proteins have been observed in vacuum extracts of root and stem pieces of blighted trees in Florida and Brazil (Beretta *et al.*, 1993) which are absent or present in much lower concentration in healthy trees. These extracts showed filamentous structures when viewed in electron microscope (Derrick *et al.*, 1991). Presence of double-stranded RNA in diseased plants also suggests viral etiology of blight disease (Nickel, 1987). However, the causal agent for blight remains undetermined. The only conclusion that can be drawn is that it is a systemic infectious agent that is limited to the root system

DIAGNOSIS AND MANAGEMENT

It has been commonly found that water transport in the xylem of blighted trees is impaired. Affected trees fail to take up or transport water, even if it is injected under pressure (Lee *et al.*, 1984). Blight disease is effectively diagnosed by reduced uptake of water when it is injected by pressure into large roots and trunks (Plate 25a) (Cohen, 1974). Xylem vessels of trunk, larger branches and roots are blocked by light-yellow amorphous plugs, or by dark-brown filamentous plugs (Plate 25b) resulting failure of water transport and development of blight symptoms due to lack of water transport to the plant canopy (Brlansky *et al.*, 1985). Depth of amorphous and filamentous plugs is an important factor for reduction in water flow but amorphous plugs play the major role in reducing water flow in blighted citrus trees (Timmer *et al.*, 1988). Nature of xylem plugging is responsible for variable water flow in infected trees on different rootstocks and their combination with different scion varieties (Beretta *et al.*, 1988). Zinc accumulates in the bark and outer xylem of the trunk of affected trees even before the formation of plugs and development of visible symptoms (Young *et al.*, 1980; Albrigo and Young, 1981). Zinc accumulation is also reported to be higher in the wood and soil of trees showing blight or declinio symptoms (Smith, 1974; Melo *et al.*, 1991; Brlansky *et al.*, 1988; Taylor *et al.*, 1988). Significance of high levels of zinc in affected trees is also useful for diagnosis of blight disease.

Antisera to 12 and 35 kDa proteins have been prepared from blight affected trees and monoclonal antibodies have also been developed against 12 kDa protein. Both these antibodies detected blight agent in immunospot and Western blot assays (Derrick *et al.*, 1990, 1993). In declinio affected trees, several proteins have been detected in plants infected naturally or inoculated experimentally but these proteins were absent in healthy trees (Nickel *et al.*, 1996). Antibodies prepared against virus particles were used to detect blight in immunogold labeling and ELISA tests (Brlansky and Hawd, 2001).

Sudden decline of citrus trees, a disease of unknown cause in Venezuela, is similar in some ways to citrus blight in Florida and citrus declinio in Brazil. Trees affected with sudden decline show blight and declinio-like decline symptoms, but affected trees usually die within 8 to 10 month after symptoms first appear. This is in contrast to trees affected with blight and declinio, which may have severe symptoms of decline for several years but seldom die. Using serological tests, trees with sudden decline were found to contain 12-KDa protein which is associated with citrus blight and declinio (Ochoa *et al.*, 1993).

Bilght-declinio diseases are detected by visual symptoms and water uptake by affected trees through syringe injection and percentage of amorphous plugs obstracting xylem vessels. Best diagnosis of blight disease in field is syringe test for water uptake (Lee *et al.*, 1984). For syringe test, a small hole is drilled into trunk of tree and a syringe is used to inject water into trunk. Blighted trees will not take up water, whereas healthy trees usually absorb water readily (Cohen and Wutscher, 1979). Zinc concentration and water soluble phenolics are high in blighted trees and may be used for diagnosis (Wutscher *et al.*, 1977). Serological tests for the 12 kDa and other proteins that are unique to blighted trees are commonly used to diagnose blighted trees (Derrick *et al.*, 1992). Disease diagnosis in young trees can be done by immunospot and Western blot techniques and other immunological techniques (Bausher and Sweeney, 1991). Root grafting can not be used for disease diagnosis as the incubation period is long (18–24 months).

For management of blight disease, blight-affected trees should be replaced with trees on tolerant rootstocks like Swingle citrumelo (*Poncirus trifoliata* × *Citrus paradisi*), Cleopatra mandarin (*C. reshni*), Empress mandarin (*C. reticulata*) or X639. Rootstock like rough lemon (*Citrus jambhiri*) should be avoided. Heavy pruning of infected trees helps in good growth of trees and normal bearing for a few years (Rossetti *et al.*, 1988).

Concave Gum – Blind Pocket

Two forms of psorosis disease were idendified: psorosis-A –the common form of the disease and psorosis-B – a less common but more severe form, characterized by early bark lesions, mature leaf symptoms and twig dieback (Facett, 1932; Fawcett and Klotz, 1938). Young leaves of concave gum-affected trees not only had young leaf symptom of psorosis-A, but also had symptoms of zonate leaf or oak leaf pattern. Psorosis young-leaf symptoms represented a common characteristic of psorosis-A, concave gum, blind pocket, crinkly leaf, infectious variegation and impietratura. Hence at one time they were thought to be members of the diseases of "psorosis family" (Wallace, 1959). However, today we know that crinkly leaf, infectious variegation and leaf rugose are members of ilarvirus group and unrelated to psorosis virus (Rana *et al.*, 1974; Gonsalves and Garnsey, 1976). The first suggestion that concave gum was unrelated to psorosis came from cross-protection experiments by Roistacher and Calavan (1965). They showed that a number of concave gum sources did not protect sweet orange seedlings against psorosis-A bark lesion. However, concave gum and blind pocket have similar symptoms and characteristics and hence both were removed from the psorosis group. Vogel and Bové (1964, 1968, 1972) always observed psorosis young-leaf symptoms, not only flecking but also oak-leaf patterns with cristacortis. Later studies showed that although concave gum and blind pocket have similar symptoms and characteristics of psorosis but are distinct from psorosis group. Concave gum disease is largely present in the Mediterranean region although it is now found worldwide. This disease can be quite severe and very debilitating to young trees (D'Onghia *et al.*, 1992).

SYMPTOMS

Field affected trees of mandarin, clementine, sweet orange, grapefruit and tangelo show mild to severe stunting. Trunk and large limbs are distorted. Concavities normally develop on the scion trunk close to bud union.Trunk symptoms are seen on the sensitive scion: mandarin, Clementine, sweet orange, grapefruit and tangelo but not on the tolerant sour orange, lemon and *P. trifoilata* rootstocks.

- ***Concave gum***: Symptoms of concave gum disease are formation of concavities of 10–20 cm or more in length and half as wide and numerous in stem and branches in addition to young leaf symptoms of vein flecking and oak-leaf pattern similar to psorosis complex (Plate 26b), these concavities interfere in normal development of wood. The cavities are found only on scion variety but not on rootstock. There is presence of gum between bark and wood at depressions. Sometimes there is cracking of bark and exudation of gum in the depressions and xylem vessels in the vicinity of concavities are plugged with gum. In cross sections of branches and limbs, concentric growth rings in the wood may be stained with brownish gum like material below the depressions. Sometimes, trees will show typical zonate leaf pattern, wood concavities and bark lesions of psorosis-A type. Such situation probably arises due to mixed infection. Concavities are common in sweet orange and mandarin trees. Concave gum disease has following characteristics:

1. It has broad concavities or depressions of various sizes on the trunk and limbs.
2. There is presence of gum beneath the concavities, in between wood and bark.
3. There is gum exudation through bark-cracks in the centre or around the margins of the depression at certain seasons.

- ***Blind pocket***: In blind pocket, concavities similar to that of concave gum are also developed on stem and branches but in this case concavities are more in number, deeper and narrower (Plate 26a). However, it is difficult to distinguish between concavities caused by concave gum or blind pocket until a person is well experienced with these diseases. Often concave gum and blind pocket depressions occur together on the same tree. Major identification of concave gum and blind pocket is based on symptoms on field trees and on the development of oak-leaf patterns on graft inoculated sweet orange, Dweet tangor or mandarin indicators which show hard gum like substance below the blind pockets. The cavities are conspicuous in sweet orange and sour lemon. Normally bark lesions are not developed due to infection by blind pocket. However, Fawcett and Bitancourt (1943) described a form of psorosis having blind pocket and bark scaling and referred it as "eruptive blind pocket". There is no further information on concave gum and blind pocket. Viruses causing leaf flecking symptoms have been compared by Timmer and Benatena (1977). Both these disorders are graft transmitted and their agents have not been identified. On the basis of trunk and leaf symptoms they are thought to be closely related and will be considered here as one single disease.
- ***Symptoms on leaves***: Young leaf flecking symptoms similar to psorosis-A and oak leaf pattern on young leaves are developed on infected trees (Plate 26b). These symptoms are visible during

new growth flushes and disappear as the leaves mature. Oak leaf pattern can be seen throughout the tree on almost all of the leaves in the young flush of growth. There are no symptoms on fruits. Wood development at the site of depressions is inhibited (Wallace, 1978). Often concave gum and blind pocket depressions occur together on the same tree

TRANSMISSION

Agent of concave gum-blind pocket is transmitted by graft inoculation of bark patches, buds or shoots. There is no report on mechanical or vector transmission.

HOST RANGE

Host range is restricted to *Citrus* species. Typical symptoms of the disease develop on mandarin, clementine, tangelo, grapefruit and a pummelo-like hyuganastsu (*Citrus tamurana*) in Japan (Leki and Ito, 1996). Infected Sour orange, lemon, and Citron show only oak leaf pattern but not trunk or branch depressions. Symptomatic hosts of concave gum-blind pocket are: Clementines, Grapefruits, Mandarins, Siamelo, Sweet oranges, Tangelos (Orlando, Sampon, Sunshine, Waft) and Dweet tangor. Mandarins and sweer oranges are highly susceptible. Symptomless hosts are: Citranges, Citrons, Lemons, Rough lemon, *P. trifoliata,* Sour orange, Sweet lime, West Indian lime and tolerant hosts are: Carrizo, Rusk Troyer, Citrons, Etrog citron, lemons, Lime (Kagzi, Key, Mexican, West Indian), *Poncirus trifoliata,* Rangpur lime, Rough lemon, Sour orange, *Severinia buxifolia* and Sweet lime.

INDICATOR HOST

Sweet orange, Dweet tangor and mandarin seedlings are used as indicator hosts where oak-leaf pattern symptoms develop on leaves within a few weeks after inoculation. Citron and lemon can also show oak-leaf pattern. For trunk and branch symptoms, Washington navel sweet orange and Orlando tangelo are used.

THE PATHOGEN

Causal agent of this disease has not been established as yet. Based on its graft transmission, this disease is considered to be caused by a virus-like pathogen.

DIAGNOSIS AND MANAGEMENT

Diagnosis of the disease can be done by observing oak leaf pattern symptom by indexing on mandarin and Dweet tangor. However, for symptoms on trunk and branches, indexing should be done on Washington navel sweet orange and Orlando tangelo. These cultivars will develop oak leaf pattern symptoms followed by cavities in stem and branches. In field trees, disease can be identified by oak leaf pattern of young leaves during spring and typical depressions on trunk and branches. For management of disease, budwood free of concave gum-blind pocket (CG-BP) should be used. The agent of CG-BP is readily eliminated by shoot tip grafting (STG). Thermotherapy can also be used to eliminate the agent of CG-BP (Calavan *et al.*, 1972; Roistacher, 1977).

Citrus Chlorotic Dwarf

A new virus-like disorder named citrus chlorotic dwarf has been reported from Eastern Mediterranian region of Turkey with an incidence of 20.3%. This disease previously believed to be a crinkly-leaf type disease is widespread in Mersin, Adana and has also begun to spread in Hatay region of Turkey (Korkmaz *et al.*, 1994). Chlorotic dwarf disease spreaded fast and reached to an epidemic level in Mersin and Adana regions. Now chlorotic dwarf disorder is being considered as one of the serious diseases of citrus in that region (Cinar *et al.*, 1993, 1994). Field symptoms consist of a V-shaped notch and chlorotic flecking on young leaves and warping, crinkling, inverted cupping and variegation on mature leaves. It has now been established that chlorotic dwarf is distinct from crinkly leaf or infectious variegation (Kersting *et al.*, 1996). Based on symptoms of the disease, it was named as "citrus chlorotic dwarf" (CCD) (Korkmaz *et al.*, 1995). Young trees infected with CCD are dwarfed and yield is reduced.

SYMPTOMS

Main symptoms of chlorotic dwarf disease on affected trees are strong leaf chlorosis and reduced leaf size (Plate 27). Young leaves show various chlorotic patterns and crinkling or distortion. Leaves on affected trees are smaller and have a notch near the apex (Korkmaz *et al.*, 1995). Field symptoms consist of a V-shaped notch and chlorotic flecking of young leaves, warping, crinkling, inverted cupping and variegation of mature leaves (Kersting *et al.*, 1996).

TRANSMISSION

The causal agent is graft-transmitted to grapefruit, lemons, mandarins and oranges. Korkmaz *et al.* (1994) reported citrus chlorotic dwarf (CCD) as a new virus-like disease which is transmitted from citrus to citrus by the whitefly *Parabemisia myricae* (Kuwana) (Homoptera; *Aleyrodidae*). It is transmitted from rough lemon to rough lemon by whitefly vector in a semi-persistent to persistent manner (Cinar *et al.*, 1994; Önelge and Çinar, 2010,). CCD is not mechanically transmitted but can be transmitted mechanically by stem slash inoculation from rough lemon to rough lemon.

HOSTRANGE

Chlorotic dwarf agent can affect most of the citrus species and cultivars. Rough lemon is highly susceptible and symptoms on rough lemon seedlings appeared 5–8 weeks after graft inoculation. Better symptoms develop at a temperature of 20–25°C but can also be developed at 30–35°C (Korkmaz *et al.*, 1994). Following varieties of citrus are commonly found infected in field:

- Aydin, Interdonata, Lamas and Kutdiken lemons (Plate 27).
- Minneola tangelo
- Fremont and Satsuma mandarins
- Marsh seedless
- Star Ruby and redblush grapefruit
- Valencia, Washington Navel and other sweet oranges
- Rough lemon
- Sour orange
- Alemow

INDICATOR HOSTS

Rough lemon, sour orange and Alemow are the best diagnostic and indicator hosts.

THE PATHOGEN

The causal agent of citrus chlorotic dwarf has not yet been characterized but preliminary purification attempts revealed that the pathogen is probably phloem restricted and has isometric particles. Martelli *et al.* (1996) isolated olive latent virus-1 from CCD infected Kutdiken lemon and Mexican lime. It could be possible that more than one virus is involved in plants with crinkling type of symptoms. CCD did not react in ELISA to antisera against citrus variegation, citrus leaf rugose, and satsuma dwarf viruses (Kersting *et al.*, 1996) suggesting involvement of an uncharacterized virus. In a purified preparation from infected sample, some bands show slightly flexuous, filamentous virus-like particles but similar particles were also observed in control extracts prepared from healthy *C. macrophylla* plants. The ultrastructure of CCD infected bark sections showed unknown filamentous virus-like particles (Howd *et al.*, 2002). However, the precise nature of these virus-like particles remains unclear.

DIAGNOSIS AND MANAGEMENT

CCD is detected by graft inoculation to diagnostic hosts such as rough lemon, Alemow and Sour orange. Management of CCD is possible by using healthy planting material and to control or minimize the whitefly population. Biological control of whitefly vector population is possible by releasing the aphelinid wasp (*Eretmocerus debachi* Rose and Rosen) (Kersting *et al.*, 1996).

Impietratura Disease

Impietratura disease of citrus is prevalent in all countries of the Mediterranean basin and also in Iran, Venezuela, South Africa and Cyprus (Reichert and Hellinger, 1930; Yedidyah, 1937; Ruggieri, 1955; Papasolomontos, 1969; Chapot, 1961; Scaramuzzi *et al.*, 1966). The name 'impietratura' has been used to designate a disease that cause skin hardening and albedo gumming of some citrus fruits (Ruggieri, 1955). In India, this disease has been reported from Kalimpong (Ahlawat *et al.*, 1984) on a local shaddock-type tree identified as *Citrus decumana*. Typical symptoms of the disease were also observed on *Soh-Shyrkhoit* and *Citrus latipes* in Northeastern region of India-the home of citrus (Bhagabati *et al.*, 1986).

SYMPTOMS

Symptoms of impietratura disease are most conspicuous on fruits of affected trees. Affected trees bear abnormal small fruits which never develop to normal size and remain small, hard like a stone and also become misshapen or lop sided (Plate 28a). Such fruits fell prematurely during June-July. Fruit bearing branches had depressed brownish spots. There are gum deposits under such spots on the wood (Ahlawat *et al.*, 1984). Fruits from affected trees show symptoms of skin hardening and show albedo gumming even when they are ¼th of size. Hardening of skin may or may not be associated with protuberances or by a halo of yellow colour surrounding the hard area (Papasolomontos, 1969). Distinct discoloration of rind appears as a circular spots on fruit skin and brown gum spot found directly under the cut surface of the spot. These discoloured spots are usually seen as raised bumps or protuberances on the rind. If such fruits are sectioned, gum spots may appear throughout the rind and albedo and a brownish gum impregnated tissue can be seen below hard area of fruit (Plate 28b). Gummy pockets can also be observed below the rind in the flesh of the fruit. In severe cases, whole fruit is deformed and hard due to gum deposits in the albedo. Sometimes, a complete ring of gum impregnated tissues can be seen in albedo of affected fruit. Hardening, malformation and albedo gumming of fruits are typical symptoms of impietratura disease in field trees. This disease can be confirmed by cutting hard fruits and see the layer of gum-impregated tissues below hard areas (Caruso *et al.*, 1993). Affected fruits rarely grow to full size and fell down prematurely. Gum impregated tissues have also observed in young green twigs of infected lemon trees in Kalimpong region of India (Ahlawat *et al.*, 1984). Seasonal variation for symptom appearance has also been noted, and strain differences are also known to occur (Catara and Scaramuzzi, 1980; Catara *et al.*, 1977; Papasolomontos, 1969).

TRANSMISSION

Impietratura disease is transmitted by grafting and no other method of transmission is known. Symptoms of the disease appear when grafted plants flower and fruit. On indicator host, symptoms may develop after 1–2 year of inoculation when they fruit. It is apparently not seed-transmissible (Bar-Joseph, 1976), and there is no evidence of mechanical or vector transmission. Impietratura disease has been transmitted by insertion of pollen from infected trees under the bark of healthy Orlando tangelo test trees (Vogel and Bové, 1980).

HOST RANGE

Sweet orange, Grapefruit, Pummelo and Volkamer lemon are highly susceptible to impietratura disease. Symptoms of disease have also been found on fruits of lemon, rough lemon, bergamot, tangelo (Orlando), citron, mandarin, clementine, Sour orange, and Dweet tangor. Chinotto, Citron and Kumquat are symptomless hosts and Chinotto, Citron, Etrog citron and Kumquats are tolerant to impietratura pathogen (Cartia and Catara, 1969). In India, symptoms of impietratura were observed on fruits of lemon, mandarin, grapefruit, sweet orange and tangelo (Mali and Chaudhuri, 1993). Specific symptoms on some hosts are described below:

- ***Volkamer lemon*:** Leaf flecking symptoms develop on new growth of inoculated plants. Albedo gumming of fruits is common when inoculated plants bear fruits. Affected fruits do not develop typical reddish colour of volkamer fruits and fruits remain pale yellow. 6–7 gum deposits can be seen in infected Volkamer lemon fruit.
- ***Rough lemon*:** Leaf flecking symptoms similar to volkamer lemon develop on inoculated plants. Fruit drop is comparatively low in this host. Gum is mostly deposited along the equatorial part of the fruit and form a round green 3–4 cm area which does not change colour even after ripening.
- ***Grapefruit and pummelo*:** Leaf flecking, heavy fruit dropping and typical fruit symptoms.
- ***Sanguinello sweet orange*:** Leaf flecking, fruit symptoms, fruit drop is less than grapefruit.
- ***Citrus decumana*:** Leaf flecking, albedo gumming, stony and malformed fruits. Gum pockets on young twigs of affected trees (Ahlawat *et al.*, 1984).

INDICATOR HOSTS

Volkamer lemon and rough lemon are good indicator hosts of impietratura disease. Alternate indicators are nucellar grapefruit or sweet orange grafted on sour orange or rough lemon seedlings. Rough lemon is a better indicator of fruit symptoms than volkamer lemon, grapefruit and sweet orange (Caruso *et al.*, 1993). Sweet orange seedlings develop vein flecking and oak-leaf patterns on young leaves after a few weeks of inoculation at low temperature. Oak leaf symptom in inoculated sweet orange seedlings was considered to be diagnostic for impietratura disease (Bar-Joseph and Loebenstein, 1970).

THE PATHOGEN

Symptoms of impietratura disease are similar to boron deficiency. However, it was ruled out by analysis of leaves from affected trees which showed adequate boron contents (Ruggieri, 1955; Papasolomontos, 1961, 1965). A normal tree should have 30–200 ppm boron in the dry matter of mature leaves, whereas a tree with boron deficiency will contain only 3–25 ppm boron in dry matter. In addition, the young leaves of boron deficient trees often show yellowish translucent spots, accompanied by marked leaf distortions or deformations. Boron deficient old or mature leaves may show corky, split veins and midribs which are absent with impietratura disease.

Evidence of transmissibility shown by several workers (Papasolomontos, 1965; Scaramuzzi *et al.*, 1966; Terranova and Scuderi, 1968; Ahlawat *et al.*, 1984) revealed that the disease is caused by a virus. Though impietratura pathogen is believed to be a virus but has not been isolated or characterized as yet. It may be related to, or part of, the concave gum family that produces a distinct oak-leaf pattern on leaves in the spring flush of growth on field trees or leaves of seedling indicators.

DIAGNOSIS AND MANAGEMENT

Diagnosis of this disease is possible by inoculating indicator plants and observing the symptoms on field trees as described above. Appearance of oak leaf pattern symptoms in young emerging leaves during first growth flush during cooler spring months in association with fruit symptoms can also be used for diagnosis of the disease in the field. Impietratura can be transmitted by grafting budwood from infected trees on to branches of healthy field trees and the symptoms reproduced in the developing fruit will indicate the presence of the disease. Any grapefruit tree, preferably on rough lemon or sour orange rootstock, is considered a superior indicator although other cultivars are also used (Papasolomontos and Economides, 1967). Bar-Joseph and Ben Shalom (1982) reported the development of symptoms on fruits in three to six months following inoculation. A short-term indexing using seedlings may be efficient for determining the presence or absence of the disease. Impietratura disease can be eliminated by certification programmes involving thermotherapy followed by shoot-tip grafting. Legislative measures and institutions are involved in the production of healthy propagating material in Greece and some other Mediterranean countries and certified material is indexed in glasshouse and also in the field (Caruso *et al.*, 1993; Bar-Joseph and Ben-Shalom, 1993).

Citrus Yellow Vein Disease

Citrus yellow vein disease was initially observed on limequat (*Citrus aurantifolia* × *Fortunella japonica*) in California (Weathers, 1957). This disease showed synergistic reaction with some other citrus viruses like psorosis and vein enation virus (Weather, 1960, 1961) and hence is one of the interesting citrus diseases. Origin of yellow vein disease is not known. Yellow vein virus is not economically important but scientifically it is important because it causes effect of synergism between two viruses.

TRANSMISSION

Yellow vein disease is transmitted only by grafting and no other means of disease spread is known.

SYMPTOMS AND HOST RANGE

Following hosts are susceptible to yellow vein disease: Citron (Etrog), *Citrus macrophylla, C. volkameriana*, Kumquats, Lemons, Limequats, Rough lemon, Tangelo (Orlando) and West Indian lime. Rough lemon and West Indian lime are highly susceptible hosts. Symptoms of yellow vein disease are similar on all susceptible citrus species and varieties but may differ in severity. Symptoms on West Indian lime, Rough lemon, Etrog citron and *Citrus macrophylla* appear as bright yellow veins of leaves and yellow blotches on stem (Plate 29). Yellowing may extend to petiole and young twigs upto a few centimetres. Occasionally vein banding effect also develops. Discoloration is sometimes restricted to half of the leaf but usually whole leaf is affected. Conspicuous symptoms develop on Florida rough lemon, Calamondin, Rangpur lime, Mandarin, and Kumquat but Citrange, Grapefruit, Lemon, sweetoranges, Sour orange and Tangelo show mild symptoms appearing on only one leaf. Subsequently these mild symptoms are masked. Trifoliate orange seems to be immune (Weather, 1960).

THE PATHOGEN

There is no information on etiology of the disease. Evidence of transmissibility revealed that the disease is caused by a virus. Though yellow vein pathogen is believed to be a virus but has not been isolated or characterized as yet. Plant infected by yellow vein virus and citrus vein enation virus will show highly synergistic effect in Lime, Sweet orange, lemon and rough lemon. Individually, each one of these viruses will show mild symptoms on these hosts but when a plant is inoculated with both viruses, severe vein yellowing symptoms develop on all leaves of the entire plant giving a look of yellow-net vein symptoms. Even the symptoms of vein enation become more conspicuous as the enations on veins become brilliant yellow and more enlarged. Doubly infected plants become stunted in growth with reduced leaf size. These studies helped to identify two strains in yellow vein virus as in combination one strain developed more yellowing and dwarfing than the second one (Weathers, 1963). Recently, it has been shown that when Mexican lime plants are inoculated both by vein enation and yellow vein, symptoms of both the diseases appeared. After shotgun cloning and sequencing, several clones showed high degree of homology with umbraviruses and luteoviruses in the case of yellow vein whereas the same luteovirus species was identified in vein enation inoculated plants. It has been proposed that the luteovirus is a major component of both diseases and the "enation" evolves to "yellow vein" as a result of the synergism between the umbra and luteo viruses (Kanrar *et al.*, 2010).

Diseases similar to yellow vein

A disease with similar symptoms on West Indian lime has been described from India named as 'yellow corky vein'. Most of the symptoms of yellow corky vein are similar to yellow vein except that the Indian disease shows vein corking on the lower side of veins and the disease is caused by the synergistic effect of two viroid species, citrus exocortis (CEVd) and hop stunt viroid (HSVd) (Rustem and Ahlawat, 1999). A yellow vein disease is also reported from Greece which is also shown due to infection of CEVd and HSVd in *Citrus bergamia* (Protopapadakis *et al.*, 2002) similar to the yellow corky vein disease reported from India (Rustem and Ahlawat, 1999). Bergamot (*Citrus bergamia*) and citron Roxane (*C. medica*) have been identified as indicator hosts of yellow vein disease in Greece. However, Etrog citron did not show yellow vein symprtoms with infection by Greece disease but showed epinasty of leaves. Better symptoms were obtained at higher temperature (32°C). To establish relationship of diseases reported from California, Greece and India needs more studies.

Yellow vein disease is detected by graft transmission to susceptible hosts. Eradication of infected trees and to avoid propagation from infected trees is the only recommended methods to contain this disease.

Leaf Curl Disease

Leaf curl disease of citrus was initially observed on a Pera orange tree at Limeira Citrus Improvement Station in Brazil. Similar disease was later found on a 12yr old Hamlin sweet orange tree at Araras (Salibe, 1959, 1965). The disease is locally known as "Crespiera citros" which means curling of leaves of citrus trees. Leaf curl disease has also been reported from India (Mali *et al.*, 1976).

SYMPTOMS

Curling and distortion of leaves similar to the one caused by heavy infestation of aphids are the major foliar symptoms (Plate 30a, b). Affected trees though bloom heavily yet set only a few fruits which remain smaller in size. Shoots become weak and can be easily broken under slight pressure. Occasionally, gum is present in xylem vessels. Heavily infected trees show dieback symptoms.

TRANSMISSION

Leaf curl disease is transmitted by graft inoculations. Pruning of inoculated plants help in quick appearance of disease symptoms. No other mode of disease spread is known.

HOST RANGE

Hosts susceptible to leaf curl disease are: Citrons, Grapefruits, Lemons, Mandarins, Pummelos, Sweet lime, Sweet oranges and Lime (Kagzi, Key, Mexican, West Indian). Sweet oranges are highly susceptible to leaf curl virus. Troyer citrange, Cleopatra mandarin, *Poncirus trifoliata* and Rangpur lime are symptomless and appears to be tolerant to leaf curl virus. The virus has been experimentally transmitted to the following hosts: Baianinha sweet orange, Hamlin and Pera sweet oranges, Maxirica Tangerine, Marsh seedless grapefruit, Doce citron (*Citrus medica*), Comprida citron, and Siciliana lemon. Dancy tangerine (*C. tangerina*), Rangpur lime and Cleopatra mandarin did not show any symptom after inoculation (Salibe, 1965). Symptoms of the disease appear in two to five months after inoculation followed by pruning in new growth of inoculated plants.

DIAGNOSIS AND MANAGEMENT

There is no information on the etiology of leaf curl disease of citrus. In Brazil, Caipira seedlings are used as indicator host. Bud inoculated seedlings of Caipira sweet orange show yellowing of leaves which are dropped prematurely. Leaves on newly developed shoots are severely curled and seedlings may die within one year of inoculation. No control measures are known but as a precaution, infected trees should be destroyed and propagation must be avoided from symptomatic trees.

Cristacortis

Cristacortis was first described on Tarocco sweetorange trees grafted on sour orange in Corsica (Vogel and Bove, 1964, 1968). Affected trees showed severe stem pitting in Sour orange and Tarocco orange. This disease is wide spread in Corsica and affects trees of Clementine on Sour orange rootsock (Vogel and Bove, 1972). They proposed the name of the disease as *Cristacortis* from Latin (crista=peg and cortis=bark) and concluded that this is an unreported disease of citrus. Cristacortis disease affects many species/varieties of citrus in Corsica, Sardinia, Sicily, Spain, Morocco, Algeria, Egypt, Turkey, Mauritius, North Africa, Mediterranean Basin, South America, Yemen and Vietnam (Scaramuzzi and Catara, 1968; Timmer *et al.*, 2000; Reuther *et al.*, 1978). This disease is also known as Sour orange stem pitting (Scaramuzzi and Catara, 1968) and Tarocco orange stem pitting (Vogel and Bove, 1964).

SYMPTOMS

Naturally infected trees are slightly stunted. Trunk, limbs and shoots show vertical depressions due to pits in the wood with corresponding pegs on cambial side of bark. Most conspicuous symptoms are numerous pits on trunk and main shoots. With certain species like tangelos, gum-like substance stains at top of peg and bottom of pit (Plate 31a). These pits have narrow longitudinal depressions which are quite distinct from blind pocket disease reported from California. Cristacortis infected trees develop pits in greater number and much earlier after inoculation than blind pocket disease. Depressions disappear with time as a consequence of radial growth. Cristacortis infected trees also show weaker branches which may bend towards ground. In Orlando and sour orange, pitting and gumming of stem develop at later stages. Verna sweet orange tree on sour orange graft inoculated with the cristacortis agent showed stem pitting on both scion and stock (Plate 31b). In cross sections of affected trunk and branches, clear brown stained lines perpendicular to wood layers are seen. Though the depressions may disappear with time but their traces can be seen in cross sections. New pits and pegs may develop at any portion on the tree. Flecking and oak-leaf pattern symptoms on young leaves similar to psorosis-concave gum-blind pocket leaf symptoms also develop on affected trees with cristacortis (Vogel and Bove, 1974). Infected trees develop comparatively larger fruits than normal but fewer in number. No fruit symptoms are known.

TRANSMISSION

Cristacortis disease is transmitted by graft and bud inoculation only to *Citrus* species. Pollens collected from cristacortis affected tree when placed under the bark of healthy Orlando tangelo bark, showed transmission and symptoms of both psorosis (young leaf symptoms) and typical cristacortis (stem pitting) symptoms. However, no natural spread of cristacortis pathogen has been observed (Vogel and Bove, 1976a, 1980). No vector of the disease is known.

HOST RANGE

Host range of cristacortis is restricted to citrus species only. Susceptible hosts to cristacortis are: Orlando tangelo, Dweet tangor, Sour orange, Sweet orange, Mandarin, Clementine, Grapefruit, Rough lemon, Satsuma, Siamelo, Sweet lime and *Citrus pectinifera*. Sour orange, Sweet oranges and Tangelos are highly susceptible to cristacortis infection. Eureka lemon, West Indian lime, Etrog citron, *C. hystrx* and *poncirus trifoilata* are tolerant or symptomless hosts. Troyer, Carrizo and Rusk citranges, Bergamot orange and Volkamer lemon did not show any symptom even after ten year of inoculation. It has not been established whether these hosts are immune or tolerant.

INDICATOR HOSTS

- ***Sweet orange seedling***: Vein flecking and oak-leaf patterns develop on young leaves within a few weeks after inoculation at cool temperature.
- ***Orlando, Webber or William's tangelo on sour orange***: Tangelo bud and two bark patches from affected tree are grafted at the same time on sour orange seedling, one bark patch above and the other below the tangelo bud. Stem pitting symptoms develop in wood and pegs on cambial side of the bark within 1–2 year after inoculation. Tangelo seedlings can also be used as test plants.

THE PATHOGEN

Nature of pathogen associated with cristacortis has not been established as yet. Based on the number of pits developed on Clementine trunk and shoots, different strains of cristacortis pathogen have been suggested (Vogel and Bove, 1976b). Cristacortis virus has been shown to be different from tristeza, cachexia and exocortis or the synergistic reaction of these pathogens (Vogel and Bove, 1972). Infected plants show leaf flecking and oak leaf pattern similar to psorosis group of viruses and therefore this disease was placed within the so called psorosis group of diseases (Vogel and Bove, 1974).

DIAGNOSIS AND MANAGEMENT

Cristacortis disease can be diagnosed by inoculating Orlando, Webber or William's tangelo on sour orange which are better indicator hosts for disease diagnosis. When susceptible species or varieties are grafted on sour orange, symptoms of depressions, pockets and stem pitting in trunk, limbs and shoots are developed. The presence of stem pitting on the sour orange stock in addition to pits on the scion is valuable for diagnosis (Plate 31b). Use of cristacortis free budwood is only method of disease management so far.

Citrus Leathery Leaf

Leathery leaf disease of citrus was observed in Sikkim orange (*Citrus reticulata*) trees in Darjeeling and Sikkim hills in 1972 (Ahlawat, 1975). Studies on this disease revealed that this is a distinct disease from tristeza, greening and crinkly leaf diseases occurring in that region. Leaves on affected trees become leathery and thick in texture. Some of these trees also showed symptoms similar to huanglongbing, infectious variegation and Satsuma dwarf diseases of citrus.

TRANSMISSION

Citrus leathery leaf virus (CLLV) is graft transmissible from Sikkim orange to several citrus species/cultivars. CLLV is readily transmitted mechanically from citrus to citrus and herbaceous hosts. It is also transmitted by aphid, *Aphis gossypii* in non-persistent manner (Ahlawat *et al.*, 1979).

HOST RANGE

Susceptible hosts of leathery leaf are: mandarin cultivars Sikkim Orange, Khasi Orange and Coorg Orange; Sweetorange cvs. Mosambi, Sathgudi and Washington navel; Eureka, Lisbon and Hill lemons; Sour orange, Rangpur lime, Grapefruit, Pummelo, Kagzi lime, Rough lemon and *C. decumana* but not trifoliate orange. In sweet orange plants, psorosis like leaf flecking symptom appears which persisted for about a week and later disappeared. In other hosts, thickening of the uppermost few leaves are the major symptoms (Fig. 23A). Leaves of inoculated plants show

Fig. 23: A. Symptom of leathery leaf on Darjeeling mandarin; B. Symptoms of leathery leaf on *Gomphrena globosa* showing mosaic symptoms

downward curling and death of apical bud. In field trees, sometimes dark green islands are also visible on these thickened leaves. Grapefruit, lemon and rough lemon are symptomless hosts of the virus (Ahlawat *et al.*, 1979).

In mechanically inoculated plants systemic mottle of leaves develops in *Gomphrena globosa* (Fig. 23B), *Phaseolus mungo, P. vulgaris* cultivars Red Kidney and Singtam local, *P. aureus* cultivars NP-23, PM-1 and ML-8. In *Sesamum indicum*, symptoms appear as chlorotic spots on inoculated leaves followed by systemic mosaic and twisting of leaves. In *Cucumis sativus*, mild mottling of leaves and in periwinkle downward curling of the leaves were the major symptoms. *Petunia hybrida, Nicotiana glutinosa and Physalis floridana* are symptomless hosts of the virus. This virus is not mechanically transmitted to 17 herbaceous hosts.

PHYSICAL PROPERTIES

Leathery leaf virus has thermal death point between 65° and 70°C, dilution end-point between 10^{-6} and 10^{-7} and longevity at room temperature (22–25°C) 96 h and at 5°C 19 days.

PATHOGEN

Nature of pathogen associated with citrus leathery leaf has not been established as yet. Studies on transmission and physical properties are suggestive of association of a virus with leathery leaf disease. Citrus leathery leaf virus resembles more closely to Satsuma dwarf (Tanaka and Imada, 1974) and citrus leaf rugose (Garnsey, 1975). However, leathery leaf virus differs from these viruses in host range and physical properties. This disease is widely present in khasi mandarin cultivated in Northeastern region of India and based on psorosis-like young leaf vein flecking symptoms, it is believed that the virus belongs to psorosis group of viruses.

DIAGNOSIS AND MANAGEMENT

Symptoms of leathery leaf disease are confusing but it can be easily diagnosed by sap inoculation to herbaceous hosts like *Gomphrena globosa* and *Sesamum indicum* which produce clear diagnostic symptoms. Use of virus-free planting material and control of aphid vectors are the measures effective for disease management.

Multiple Sprouting

A disease with multiple bud sprouts in sweet orange trees was described from South Africa (Schwartz, 1970). Symptoms of multiple sprouting were distinct from copper deficiency, bud mite (*Aceria sheldoni*) infestation or genetic factors. Two types of symptoms have been described for multiple sprouting disease (1) Infected trees showed better growth than healthy trees of same age but bushy with multiple branching having deformed leaves. Many shoots develop from a single bud giving a look of witches' broom symptoms (Fig. 24). Branches with multiple sprouts may show negatively geotropic habit (Searle, 1969). With the growth habit of affected plants, they can be easily idendified in the field. Trunk of affected trees on rough lemon roots show extremely reduced trunk diameter. A disease with multiple sprouting symptoms was observed in kagzi lime trees (*Citrus aurantifolia)* from Experimental farm of Indian Agricultural Research Institute, Pune in India (Ahlawat and Chenulu, 1985). A bud transmissible disease with almost similar symptoms has also been reported from India with the name of "Blastomania" (Mali *et al.*, 1975). However, Blastomania disease has been shown to be associated with phytoplasma-like bodies (Sharma and Singh, 1988).

Fig. 24: Twigs of key lime showing multiple sprouting-Photo by author from IARI, Regional Station, Pune, India

SYMPTOMS

The growth of field infected trees is bushy with multiple branching having deformed leaves. Multiple sprouts will show negatively geotropic growth. Experimentally inoculated citrus species showed only shortening of internodes, leaf abscissions and dieback symptoms (Schwarz, 1970). Field symptoms in Key lime trees in India were extremely reduced size of leaves on multiple sprouts. Multiple branches remained green in colour for quite a long time or until they are removed.

TRANSMISSION

Multiple sprouting disease agent is graft transmitted to several citrus hosts. It is also transmitted by mechanical inoculations. Multiple sprout disease is also sap transmitted in India (Ahlawat and Chenulu, 1985). The Indian isolate was transmitted by mechanical inoculation from citrus to *Chenopodium quinoa.*

HOST RANGE

Following hosts are susceptible to multiple sprouting: Bergamot and Grapefruit are symptomless hosts. Citranges (Carrizo, Rusk, Troyer), Citrumelos and Trifoliate orange are tolerant hosts of multiple sprouting.Sweetorange plants showed milder symptoms than sour orange, lemon and grapefruit. Multiple sprouting has been shown to be sap transmitted from citrus to *Chenopodium quinoa, C. amaranticolor, Nicotiana glutinosa and Vigna sinensis* showing virus-like symptoms (Majorana and Schwarz, 1972). *Celosia argentea* and *Gomphrena globosa* are symptomless hosts. Causal agent of multiple sprouting in India is also easily sap transmitted from affected leaves of multiple sprouts to *C. quinoa* causing local lesions. However, the multiple sprouting agent could not be transmitted back to citrus or any other host from local lesions of *C. quinoa.*

THE PATHOGEN

Repeated attempts to identify virus-like particles from citrus and sap inoculated *C. quinoa* were unsuccessful (Ahlawat and Chenulu, 1985). The cause of this disease is yet to be ascertained from Indian citrus and also from South Africa.

DIAGNOSIS AND MANAGEMENT

Diagnosis of multiple sprouting (MS) is easy in the field by observing typical multiple sprouts on affected trees. MS agent can also be detected by sap transmission to *C. quinoa* where it produces local lesions on inoculated leaves. Removal of affected branches and propagation from MS-free trees are the best ways to manage this disease effectively.

Citrus Dieback Disease in India

Dieback is symptoms and not a disease but it is commonly called as a dieback syndrome. Dieback is a serious disorder of citrus species and cultivars in India and is responsible for decline of citrus industry in the country. This disease complex can be identified from distances as it causes defoliation of young branches and dieback of twigs from tips. Affected trees normally do not die but vigour, general health and production is adversely affected (Raychaudhuri *et al.*, 1969, 1977). Dieback or decline of citrus trees is also common in Pakistan (Akhtar and Ahmad, 1999).

SYMPTOMS

In dieback syndrome, leaves of affected trees fell down from twigs and branches resuting symptoms of defoliation and dieback of twigs (Plate 32). This is followed by secondary growth of axillary buds which consist of short, upright, small weak shoots and smaller leaves, sometime showing bushy appearance. Leaves on these shoots may show typical symptoms of zinc deficiency such as green blotches or small green spots on yellow tissue or greening of veins with chlorosis of interveinal tissue. These branches may develop unusual and excessive flowering. Dieback of twigs moves downward and affected twigs may be attacked by weak wood rotting fungi. Initially, symptom starts on one or two branches but subsequently the whole tree is involved. Affected trees bear excessive flowering which produces sun blotched fruits that seldom mature. Feeder roots become dark and show rotting.

Citrus dieback syndrome has been studied by several scientists of different disciplines in India. Hence, different causes of dieback have been suggested such as soil types and conditions, high water table, nutritional deficiencies and infection by fungi, viruses, phytoplasmas, viroids and nematode infestation (Raychaudhuri *et al.*, 1969, 1977). Citrus dieback symptoms have been attributed by citrus tristeza virus (Capoor, 1963; Vasudeva and Capoor, 1958). Greening disease now known as Huanglongbing was also suggested the cause of dieback in India (Fraser and Daljit Singh, 1966, 1968; Fraser *et al.*, 1966; Nariani *et al.*, 1967). Further investigations revealed that the disease is caused by a complex of the Huanglongbing organism and certain fungi (Raychaudhuri *et al.*, 1967). With the development of diagnostic technology and increasing knowledge of new pathogens, cause of dieback syndrome is now known due to infection of certain viruses (Tristeza, Indian citrus ringspot and citrus yellow mosaic badnavirus), Huanglongbing bacterium and exocortis and hop stunt viroid in two or more combinations in addition of huanglongbing (Ahlawat, 1997). In Pakistan also, dieback disease has been attributed to both biological (Huaglongbing, CTV, Exocortis viroid, *Phytophthora gummosis*, Bud-union crease) and abiological (Over-irrigation, poor drainage, excess salt and/or heavy ploughing for clean cultivation) factors. Huanglongbing is probably one of the major causes of citrus decline in Pakistan. Fungi similar to isolated from dieback-affected citrus trees (Raychaudhari *et al.*, 1977; Ahlawat and Raychaudhuri, 1988) in India have also been isolated from Pakistan (Girmaldi and Catara, 1989). The role of these fungi has probably been overemphasized, since huanglongbing in itself is enough to induce decline and dieback. Dieback disease has also been described in Australian citrus which is probably caused by Huanglongbing bacterium (Broadbent *et al.*, 1976).

Development of new diagnostic methodology and technology for freeing viruses and virus-like pathogens through techniques like shoot tip grafting (STG) became extremely helpful to minimize dieback incidence. STG and recent diagnostics like ELISA and PCR are of great advantage to raise new plantings free from pathogens and will hopefully minimize the century old dieback in India. Diagnostic methods and management practices for individual pathogen involved in dieback syndrome have been described in this book.

Bud-Union Crease

Bud-union abnormalities became apparent as soon as the practice of growing seedling trees was shifted to grow grafted trees to maintain true-to type qualities of citrus cultivars. Bud union disorders affecting citrus trees on certain rootstock have been reported from several countries. Some of these abnormalities are considered due to genetical incompatibility, others are graft transmitted and caused by pathogens. The best known example of this type of diseases is tatter leaf which induces bud union crease in different citrus species propagated on trifoliate rootstock (See description of Tatter leaf-Citrange stunt). Other graft-transmissible bud union disorders are: A) Sweet orange on rough lemon, B) Nagami kumquat on trifoliate, and C) Parson's special mandarin on Volkamer lemon rootstocks. Bud union crease was first described in Palestine on Shamaouti sweet orange trees grafted on rough lemon rootstock (Yedidyah, 1937), and later in Florida (Grimm *et al.*, 1955), South Africa (McClean *et al.*, 1958; McClean, 1974), Brazil (Grant *et al.*, 1957) and Spain (Navarro *et al.*, 1993) affecting different sweet orange cultivars propagated on rough lemon. Usually bud union disorder appears to be a non-destructive disease as data on damage caused by budunion are not available. Sweet orange trees grafted on rough lemon are susceptible to bud-union crease in South Africa and Spain and Parson's Special mandarins grafted on Volkamer lemon are susceptible in Corsica. Bud-union crease has also been noticed on Bloodred, Mosambi, Valencia late and Pineapple sweet orange trees on rough lemon rootstock in Pakistan (Bove, 1995). In Bloodred and Mosambi sweet orange old trees, symptoms have been described in India (Bhutani *et al.*, 1972)

SYMPTOMS

Symptoms of bud union crease appear at a point of bud union as continuous depression in the wood with corresponding projections from the inner face of the bark. Sometimes this depressed line is continuous (Plate 33a), but in most cases indentations with gum impregnation are observed. Necrosis of bark at the bud union also develops in susceptible cultivars of sweet oranges. In some cases bark appear normal but if the bark is removed at the bud union, clear symptoms of the disease are visible (Plate 33b). Bark breaks easily at the constricted bud-union line, revealing gum deposits at the union. Some sweet orange varieties or selections on rough lemon rootstock show conspicuous bud union crease but other clones of the same variety grow normal such as nucellar clones of Pineapple, Hamlin and Valencia sweet oranges. In Pakistan, Valencia late and Pineapple sweet oranges showed an irregular and almost continuous ring of gum-impregnated projections on the cambial face of the bark at the bud-union line, which fits in the crease in the wood. In India, bud-union crease has been frequently observed by the author on Mosambi sweet orange and Nagpur mandarin trees grafted on rough lemon (Plate 33b). Affected trees are slightly stunted and sprouts from rootstock were observed. There are no symptoms on leaves and fruits. The layer of functional phloem at the union is very thin and seems to be non-functional.

TRANSMISSION

Transmission is normally through infected budwood or bark patches. No vector transmission is known.

HOST RANGE

Blood oranges, Shamouti, Jaffa, Med-sweet, Malta Egg are susceptible hosts. In South Africa, Tomango, Hall's midseason, Letaba midseason, Mid-knight, and Premier oranges are susceptible (McClean, 1974). Whereas, in Spain susceptible cultivars on rough lemon rootstock are: Valencia temprana, Royal temprana, Doblefina, *Cadenera punxosa*, Santomera, Navel sangre, Ricalate, Parson Brown and Taroco. Parson Brown and Taroco were imported from Florida and Italy, respectively (Navarro *et al.*, 1993). In Spain, *Cadenera punxosa* plants propagated on rough lemon are susceptible and showed symptoms of stunting and bud union crease two years after graft inoculation but inoculated plants of Marsh grapefruit, Nules clementine or Eureka lemon grew normally and did not show bud union crease. Symptoms shown by Bloodred and Mosambi sweet oranges correspond to a severe form of the disease, while those seen on the other two sweet oranges represent a milder form (McClean, 1974).

THE PATHOGEN

The causal agent is graft-transmissible, but its nature is unknown. Bud-union crease disorder has been considered due to a virus or a genetic factor (Reuther *et al.*, 1978). Bove (1995) has provided evidence that the bud-union crease of sweet orange trees on rough lemon rootstock is caused by a graft-transmissible pathogen.

DIAGNOSIS AND MANAGEMENT

Trees indexed positively for gummy bark viroid were also positive for bud-union constriction (Nour-Eldin, 1968). Nucellar sweet orange clones on rough lemon roots grew healthy upto eight years of age but develop symptoms within four years after inoculation (Reuther *et al.*, 1978). Plants of sensitive sweet orange cultivars on rough lemon will show external crease symptoms within two years after graft inoculation with bark patches from affected trees. Symptoms of bud union disorder appear earlier under field conditions than in the greenhouse. Selection of disease-free scion budwood and avoidable susceptible stionic combinations are the advisable approaches to contain the disease. The disease is controlled by using new-line sweet orange on rough lemon where old lines are found to make abnormal unions. The disease can also be eliminated by shoot tip grafting *in vitro* (Navarro *et al.*, 1993).

Citrus Measles

Measles in citrus is a disorder that occasionally affects a few trees in a grove. The disease is rarely found in Florida and Brazil for a long time but it has also been recorded in California (near Farmersville) for the first time. Symptoms of measles appear as numerous pale yellow spots on leaves of affected trees. These spots are seen as yellow from upper surface and tan to brown and raised on the lower surface of leaves. Once a tree is affected there is no masking of symptoms and they reappear every year throughout the tree canopy. Sometimes only one branch or limb is affected, but the symptoms can appear throughout the tree canopy. Varieties of sweet orange such as Hamlin, Pineapple, Valencia, and Navel orange trees and several other citrus species develop symptoms of measles. Cause of measles disorder is not known and has been considered as a genetic disorder (Whiteside *et al.*, 1988). However, this disorder has been shown to be graft transmitted to sweet orange using blind buds from symptomatic field trees (Lee *et al.*, 1993). Faint translucent spots develop on young leaves of inoculated plants which subsequently develop into typical measles spots as the leaves matured. Thus graft transmissible nature of citrus measles was confirmed. Symptoms of measles disorder develop at day temperature of 38°C and night temperature of 18°C. There is no information on the causal agent. Graft transmission of measles disorder to sweet orange indicator plants suggests that the disease is caused by an infectious agent similar to viruses (Lee *et al.*, 1993). No virus-like particles or double stranded RNA species were observed in measles affected tissues. Budwood from contaminated trees should be avoided to manage the disease.

Kassala Disease

Kassala disease was first observed in 1987 from a place known as Kassala in Eastern Sudan (Bove, 1995). Symptoms of Kassala disease have also been observed in Southern Yemen. Gum impregnation in bark and mild stem pitting on affected grapefruit trees are the main symptoms of the disease. Mild stem pitting of kassala disorder resembles to those of cachexia in mandarin and gummy bark in sweet orange.

Symptoms of cassala disease can be easily observed after removing the outer bark of grapefruit scion. Whitish bark tissues below the bark show impregated streak of reddish-brown gum like material near the bud union (Plate 34). Pits in xylem of grapefruit trees are also seen after removal of outer bark. Bark gumming can stand upto 2 m above the bud union. The nature of the agent is unknown. Leaves and fruits on affected trees do not show any symptom. Graft-transmissibility of kassala disease agent has not been studied. Natural transmission through budwood is possible but not demonstrated.

Host range of this virus is restricted to *Citrus* species only. Cassala disease was also seen on the scions of Foster and Marsh seedless grapefruit trees grafted on Sour orange. Sour orange is a symptomless or tolerant host. This disease is diagnosed by Scraping the bark of grapefruit trunk above the bud union which reveals gum deposits. Bark removal across the bud-union shows stem pitting on the grapefruit scion. Symptoms of this disease such as stem pitting resemble to those of cachexia on cachexia sensitive hosts and those of gummy bark on sweet orange. However, the amount of gum in the bark seems to be smaller than that seen in the case of cachexia and gummy bark. In field trees, the disease is diagnosed by scraping the bark of the grapefruit trunk above the bud union which will show gum deposits. Bark removal across the bud-union shows stem pitting on the grapefruit scion. Cassala disease free budwood should be used for disease management.

Rio Grande Gummosis (RGG)

A disease with the name of ferment gum disease (Rio Grande gummosis) was first reported on a grapefruit tree in California (Calavan, 1961). This disease was further studied and reviewed by Childs (1978a, b) in two parts. Rio Grande gummosis (RGG) is a disorder of the trunk and limbs of mature citrus trees. It occurs primarily in the Caribbean Basin, Somalia and Morocco but has also been reported sporadically from some other citrus areas. Grapefruit and lemon are the most commonly affected species. Narrow cracks (lesions) exuding yellow gum are seen in bark of affected trees. Severely affected trees produce more gum and their branches and trunk become black upto 100 cm of lesions. The cambial tissues below the bark become pink or orange in colour which may be seen after removal of bark (Plate 35). Gum is water soluble and hence washed away during rains. As the symptoms advance, lesions become darker and wood tissue become impregnated with gum followed the death of bark and decay of wood. If limbs are girdled by bark necrosis, death of the entire limb occurs. The disease is common to older trees but may also appear in young trees. Transmission of the disease is not known.

The cause of the disease is not known. It has been variously attributed to soil conditions and fertilization, viruses and wood-attacking fungi (Davis, 1980, Sonoda and Pelosi, 1990). In Florida, the disease has been associated with an uncharacterized, graft-transmissible agent, unrelated to psorosis (Powell *et al.*, 1998). The disease does not appear to be spreading in nature. The disease is aggravated by application of chlorides and other fertilizer salts and occurs more commonly on calcareous soils or those with calcareous subsurface layers. Symptoms are the only method of disease diagnosis. No control measures are known for the disease except to avoid making large pruning wounds near the trunk (Timmer and Powell, 2006).

Citrus Sudden Death Disease

Citrus Sudden Death (CSD) is a new disease that was first reported in 1999 at the Triangulo mineiro and Sao Paulo in Brazil and was considered to be a new threat to Brazilian citrus. Citrus Sudden Death (CSD) is a new epidemic disease of alarming proportions that already has caused the death of more than 1 million sweet orange trees grafted on Rangpur lime in Brazil (Maccheroni *et al.*, 2005). CSD disease causes serious economic losses in sweet orange production in the States of São Paulo and Minas Gerais. Currently, about three million orange trees have been affected by this disease. Graft-transmission tests reproduced the symptoms of the disease, but search for endogenous and exogenous bacteria as well as for viroids gave negative results (Barros *et al.*, 2005). The death of the plants was due to disorganization and death of phloem vessels at the graft union (Kitajima, 2010). Natural transmission of CSD has been suggested by an aerial vector (Yamamoto *et al.*, 2010). Recently, sudden death virus has been transmitted to *Chenopodium quinoa, C. amaranticolor, Nicotiana clevelandii* and *Nicotiana* sp., which are not the host of CTV (Santos *et al.*, 2010). The casual agent of CSD disease is not yet known but based on transmission results, a virus seems to be the most plausible candidate to cause the disease. There is a hypothesis that correlates CSD and a mutated *Citrus tristeza virus* (CTV). Epidemiological studies have suggested that CSD is probably caused by a new variant of *Citrus tristeza virus* (CTV) (Coletta-Filho *et al.*, 2005; Maccheroni *et al.*, 2005). Investigations are under way to find out whether more virulent CTV variants might be associated with CSD. Analysis of CSD affected trees as well as asymptomatic trees were found to be infected with *Citrus tristeza virus* (CTV). However, a newly isolated and identified tymo-like virus (TlV) was detected from mainly CSD affected orange trees (Barros *et al.*, 2005).

All attempts to associate the disease with a specific CTV isolate so far have failed. The putative variability found was due to geographic or environmental differences. However, the presence of a *Tymoviridae* sequence was observed only in samples from CSD affected trees (Maccheroni *et al.*, 2005). They named this new virus as Citrus sudden death-associated virus (CSDaV). EM preparations showed the presence of two types of isometric particles, typical of *Tymoviridae* (Maccheroni *et al.*, 2005). A tymo-like virus isolated from CSD-trees was closely related to the tymovirus group that possesses a poly-A tail, but not to a tymovirus with a t-RNA like structure in the 3′ end of genome (Barros *et al.*, 2005). Preparations from partially purified virus were examined by electronic microscopy and tymo-like particles were observed, although at a very low concentration. However, the results of some studies agree with the hypothesis that CSD could be associated with different populations of CTV

(Coletta-Filho *et al.*, 2005). Further studies are necessary to verify if the isolated tymo-like virus, or more virulent CTV variants or a synergistic interaction of both is the cause of CSD (de Jesus Junior *et al.*, 2005). The complete genomic sequence of this virus was obtained and phylogenetic analysis showed a close relationship with viruses of the *Marafivirus* genus. In all affected trees, Marafivirus was constantly detected in association with citrus tristeza virus. However, citrus sudden death virus was detected in all symptomatic leaves of inoculated *Chenopodium quinoa, C. amaranticolor, Nicotiana clevelandii* and *Nicotiana* sp., while CTV was never detected (Santos *et al.*, 2010) suggesting that the disease is caused by *Marafivirus* genus.

REFERENCES

Ahlawat, Y.S. 1975. Leathery leaf – a new citrus disease isolated from greening affected mandarin trees. *Indian Phytopath.* 28: 146 (Abs).

Ahlawat, Y.S., Nariani, T.K. and Sardar, K.K. 1979. Leathery leaf: A new virus disease of citrus. *Indian Phytopath.* 32: 198–201.

Ahlawat, Y.S., Chenulu, V.V., Chakraborty, N.K. and Viswanath, S.M. 1984. Occurrence of impietratura disease of citrus in India. *Curr. Sci.* 53: 384–385.

Ahlawat, Y.S. and Chenulu, V.V. 1985. Rubberywood and multiple sprouting-two newly discovered diseases of citrus. *Indian Phytopath.* 38: 625.

Ahlawat,Y.S. and Raychaudhary, S.P. 1988. Status of *Citrus tristeza* and dieback diseases in India and their diagnosis. pp. 871–879. *In*: Proc. 6th Internl. Citrus Congress, (*Eds.*) Goren, R. and Mendel, K.,Tel Aviv, Israel.

Ahlawat, Y.S. 1997. Viruses, greening bacterium and viroids associated with citrus (*Citrus* spp.). *Indian J. Agric. Sci.*, 67: 51–57.

Akhtar M.A. and Ahmad, I. 1999. Incidence of citrus greening disease in Pakistan. *Pakistan Journal of Phytopathology*, 11: 1–5.

Albrigo, L.G. and Young, R.H. 1981. Phloem zinc accumulation in citrus trees affected with blight. *HortScience* 16: 158–160.

Albrigo, L.G., Syvertsen, J.P. and Young, R.H. 1986. Stress symptom of citrus trees in successive stages of decline due to blight. *J. American Soc. Hort. Sci.* 111: 465–470.

Albrigo, L.G., derrick, K.S., Timmer, L.W., Tucker, D.P.H. and Graham, J.H. 1992. Inability to transmit citrus blight by limb grafts. pp. 131–138. *In*: Proc. 12th Conf. IOCV, IOCV, Riverside, CA.

Bar-Joseph, M. 1976. Some effects of temperature on symptom appearance and therapy of citrus impietratura disease. pp. 105–108. *In*: Proc. 7th Conf. IOCV, IOCV, Riverside, CA.

Bar-Joseph, M. and Loebenstein, G. 1970. Leaf flecking on indicator seedlings with citrus in Israel. A possible indexing method. *Plant Dis. Rep.*, 54: 643–646.

Bar-Joseph, M. and Ben-Shalom, J. 1982. Limited systemic spread of impietratura and psorosis-A in graft-inoculated grapefruit trees. *Plant Disease* 66: 820–821.

Bar-Joseph, M. and Ben-Shalom, J. 1993. A combined indexing and budwood propagation method for citrus. *Phytoparasitica* 21: 269–272.

Barros, C.C.P., de Jesus Junior, W.C., Teixeira, D.C., Gomes, C.C., Dutra, L.S. and Nagata, T. 2005. Molecular Characterization of Tymo-Like Virus Isolated from Orange Trees with Sudden Death Symptoms. Proc. 16th Conf. IOCV, IOCV, Riverside, CA (Abs).

Bausher, M.G. and Sweeney, M.J. 1991. Field detection of citrus blight using immunological techniques. *Plant Disease* 74: 168–170.

Beretta, M.J.G., Brlansky, R.H. and Lee, R.F. 1988. A comparison of histochemical staining reactions of the xylem occlusions in trees affected by citrus blight and declinio. *Plant Disease* 72: 1058–1060.

Berreta, M.J.G., Rossetti, V., Teixeira, A.R.R. and Sempionato, O. 1991. Positive diagnostic tests for declinio on plants root-graft inoculated in Brazil. pp. 256–260. *In*: Proc. 11th Conf. IOCV. IOCV, Riverside, CA.

Beretta, M.J., Derrick, K.S., Lee, R.F., Barthe, G. and Rosetti, V. 1993. Citrus declinio in Brazil: Rate of spread and serological comparison with other declines. pp. 113–115. *In*: Proc. 12th Conf. IOCV, IOCV, Riverside, CA.

Bhagabati, N.K., Borthakur, B.C. and Ahlawat, Y.S. 1986. Occurrence of impietratura disease of citrus in northeastern India. *Indian Phytopath*. 39: 110–111.

Bhutani, V.P., Bakhshi, J.C. and Knorr, L.C. 1972. Biochemical changes in healthy and decline sweet orange trees associated with bud union crease. pp. 229–233. *In*: Proc. 5th Conf. IOCV, IOCV, Riverside, CA.

Bové, J.M. 1995. Virus and virus-like diseases of citrus in the Near East region. Chapter 20, The Sudan. FAO, Rome, pp. 303–318.

Brlansky, R.H., Lee, R.F. and Collins, M.H. 1985. Structural comparison of xylem occlusions in the trunks of citrus trees with blight and other decline diseases. *Phytopathology* 75: 145–150.

Brlansky, R.H., Lee, R.F. and Agostini, J.P. 1988. The occurrence of citrus blight on sweetorange and *Cleopatra mandarin* rootstocks in Northwest Argentina. pp. 381–383. *In*: Proc. 10th Conf. IOCV, IOCV, Riverside, CA.

Brlansky, R.H. and Howd, D.S. 2001. Purification of virus-like particles from blight-affected citrus trees. pp. 297–305. *In*: 15th Conf. IOCV, IOCV, Riverside,

Broadbent, P., Fraser, L.R. and McGechan, R. 1976. Australian citrus dieback. pp. 141–146. Proc. 7th Conf. IOCV, IOCV, Riverside, CA.

Calavan, E.C. 1961. Ferment gum disease (Rio Grande gummosis) of grapefruit. *Calif. Citrogr*. 46: 231–232.

Calavan, E.C., Roistacher, C.N. and Nauer, E.M. 1972. Thermotherapy of citrus for inactivation of certain viruses. *Plant Dis. Reptr*. 56: 976–980.

Capoor, S.P. 1963. Decline of citrus trees in India. Symposium on plant and animal viruses. *Nat. Inst. Sci. India Bull*. 24: 48–64.

Cartia, G. and Catara, A. 1969. Susceptibility of citrus cultivars to impietratura. pp. 184–187. *In:* 5th Conf. IOCV, IOCv, Riverside, CA.

Caruso, A., Davino, M. and Terranova, G. 1993. Observations on impietratura disease symptoms in four citrus species. pp. 379–382. *In*: Proc. 12th Conf. IOCV, IOCV, Riverside, CA.

Catara, A. and Scaramuzzi, G. 1980. Impietratura. *In*: Bové, J.M. and Vogel, R. *(Eds.)*. Description and illustration of virus and virus-like diseases of citrus. A collection of colour slides. Paris, I.R.F.A. SETCO-FRUITS.

Catara, A., Bassi, M., Perrotta, G., Magnano di san Lio, G. and La, Ferlita 1977. Present status of impietratura disease. *In*: *Proc. Int. Soc. Citriculture*, 3: 946–953.

Chapot, H. 1961. Impietratura in Mediterranean countries, pp. 177–181. *In*: Proc. 2nd Conf. IOCV, IOCV, Riverside, CA.

Childs, J.F.L. 1978a. Rio Grande gummosis of citrus trees. Part I. A brief review of the history and occurrence of Rio Grande gummosis. *Plant Dis. Rep*. 62: 390–394.

Childs, J.F.L. 1978b. Rio Grande gummosis of citrus trees. Part II. The causal relations of Rio Grande gummosis. *Plant Dis. Reptr*. 62: 395–399.

Childs, J.F.L. 1979. Florida *Citrus* Blight, Part I. Some causal aspects of citrus blight, Part II. Occurrence of citrus blight outside Florida. *Plant Dis. Reptr*. 63: 565–569.

Cinar, A., Korkmaz, S. and Kersting, U. 1994. Presence of new whitefly-born citrus disease of possible viral aetiology in Turkey. *FAO Plant Prot. Bull*. 42: 73–75.

Cinar, A., Kersting, U., Onelge, N., Korkmaz, S. and Sas, G. 1993. Citrus virus and virus-like diseases in eastern Mediterranean region of Turkey. pp. 397–400. *In*: Proc. 12th Conf. IOCV, IOCV, Riverside, CA.

Cohen, M. 1974. Diagnosis of young tree decline, blight and sand hill decline of citrus by measurement of water uptake using gravity injection. *Plant Dis. Reptr*. 58: 801–805.

Cohen, M. and Wutscher, H.K. 1979. Diagnosis of trees with citrus blight (YTD). *In*: Proceedings of the International Society of Citriculture 1977, Vol. 3, pp. 884–886. ISC, Lake Alfred, Florida, USA.

Coletta-Filho, H.D., Targon, M.L.P.N., Takita, M.A., Muller, G.W., Santos, F.A., Dorta, S.O., de Souza, A.A., Astua-Monge, G., Frietas-astua, J. and Machado, M.A. 2005. *Citrus tristeza* virus Variant Associated with Citrus Sudden Death and Its Specific Detection by RT-PCR. Proc. 16th Conf. IOCV, IOCV, Riverside, CA (Abs).

Davis, R.M. 1980. Evidence for the involvement of *Diplodia natalensis* in Rio Grande gummosis of citrus. *J. Rio Grande Val. Hortic. Soc*. 34: 55–60.

De jesus Junior, W.C., Teixeira, D.C., Tanaka, F.A.O., Martins, E.C., Barros, C.C.P., Nagata, T., and Rezende, J.A.M. 2005. Purification of Tymo-Like Virus Particles Isolated from Orange Trees with Sudden Death. Proc. 16th Conf. IOCV, IOCV, Riverside, CA (Abs).

Derrick, K.S., Lee, R.F., Brlansky, R.H., Timmer, L.W., Hewitt, B.G. and Barthe, G.A. 1990. Proteins associated with citrus blight. *Plant Disease* 74: 168–170.

Derrick, K.S., Lee, R.F., Brlansky, R.H., Timmer, L.W., Hewitt, B.G. and Barthe, G.A. 1991. Proteins and filamentous structures associated with citrus blight. pp. 265–268. *In*: Proc. 11th. Conf. IOCV. IOCV, Riverside, CA.

Derrick, K.S., Barthe, G.A., Hewitt, B.G., Lee, R.F. and Albrigo, L.G. 1992. Detection of citrus blight by serological assays. *Proc. Fla. State Hortic. Soc*. 105: 26–28.

Derrick, K.S., Barthe, G.A., Hewitt, B.G. and Lee, R.F. 1993. Serological tests for citrus blight. pp. 121–126 *In*: Proc. 12th Conf. IOCV, IOCV Riverside, CA.

Derrick, K.S., Beretta, M.J. and Barthe, G.A. 2006. Detection of an Idaeovirus in citrus with implications as to the cause of citrus blight. *Proc. Fla. State Hort. Soc*. 119: 69–72.

Donadio, L.C. and Banzato, D.A. 1988. Relationship of citrus cultivars and declinio. pp. 384–387. *In*: Proc. 10th IOCV, IOCV, Riverside.

D'Onghia, A.M., de Marco, P. and Savino, V. 1992. Gravi casi di concavita' gommose su navalina in Puglia. (Severe cases of concave gum on Navalina orange in Apulia. *Informaatore Fitopatologico* 3: 39–41.

Fawcett, H.S. 1932. New angles on treatment of bark diseases of citrus. *California Citograph* 17: 406–408.

Fawcet, H.S. and Klotz, L.J. 1938. Types of symptoms of psorosis and psorosis-like diseases of citrus. (Abstr.) *Phytopathology* 28: 670.

Fawcett, H.S. and Bitancourt, A.A. 1943. Comparative symptomatology of psorosis varieties of citrus in California. *Phytopathology* 33: 837–864.

Fraser, L.R. and Singh, Daljit 1966. Greening virus – new threat to citrus industry. *Indian Hort.* 10: 21–22.

Fraser, L.R., Singh, Daljit, Capoor, S.P. and Nariani, T.K. 1966. Greening virus, the likely cause of citrus dieback in India. *FAO Plant Prot. Bull.* 14: 127–130.

Fraser, L.R. and Singh, D. 1968. Citrus dieback in India – the contribution of greening virus. pp. 141–144. *In*: Proc. 4th Conf: IOCV, IOCV, Riverside, CA.

Garnsey, S.M. 1975. Purification and properties of citrus leaf rugose virus. *Phytopathology* 65: 50–57.

Girmaldi, V. and Catara, A. 1989. Detection of citrus tristeza and greening in Pakistan through electron microscopy. *Journal of Phytopathology* 162: 17–21.

Grant, T.J., Moreira, S. and Childs, J.F.L. 1957. Observations on abnormal citrus rootstock reaction in Brasil. *Plant. Dis. Reptr.* 31: 743–748.

Grimm, G.R., Grant, T.J., Childs, J.F.L. 1955. A bud union abnormality of rough lemon rootstock with sweet orange scions. *Plant Dis. Reptr.* 29: 810–811.

Gonsalves, D. and Garnsey, S.M. 1976. Association of particle size with sedimentation velocity of the nucleoprotein components of citrus variegation and citrus leaf rugose viruses. pp. 109–115. *In*: Proc. 7th Conf. IOCV, IOCV, Riverside, CA.

Hopkins, D.L., Bistline, F.W., Russo, L.W. and Thompson, C.M. 1991. Seasonal fluctuation in the occurrence of *Xylella fastidiosa* in root and stem extracts from citrus with blight. *Plant Disease* 75: 145–147.

Howd, D.S., Hartung, J.S., and Brlansky, R.H. 2002. Ultrastructure of citrus cholorotic dwarf-infected leaves and bark. pp. 373–377. *In*: Proc. 14th Conf. IOCV, Riverside, CA.

Kanrar, S., Afunian, M., Loannis, G.G., Tzanetakis, E., and Vidalakis, G. 2010. Molecular characterization of viruses associated with yellow vein and vein enation disease of citrus. p. 47. *In*: (Abs) 18th Conf. IOCV held in Campinas/S.P., Brazil, Nov. 7–12, 2010.

Kitajima, E.W. 2010. Citrus viruses and bud transmitted bacteria in Brazil: History and perspectives. p. 2. *In*: 18th Conf. IOCV held at Campinas/SP, Brazil 7–12 Nov. 2010.

Korkmaz, S., Cinar, A., Bozan, O., and Kersting, U. 1994. Distribution and natural transmission of a new whitefly-borne virus disease of citrus in the eastern Mediterranean region of Turkey. pp. 437–439. *In*: Proc. 9th Congr. Mediterr. Phytopathol.Union. Aydin/Turkey.

Korkmaz, S., Cinar, A., Demirer, E. and Onelge, N. 1994. Greenhouse observation on the susceptibility of 36 citrus varieties to a new whitefly-borne virus. pp. 305–306. *In*: Proc. 9th Cong. Mediterr. Phytopathol. Union, Kusadasi-Aydin, Turkey.

Korkmaz, S., Cinar, A., Kersting, U. and Garnsey, S.M. 1995. Citrus chlorotic dwarf: A whitefly transmitted viruslike disease of citrus in Turkey. *Plant Dis.* 79: 1074.

Kersting, U., Korkmaz, S., Cinar, A., Ertugrul, B., Onelge, N. and Garnsey, S.M. 1996. Citrus chlorotic dwarf, a new whitefly transmitted disease of eastern Mediterranean region of Turkey. pp. 220–225. *In*: Proc. 13th Conf. IOCV, IOCV, Riverside.

Lee, R.F., Marais, L.J., Timmer, L.W. and Graham, J.H. 1984. Syringe injection of water into the trunk: a rapid diagnostic test for citrus blight. *Plant Disease* 68: 511–513.

Lee, R.F., Derrick, K.S., Futch, S.H. and Tucker, D.P.H. 1993. Graft transmission of citrus measles. pp. 371–374. *In*: Proc. 12th Conf. IOCV, IOCV, Riverside, CA.

Ieki, H. and Ito, T. 1996. Occurrence of concave gum on Hyuganatsu (*Citrus tamitrana*) in Japan, pp. 346–348. *In*: Proc. 13th Conf. IOCV, IOCV, Riverside, CA.

Lima, J.E.O. 1982. Observation on citrus blight in Sao Paulo, Brazil. *Proc. Fla, State Hort. Soc.* 95: 72–75.

Maccheroni, W., Alegria, M.C., Greggio, C.C., Piazza, J.P., Kamla, R.F., Zacharias, P.R., Bar- Joseph, M., Kitajima, E.W., Assumpção, L.C., Camarotte, G., Cardozo, J., Casagrande, EC., Ferrari, F., Franco, S.F., Giachetto, P.F., Girasol, A., Jordão, H. Jr., Silva, V.H., Souza, LC., Aguilar-Vildoso, C.I., Zanca, A.S., Arruda, P., Kitajima, J.P., Reinach, F.C., Ferro, J.A. and da Silva, A.C. 2005. Identification and genomic characterization of a new virus (Tymoviridae family) associated with citrus sudden death disease. *J.Virol* 79: 3028–3037.

Maccheroni, W., Greggio, C.C., Piazza, J.P., Kamla, R.F., Zacharias, P.R.A., Kitajima, J.P., Ferro, J.A. and da Silva, A.C.R. 2005. *Citrus tristeza* virus genome variability in Citrus sudden death affected areas in Brazil. Proc. 16th Conf. IOCV, IOCV, Riverside, CA (Abs).

Maccheroni, W., Alegria, M.C., Greggio, C.C., Piazza, J.P., Kamla, R.F., Zacharias, P.R.A., Bar-Joseph, M., Ferro, J.A. and da Silva, A.C.R. 2005. A new tymoviridae virus associated with citrus sudden death disease in Brazil. Proc. 16th Conf. IOCV, IOCV, Riverside, CA (Abs).

Majorana, G. and Schwarz, R.E. 1972. Studies on two mechanically transmissible citrus viruses. pp. 188–192. *In*: Proc. 5th Conf. IOCV, IOCV, Riverside, CA.

Mali, V.R., Choudhury, K.G. and Rane, S.D. 1975. Blastomania – A new bud transmissible disease of Citrus. *Curr. Sci.* 44: 627–628.

Mali V.R., Chaudhury, K.G. and Rane, S.D. 1976. Leaf curl virus disease of Citrus in India. *Sci. and Cult.* 42: 525–527.

Mali, V.R. and Chaudhuri, K.G. 1993. Occurrence of citrus virus and virus-like diseases in Maharashtra, India. p. 466. *In*: Proc. 12th Conf. IOCV, IOCV, Riverside (Abs).

Marais, L.J. and Lee, R.F. 1991. Experimental transmission of citrus blight in South Africa. pp. 261–264. *In*: Proc. 11th Conf. IOCV. IOCV, Riverside, CA.

Martelli, G.P., Yilmaz, M.A., Savino, V. and Balogu, S. 1996. Properties of citrus isolate of olive latent virus-1. A new necrovirus. *European J. Plant Pathol.* 102: 527–536.

McClean, A.P.D. and Engelbrecht, A.P.H. 1958. Xyloporosis, cachexia and abnormal bud unions in South African citrus trees. *S. African J. Agr. Sci.* 4: 389–431.

McClean, A.P.D. 1974. Abnormal bud union between some sweet oranges and rough lemon rootstock: evidence that it is caused by a transmissible pathogen. pp. 203–210. *In*: Proc. 6th Conf. IOCV, IOCV, Riverside, CA.

Melo, W.J., Oliveira, H., Beretta, M.J.G., Chelli, R.A. and Leite, S.A.S. 1991. Zinc in soil and citrus trees showing declinio symptoms. pp. 283–288. *In*: Proc. 11th Conf. IOCV, IOCV, Riverside, CA.

Nariani, T.K., Raychaudhuri, S.P. and Bhalla, R.P. 1967. Greening virus of citrus in India. *Indian Phytopath.* 20: 146–150

Navarro, L, Pina, J.A., Juárez, J. and Ballester, J.F. 1993. Elimination of a bud union abnormality of sweet orange grafted on rough lemon by shot-tip grafting *in vitro*. pp. 375–378. *In*: Proc. 12th Conf. IOCV, IOCV, Riverside, CA.

Nickel, O. 1987. Double-stranded RNA isolated from "declinio"-affected 'Pera' sweet orange. *Fitopatologia Brasileira* 12: 398–399.

Nickel, O., Couto, R.C.S., Coelho, Y.S., Vilarinhos and Oliveira, R.P. 1996. Proteins associated with transmission of citrus declinio in Bahia, Brazil. pp. 286–290. *In*: Proc. 13th Conf. IOCV, IOCV, Riverside, CA.

Nour-Eldin, F. 1968. Gummy bark of sweet orange. *In*: Childs, J.F.L. (*ed*). Indexing procedure for 15 virus diseases of citrus trees. USDA Handbook 333:50-53

Ochoa, F., Beretta, M.J.G., Derrick, K.S., Lee, R.F. and Rangel, E. 1993. Comparison of Citrus Sudden Decline from Venezuela with citrus blight from Florida and Delinio from Brazil. pp. 116–120. *In*: Proc. 12th Conf. IOCV, IOCV, Riverside.

Onelge, N. and Çinar, A. 2010. Virus and Virus-like Diseases in Turkey Citriculture. *In*: Proc. 17th Conf. IOCV, IOCV, Riverside,CA.

Powell, C.A., Pelosi, R.R., Sonoda, R.M. and Lee, R.F. 1998. A psorosis-like agent prevalent in Florida's grapefruit groves and budwood sources. *Plant Dis*. 82: 208–209.

Papasolomontos, A. 1961. Preliminary report on a new disorder of grapefruit. Countryman [Nicosia] Nov.-Dec. pp. 15–16.

Papasolomontos, A. 1965. The present status of impietratura, a citrus disease in Cyprus. *Plant Dis. Reptr*. 49: 111–113.

Papasolomontos, A. and Economides, C.V. 1967. Effect of rootstocks on the incidence of impietratura-diseased grapefruit fruits. *Plant Dis. Rep*., 51: 684–686.

Papasolomontos, A. 1969. A report on impietratura disease of citrus; its distribution and importance. pp. 1457–1462. *In*: Proc. 1st Int. Citrus Symp.

Protopapadakis, E., Tzortzakaki, S., Kasapakis, J. and Tsagris, M. 2002. A new graft transmissible disease of Bergamot in Greece. pp. 382. *In*: Proc. 15th Conf. IOCV, IOCV, Riverside, CA.

Rana, G.L., Quacquarelli, A. and Martelli, G.P. 1974. Some properties of citrus variegation virus. pp. 165–168. *In*: Proc. 6th Conf. IOCV, Riverside, CA.

Raychaudhuri, S.P., Nariani, T.K. and Lele, V.C. 1967. Citrus dieback complex – a serious threat to citrus industry in India. *In*: Proc.Internl. Symp. On sub tropical and tropical Horticulture, New Delhi.

Raychudhuri, S.P., Nariani, T.K. and Lele, V.C. 1969. Citrus dieback problem in India. Proc. First Internl. *Cirus Symp*. 3: 1433–1437.

Raychaudhuri, S.P., Nariani, T.K. and Ahlawat, Y.S. 1977. Dieback of citrus in India. Proc. 3rd Symp. Internl. Soc. Citriculture 3: 914–918.

Reichert, I. and Hellinger, E. 1930. Internal decline. A new physiological disease of citrus fruits in Palestine. *Hadar* 3: 220–224.

Reuther, W., Calavan, E.C. and Carman, G.E. (*Eds*.) 1978. pp. 170–171. *In*: The citrus Industry. Dept. of Agric. Sci., Univ. California, Riverside.

Rhoads, A.S. 1936. Blight-a new parasitic disease of citrus trees. *Florida Agric. Exp. Stan. Bull*. 296: 64.

Roistacher C.N. and Calavan, E.C. 1965. Cross protection studies with strains of concave gum and psorosis viruses. pp. 154–161. Proc. 3rd Conf. IOCV, IOCV, Riverside, CA.

Roistacher, C.N. 1977. Elimination of citrus pathogens in propagative budwood. I. Budwood selection, indexing and thermotherapy. *Proc. Int. Soc. Citriculture* 3: 965–972.

Rossetti, V., Beretta, M.J.G., Lefevre, A.F.V. and Trindade, M.L.B. 1988. Evaluation of possible recovery of declinio affected citrus trees. pp. 396–399. *In*: Proc. 10th Conf. IOCV, IOCV, Riverside, CA.

Rossetti, V., Berreta, M.J.G. and Teixeira, A.R.R. 1991. Experimental transmission of declinio by approach root grafting in sao Paulo State, Brazil. pp. 250–255. *In*: Proc. 11th Conf. IOCV, IOCV, Riverside, CA.

Ruggieri, G. 1955. Le arance impietrate. *Riv. Agrumic.* 1: 65–69.

Rustem, F.A. and Ahlawat, Y.S. 1999. Yellow corky vein: a viroid disease of citrus in India. *Indian J. Agric.Sci.* 69: 789–94.

Salibe, A.A. 1959. Leaf curl – a transmissible virus disease of citrus. *Plant dis. Reptr.* 43: 1081–1083.

Salibe, A.A. 1965. Susceptibility of citrus varieties to leaf curl virus. pp. 175–178. *In*: Proc. 3rd Conf. IOCV, IOCV, Riverside, CA.

Salibe, A.A. and Jahen, A. 1984. Consideration on the possible nature of citrus decline. pp. 316–321. *In*: Proc. 9th Conf. IOCV, IOCV, Riverside, CA.

Santos, M.A., Wulff, N.A., Toloy, R.S., Ayres, A.J. and Rezende, J.A.M. 2010. Experimental transmission of citrus sudden death associated virus CSDaV to herbaceous plants. p. 104. *In*: Abs.18th Conf. IOCV held in ampinas/SP,Brazil, Nov. 7–12, 2010.

Scaramuzzi, G., Catara, A. and Cartia, G. 1966. Investigations on impietratura disease, pp. 197–200. *In*: Proc. 4th Conf. IOCV, IOCV, Riverside, CA.

Scaramuzzi, G. and Catara, A.A. 1968. Studies on sour orange stem pitting in Sicily. pp. 201–205. *In*: Proc. 4th Conf. IOCV, IOCV, Riverside, CA.

Schwartz, R.E. 1970. Multiple sprouting disease of citrus. *Plant Dis. Reptr.* 54: 1003–1007.

Searle, C.M. 1969. A preliminary report on off- type trees. Mazoe Citrus Estates, (Rhodesia), p. 15.

Sharma, S.R. and Singh, S.J. 1988. Association of mycoplasma-like organisms with *Citrus blastomania* disease. *Indian J. Pl. Pathol.* 6: 86–92.

Smith, P.F. 1974. Zinc accumulation in the wood of citrus trees affected with blight. *Proc. Fla. State Hort. Soc.* 87: 91–95.

Smith, P.F. and Reitz, F.J. 1977. A review of the nature and history of citrus blight in Florida. *Proc. Interl. Soc. Citric.* 3: 881–884.

Sonoda, R.M. and Pelosi, R.R. 1990. Fungi in Rio Grande gummosis lesions and patterns of gummosis-affected grapefruit trees in Indian River area groves. *Proc. Fla. State Hortic. Soc.* 103: 59–61.

Tanaka, S. and Imada, J. 1974. Mechanical transmission of viruses of Satsuma dwarf, citrus mosaic, navel infectious mottling and natsudaidai dwarf to herbaceous plants. pp. 141–145. *In*: Proc. 6th Conf. IOCV, IOCV, Riverside, CA.

Taylor, K.C., Albrigo, L.G. and Chase, C.D. 1988. Zinc complexation in the phloem of blight affected citrus. *J. American Soc. Hort. Sci.* 113: 407–411.

Terranova, G. and Scuderi, A. 1968. Further research on impietratura disease, pp. 242–247. *In*: Proc. 4th Conf. IOCV, IOCV, Riverside, CA.

Timmer, L.W. and Benatena, H.N. 1977. Comparison of psorosis and other viruses causing leaf flecking in citrus. *Proc. Intl. Soc. Citriculture* 3: 930–935.

Timmer, L., Brlansky, R.H. and Lee, R.F., Graham, J.H., Agostini, J.P., Fischer, H.U. and Casafus, C. 1984. Characteristics of citrus trees affected with blight in Florida, declinamiento in Aregentina and by declinio in Brazil. *Proc. Internl. Soc. Citriculture* 1: 371–374.

Timmer, L., Brlansky, R.H., Agostini, J.P. and Lee, R.F. 1988. Effect of amorphous plugs, filamentous plugs and depth into trunk xylem on the water flow in declining and healthy citrus trees. pp. 370–376. *In*: Proc. 10th Conf. IOCV, IOCV, Riverside, CA.

Timmer, L.W., Brlansky, R.H., Derrick, K.S. and Lee, R.F. 1991. Transmission of citrus blight by root graft inoculation. pp. 277–282. *In*: Proc. 11th Conf. IOCV, IOCV, Riverside, CA.

Timmer, L.W., Lee, R.F., Brlansky, R.H., Graham, J.H., Albrigo, L.G., Derrick, K.S. and Tucker, D.P.H. 1992. The infectious nature of citrus blight. *Proc. Fla. State Hortic. Soc.* 105: 21–26.

Timmer, L.W., Garnsey, S.M. and Graham, J.H. (*Eds.*) 2000. Compendium of citrus diseases, Second Edition. APS Press, st. Paul, MN. pp. 65–66.

Timmer, L.W., Garnsey, S.M. and Graham, J.H. (*Eds.*) 2000. 2nd Ed. Cristacortis. *In*: Compendium of citrus diseases. APS Press, St. Paul, Minnesota, USA.

Timmer, L.W. and Powell, C.A. 2006. Rio Grande gummosis. www.ivia.es/iocv.

Tucker, D.P.H., Lee, R.F., Timmer, L.W., Albrigo, L.G., Brlansky, R.H. 1984. Experimental transmission of citrus blight. *Plant Dis.* 68: 979–980.

Vasudeva, R.S. and Capoor, S.P. 1958. Citrus decline in Bombay state. *FAO Plant Prot. Bull.* 6: 91.

Villalobos, W., Moreira, L., Derrick, K.S., Beretta, M.J.G., Lee, R.F. and Rivera, C. 2005. Occurrence of citrus blight in Costa Rica. p. 509. *In*: Proc. 16th. Conf. IOCV. IOCV, Riverside, CA.

Vogel, R. and Bové, J.M. 1964. Stem pitting sur bigaradier et sur oranger Tarocco en Corse : une maladie à virus. *Fruits* 18: 269–274.

Vogel, R. and Bové, J.M. 1964. Stem pitting sur bigaradier et sur oranger Tarocco en Corse: une maladie à virus. *Fruits* 19: 264–274.

Vogel R. and Bové, J.M. 1968. Cristacortis, a virus disease inducing stem pitting on sour orange and other citrus species. pp. 221–228. *In*: Proc. 4th Conf. IOCV, IOCV, Riverside, CA.

Vogel R. And J. M. Bove.1972. Relation of cristacortis virus to other citrus viruses, p. 178-184. In Proc. 5th Conf. IOCV, IOCV, Riverside.

Vogel, R. and Bove, J.M. 1974. Studies on the cause of leaf symptoms associated with cristacortis disease of citrus. pp. 131–134. *In*: Proc. 6th Conf. IOCV, IOCV, Riverside, CA.

Vogel, R. and Bové, J.M. 1976a. Transmission de maladies infectieuses d'agrumes à agrumes par le pollen d'arbres appliqué sous l'écorce de plantes saines. C.R. Acd. Sci. Paris, série D, 283: 1409–1412.

Vogel, R. and Bove, J.M. 1976b. Evidence of the existence of strains of the cristacortis pathogen. pp. 101–104. *In*: Proc. 7th Conf. IOCV, IOCV, Riverside, CA.

Vogel, R. and Bove, J.M. 1980. Pollen transmission to citrus of the agent inducing cristacortis stem pitting and psorosis young leaf symptoms. pp. 188–190. *In*: Proc. 8th Conf. IOCV, IOCV, Riverside, CA.

Wallace, J.M. 1959. A half century of research on psorosis. pp. 5–21. *In*: Wallace, J.M. (*Ed*.) Citrus virus diseases. Univ. Calif. Div. Agr. Sci., Berkeley, Calif.

Wallace, J.M. 1978. The psorosis diseases. pp. 69–87. *In*: Reuther, W., Calavan, E.C. and Carman, G.E. (*Eds*.). The Citrus Industry, vol. IV, Univ. Calif., Div. Agr. Sci.

Weathers, L.G. 1957. A vein yellowing disease of citrus caused by a graft transmissible virus. *Plant Dis. Reptr*. 41: 741–742.

Weathers, L.G. 1960. Yellow vein disease of citrus and studies of interactions between yellow vein and other viruses of citrus. *Virology* 11: 753–764.

Weathers, L.G. 1961. Responses of citrus to concurrent infection with two or more unrelated viruses. pp. 187–196. *In*: Proc. 2nd Conf. IOCV, IOCV, Riverside, CA.

Weathers, L.G. 1963. Use of synergy in the identification of strains of citrus yellow vein virus. *Nature* 200: 812–813.

Whiteside, J.O., Timmer, L.W. and Graham, J.H. 1988. pp. 62–63. *In*: Compendium of citrus diseases (Whiteside, J.O., Garnsey, S.M. and Timmer, L.W. *Eds*.) APS Press st. Paul, MN.

Wutscher, H.K., Cohen, M. and Young, R.H. 1977. Zinc and water-soluble phenolic levels in the wood for the diagnosis of citrus blight. *Plant Dis. Rep*. 6: 572–576.

Yedidyah, S. 1937. Citrus growing. Hassadeh Publ. Co. Tel Aviv, p. 478 (*In*: Hebrew).

Young, R.H., Wutscher, H.E. and Albrigo, L.G. 1980. Relationship between water translocation and zinc accumulation in citrus trees with and without blight. *Journal of the American Society for Horticultural Science* 105: 444–447.

Young, R.H., Albrigo, L.G., Tucker, D.P.H. and Williams, G. 1980. Incidence of citrus blight on Carrizo citrange and some other rootstocks. *Proc. Fla. State Hort. Soc*. 93: 14–17.

Yamamoto, P.T., Bassanezi, R.B., Wulff, N.A., Santos, A.M., Sanches, A.L., Toloy, R.S., Fernandes, N.G., Ayres, A.J., Tatsuya Nagata, W.C.J. Jr., Tanaka, F.A.O., Kitajima, E.W. and Bove, J.M. 2010. Studies on citrus sudden death transmission. p. 27. *In*: Abs. 18th Conf. IOCV held in Campinas/SP,Brazil, Nov. 7–12, 2010.

Yokomi, R.K., Garnsey, S.M., Young, R.H. and Grimm, G.R. 1984. Spatial and temporal analysis of citrus blight incidence in 'Valensia' orange groves in central Florida. pp. 260–269. *In*: Proc. 9th Conf. IOCV, IOCV, Riverside, CA.

4

OTHER GRAFT TRANSMISSIBLE DISEASES

There are certain diseases where limited information is available. These diseases have been tabulated in Table 8.

Table 8: Other graft transmissible diseases of citrus

Name of disease	*Symptoms*	*Transmission & other properties*
Algerian navel orange virus	There are no symptom in citrus hosts	Mechanically transmitted to *Chenopodium quinoa* and back to citrus. Flexuous virus-like particles in extract of infected citrus and herbaceous hosts (Garnsey, 1975)
Brittle twig yellows	Twigs become brittle with yellowing of leaves	Graft transmitted,Some similarities to stubborn disease (Samadi *et al.*, 1977)
Yellow mottle	Vein clearing and ring spot symptoms in Satsuma mandarin	Graft transmitted, rod shaped virus particles measuring 690–740 nm (Ushiyama *et al.*, 1984)
Fatal yellows	Vein clearing, vein corking, leaf epinasty and stunting in lemon on *C. macrophylla* rootstock and seedlings of *C. macrophylla*	Graft transmissible (Schneider, 1984)
Fovea	Tree decline and inverse stem pitting in trunk. Found on Murcott and other mandarin hybrids	Similar stem pitting symptoms in Murcott reported due to cachexia viroid (Olson, 1960)

Table 8: (*Contd...*)

Table 8: (*Contd...*)

Name of disease	*Symptoms*	*Transmission & other properties*
Grapefruit bark scaling	Excessive bark scaling in trunk of grapefruit trees similar to exocortis viroid on rootstocks.	Symptoms are similar to shell bark viroid but no relationship established. No method of transmission is known (Timmer *et al.*, 2000).
Milam lemon stem pitting	Severe stem pitting to trunk and major limbs in Milam lemon and some mandarin hybrids. Affected trees grow vigorously.	Method of transmission not known (Garnsey *et al.*, 1979)
Nagami Kumquat disease	Tristeza-like vein clearing on sweetorange and other citrus varieties. Associated with graft incompatibility of Nagami Kumquat on Troyer citrange.	Graft transmissible agent (Navarro *et al.*, 1984)

REFERENCES

Garnsey, S.M. 1975. Two mechanically transmissible viruses in navel orange selections introduced from Algeria. *Plant Dis. Reptr*. 59: 689–693.

Garnsey, S.M., Youtsey, C.O. and Bridges, G.D. 1979. Reaction of eight citrus varieties topworked on Milam trees with severe stam pitting symptoms. *Proc. Fla. State Hort. Soc*. 92: 18–20.

Navarro, L., Pina, J.A., Ballester, J.F., Moreno, P. and Cambra, M. 1984. A new graft transmissible disease found in Nagami kumquat. pp. 234–240. *In*: Proc. 9th Conf. IOCV, IOCV, Riverside, CA.

Olson, E.O. 1960. Xyloporosis (cachexia or fovea) disease of Murcott honey "orange" in Texas. *J. Rio Grande Val. Hortic Soc*. 14: 26–28.

Samadi, M., Cochran, L.C., and Martinez, A.L. 1977. Brittle twig yellows, a new disorder to sweet orange in Southern Iran. Proc. Intl. Soc. Citriculture 3: 900

Schneider, H. 1984. Host range of the fatal yellows disease. pp. 222–228. *In*: Pro. 9th Conf. IOCV, IOCV, Riverside, CA.

Timmer, L.W., Garnsey, S.M. and Graham, J.H. (*Eds*.) 2000. Compendium of citrus diseases, Second Edition. APS Press, st. Paul, MN. pp. 65–66.

Ushiyama, K., Usigi, T. and Hibano, H. 1984. A new citrus virus disease: citrus yellow mottle. pp. 204–210. *In*: Pro. 9th Conf. IOCV, IOCV, Riverside, CA.

5

DISEASES CAUSED BY PROKARYOTES

Following two diseases are commonly observed in citrus which are caused by prokaryotes:

1. **Huanglongbing (Greening) Disease**
2. **Citrus Variegated Chlorosis**

In citrus, prokaryotes cause some serious and economically important diseases such as huanglongbing, commonly called as citrus greening disease and citrus variegated chlorosis. Huanglongbing or greening disease is caused by an insect-borne bacterium which is restricted to sieve tubes of phloem tissue. Huanglongbing disease is widely distributed in several countries in Asia, Africa, part of Arabian Peninsula, Brazil and United States of America. Official name of the disease was changed from "greening" to "huanglongbing" in 1995 during the 13th Conference of International Organization of Citrus Virologists (IOCV) held in China because this disease was first shown to be a graft transmissible disease in China. In literature, this disease has been variably spelled by various workers like Huanglungbin, Huanglongbing and huanglongbin. Although the name 'greening' is commonly used name but changed name 'huanglongbing' will be used in this chapter.

Citrus variegated chlorosis is caused by a xylem-limited bacterium, *Xylella fastidiosa* and is found in Brazil. Other diseases like witches' broom disease of lime is caused by '*Candidus* Phytoplasma aurantifolia and is prevalent in Sultanate of Oman, United Arab Emirates, Iran and may be in some other Arabian countries where citrus is grown. Witches' broom and rubberywood –two phytoplasma diseases are known to occur in India. Other prokaryotic disease is stubborn. Stubborn name was first used by a propagator in USA to describe the poor response of trees top worked with selected buds in 1921 and was used as a disease namc by J.C. Perry in 1929 (Fawcett and Klotz, 1948). Stubborn disease is found in USA, Eastern Mediterranean Basin, Middle East, Israel and Corsica (France). These diseases will be discussed in detail in following pages.

Huanglongbing (Greening) Disease

A disease with the name of yellow shoot was reported from China as early as 1919 (Reinking, 1919). In China, this disease was of minor importance at that time but by 1936, it spread to major citrus growing regions of the whole country and became a major disease of citrus in China. Symptoms of this disease were initially considered to be caused by mineral or micronutrient deficiency or due to water logging conditions. A disease with similar symptoms was described with the name of Huanglongbing by Professor Lin Kung Hsiang from China and was shown to be graft transmitted (Lin, 1944, 1956a, b). The new flush on a tree is called huang lung in the Chaozhou district of China from where the name of the disease as Huanglongbing (yellow shoot or yellow dragon) originated. Later studies revealed that yellow shoot disease earlier reported by Reinking was similar to huanglongbing disease reported by Lin. In China, this disease is called by several names such as Wei Huang (withering yellow), Ji Tou Huang (yellow chicken head), Huang Pi Tou (yellow surface disease), Shuo Ding (weathering head), Huang Qi Bin (yellow crab disease), Jin Ban Yie (golden board leaf) and La Jiao Yie (pepper leaf). Later, a disease with similar symptoms was described from South Africa by the name of greening disease (Van de Merwe and Andersen, 1937). Transmission of South African greening by graft-inoculation was reported in 1965 (McClean and Oberholzer, 1965a) much later than huanglongbing in China. Transmission of the greening agent by the African citrus psyllid vector, *Trioza erytreae*, was also published in 1965 (McClean and Oberholzer, 1965b). Transmission of the HLB agent by the Asian citrus psyllid vector, *Diaphorina citri*, was reported simultaneously in 1967 in India (Capoor *et al.*, 1967) and in the Philippines (Martinez and Wallace, 1967).

Huanglongbing disease is known by different names in different countries. It is called as likubin in Taiwan, leaf mottle yellows in the Philippines, citrus dieback in India, Huanglongbing or yellow shoot in China and citrus vein phloem degeneration in Indonesia. There are three forms of Huanglongbing (HLB) – the Asian form, African form and the American form. HLB is a very destructive disease for sweet orange, mandarin, tangelo and grapefruit trees, irrespective of rootstocks. Seedling trees of these cultivars are equally affected. HLB causes 30–100% losses and citrus cultivation in many countries due to HLB has been eliminated. Severe epidemics of this disease occurred in 1960s especially in Java and Sumatra where about 3 million trees were destroyed (Tirtawidjaja, 1980). About 100 million trees are affected with HLB throughout the world each year. Once a tree is infected with HLB, it will soon decline and become uneconomical. Since HLB is a vector borne disease, it spreads very fast in citrus groves. For example, in Brazil, HLB was first recorded in 2004 and since then about three million trees have been eradicated under HLB control programme. In India, it has been estimated that HLB causes loss of about Rs. 116000 (US$ 2600) if a tree is declined after 8 years of planting in an orchard of 1000 trees.

Asian form of HLB and its psyllid vector, *Diaphorina citri* is present in about 20 countries in Asia, Southeast Asia and Oceania (India, Pakistan, Nepal, Bhutan, Bangladesh, Sri Lanka, Myanmar, Thailand, Malaysia,, Combodia, Laos, Vietnam, China, Taiwan, Japan, Philippines, Indonesia, East Timor, Papua Guinea and Iran. In Iran, *D. citri* was reported in 2000 and later in 2008. *Candidus* Liberibacter asiaticus was detected in vector and plants in 2009 by nested PCR (Golmohammadi *et al.*, 2010). Both Psyllid vector and Asian HLB are temperature tolerant (32–35°C.). African Form of HLB and its psyllid vector, *Trioza erytreae* are present in South Africa, Zimbabwe, Swaziland, Malawi, Burundi, Tanzania, Kenya, Somalia, Ethiopia, Cameroon, Nigeria and Central African Republic as well as Madagascar Island. Both psyllid vector and African HLB survive well at low Temp. below 30°C. HLB has also been reported from Sao Paulo, Brazil in 2004 (Coletta-Filho *et al.*, 2004; Teixeira *et al.*, 2005a) though the psyllid vector *Diaphorina citri* is known to occur there since 1942. Asian type HLB has been reported from Florida in 2005 (Halbert, 2005) and in Louisiana in 2008. In Cuba Asian form of HLB was reported in 2006 (Luis Pantoja *et al.*, 2008) but the psyllid vector is known to occur there since 1999. In Brazil, both Asian and American forms are present. However, there are areas where both Asiatic and African forms and their vectors are present. These regions are islands in Indian Ocean, Reunion and Mauritius and border of Saudi Arabia and Yemen. Interestingly, there are places which are free from HLB and its vector(s) as yet. These places are: Countries around the Mediterranean basin, Australia, New Zealand and North-and south-Pacific islands except Hawaii and Maui islands where *D. citri* is present. Huanglongbing and its vectors have been reviewed time to time (Bové and Saglio 1974; Bové, 2006; da Graça 1991; da Graça and Korsten, 2004; Garnier and Bové, 1993; Gottwald *et al.*, 2007; Halbert and Manjunath, 2004).

Dieback of citrus trees in India is known to occur since 18th century. It was reported from Bonavia (Assam) in 1888 and was also observed by Roghoji Bhonsale in 18th century (Capoor, 1963). Dieback in India had been a baffling problem since 1920 and was considered not to be a specific disease (Asana, 1958). Dieback was attributed by several factors, including soil disorders, nutritional deficiencies, twig fungi, and viruses such as citrus tristeza virus (Capoor, 1963; Cheema and Bhat, 1929). In view of serious nature of huanglongbing disease and its economic importance to Indian citriculture, Government of India in 1965 invited Lilian Fraser-a scientist from Australia under Colombo plan to advise the Indian Government about the possible cause of citrus dieback in India. She visited major citrus growing areas of the country and concluded that the dieback affected citrus trees in India carry the symptoms of citrus greening (huanglongbing) disease similar to the one known to occur in South Africa and suggested that greening appears to be the cause of citrus dieback disease in India (Fraser and Singh, 1966; Fraser *et al.*, 1966). The experimental proof for the presence of huanglongbing disease came in 1967 when successful transmission of HLB agent was obtained by the psyllid vector (Capoor *et al.*, 1967).

Huanglongbing disease was considered to be a virus disease till 1970 but it is now known that a bacterium, '*Candidus* Liberobacter species' is associated with it. Huanglongbing (HLB) is adversely affecting production of citrus in the Indian subcontinent and also in major citrus growing countries in Asia. In 1970s it was established that dieback in India is associated with HLB and some fungi (Raychaudhuri *et al.*, 1972; Raychauduri *et al.*, 1977). However further studies revealed that dieback symptoms are caused by viruses like citrus tristeza, Indian citrus ring spot, Citrus yellow mosaic badnavirus, HLB and citrus yellow corky vein viroid (Exocortis+hop stunt) in two or more combinations (Ahlawat, 1997, 2007). However, HLB remains a major component in citrus dieback syndrome in India. Dieback disease is characterized by defoliation of young branches and dieback of twigs from the tip resulting in loss of vigour, general health and ultimately death of tree. Northeast region of India is known as original home of citrus species where Khasi mandarin (*Citrus reticulata)* seedling trees are grown since centuries. During a survey to this region, it was observed that HLB is widely distributed in this region except a place known as Digaru (Assam) because Digaru is surrounded by hills which act as a natural barrier for vector movement (Bhagabati *et al.*, 1989). HLB is widely distributed in Darjeeling and Sikkim, an area bordering Nepal and Bhutan. In this region vector, *Diaphorina citri* is not found at higher altitudes but HLB was recorded from 18 places out of 32 surveyed in Darjeeling district suggesting that presence of HLB disease in this region is a result of inadvertent propagation from contaminated trees as no spread of HLB was observed in this region (Ahlawat and Sardar, 1976). In India, HLB is widely distributed in all citrus growing states including North-eastern region, Bihar, West Bengal, Orissa, Maharashtra, Punjab, Andhra Pradesh, Kerala and several other areas where citrus is grown (Nariani and Raychaudhuri, 1968; Nariani *et al.*, 1967, 1970; Gupta *et al.*, 1972; Raychaudhuri *et al.*, 1972). HLB is also widely distributed and destructive disease of citrus trees in Pakistan (Cochran, 1976; Arif *et al.*, 2005).

Huanglongbing (HLB) disease is probably one of the most destructive diseases of citrus. In addition to Asia and Africa where HLB is present for several decades, this has also been reported in the Western hemisphere from Brazil in 2004 (Coletta-Filho *et al.*, 2004; Teixeira *et al.*, 2005a, b, c, d, e), Florida in 2005 (Halbert, 2005) and in Cuba in 2006 (Cordero *et al.*, 2010). All infected commercial plantations continue to decline because adequate control measures are not available. *Diaphorina citri,* the psyllid vector of HLB, has been present in Brazil for over sixty years. This disease is spreading at an alarming rate in both areas. In Florida, large scale distribution of citrus huanglongbing and its vector, *Diaphorina citri* has been observed (Halbert *et al.*, 2008) although the vector was established earlier (1998) than the disease (Halbert *et al.*, 2003). *Diaphorina citri,* the psyllid vector of the disease, is already present in Texas, Mexico, and Hawaii and have entered in California in 2008 (Luque-Williams and Polek, 2010) posing an immediate threat of introduction of HLB into California. HLB is caused by phloem limited gram negative bacteria (*Candidus* Liberibacter species). At present, three liberibacter species, *Candidus* Liberibacter asiaticus, *Candidus* Liberibacter africanus and *Candidus* Liberibacter americanus infecting citrus are known.

HLB epidemics take several years to reach high incidence levels. Temporal progress of HLB incidence depends on 1) vector population 2) extent of inoculums reservoir and age of grove at first infection. Disease progression in the orchard can be relatively fast, reaching more than 0.95% incidence in 3 to 13 years after onset of the first symptoms (Aubert *et al.*, 1984; Gottwald *et al.*, 2010). Gottwald *et al.* (2007) have recently reviewed the literature on HLB epidemiology and pointed out the difficulties of conducting quantitative epidemiological studies on HLB. Usually, higher concentrations of symptomatic trees are initially found at the borders of blocks probably due to first settlement of the psyllid vector.

SYMPTOMS

Citrus trees can be infected by HLB bacterium at any stage of the tree right from nursery plants to orchard trees. Generally, young trees are more susceptible than older ones. Affected trees are stunted, have spreaded growth and sparse yellow foliage and show twig dieback. HLB symptoms first appear on one or two branches but finally the whole tree is affected (Plate 36a). Leaf symptoms of HLB disease remain confusing with zinc deficiency but through experiments, it has been established that the most characteristic symptom of greening disease is mottling of leaves (Plate 36b, c). Mottling symptoms on leaves normally appear on lower canopy (Plate 36b). Defoliation followed by dieback of twigs is a common symptom of HLB disease in India. Twig dieback depends on strain of the pathogen and cultivar. Mottled leaf symptoms are conspicuous on sweetorange leaves (Plate 36b) but less prominent in mandarin leaves. In mottled leaves, yellow, pale green and dark green areas are normally seen and such symptoms are known as 'Blotchy mottle'. Mottled symptoms are characterized as discolouration of inter-veinal areas, alternating with green corresponding islands similar to mineral deficiency symptoms (zinc deficiency). Midribs and lateral veins of mature leaves of affected trees often show yellowing. Whole leaf may become chlorotic with dark green small circular islands over it without any definite arrangement. Most leaves on affected branches fell off with the onset of summer resulting dieback of twigs. This condition is followed by secondary growth from axillary buds which consist of short, upright, small weak shoots which may be bushy. New leaves develop on these shoots are small with upright growth. These leaves show different types of discolouration resembling zinc, manganese or iron deficiency. Initially, symptoms develop on one or two branches but ultimately the whole tree is involved. Seedling mandarin trees in Sikkim, Darjeeling hills, Nepal, Bhutan show yellowing of leaves on affected trees (Plate 36a). African form of HLB develops leaf symptoms under cool conditions (20–25°C) but Asian form of HLB shows symptoms under both cool and warm temperatures (up to 35°C). Symptoms of zinc deficiency are often confused with greening symptoms. Therefore, mottling and non-mottling (Zn deficiency-type) leaf samples from different locations in India were analysed by electron microscopy and DNA-DNA hybridization tests. Only leaves with mottled symptoms showed positive reaction in these tests suggesting that leaf mottling is the main diagnostic symptom of HLB disease (Varma *et al.*, 1993). Sometimes, leaves on affected trees or in glasshouse inoculated plants show curling and corking of leaf veins and a few leaves also become leathery in texture (Plate 37a).

Affected trees show excessive and early flowering. Fruit set is very poor and off-season blooming is common on affected trees. Fruits on HLB affected trees are smaller, lop sided, drop easily and poorly coloured (Plate 37b) and remain dull green, hence the name greening was given to the disease in South Africa. In affected fruits, orange colour starts first at the peduncular end, at a time when the stylar end is still green. Such fruits contain brown aborted seeds (Plate 37c). Fruit drop is very common. Smaller and lopsided fruits with aborted seeds are most characteristic symptoms of HLB. Fruits from HLB affected trees are juiceless and bitter in taste, low in soluble solids and high in acid contents. Rootlets or feeder roots show rotting at advanced stages of the disease. There is no symptom on trunk or twigs.

TRANSMISSION

HLB is transmitted by bud, shoot, bark or leaf patch grafting but the transmission is poor as the causal agent is unevenly distributed in infected citrus species. HLB disease is also transmitted experimentally by dodder (*Cuscuta campestris* or other species) to periwinkle (*Catharanthus roseus)* and tobacco (Garnier and Bove, 1983) (Plate 38b), and tomato (Duan *et al.*, 2008a). HLB pathogen is transmitted to periwinkle through dodder initially but once the periwinkle got infected, HLB pathogen can be transmitted from infected periwinkle to healthy periwinkles through grafting.

HLB is seed transmitted in periwinkle and dodder seeds. Seed transmission in periwinkle and dodder resulted in low bacterial titre and very mild disease in periwinkle (Zhou *et al.*, 2008). However, transmission of the HLB pathogen through citrus seed was not confirmed (Hartung *et al.*, 2010; Van Varun *et al.*, 2010). Transmission of HLB has also been reported to cucumber (Schwarz in 1968). Natural transmission of HLBs is through psyllid vectors in persistent manner. South African HLB is transmitted by psyllid, *Trioza erytreae* Del Guercio (McLean and Oberholzer, 1965a, b). Asian HLB is transmitted by another psyllid, *Diaphorina citri* Kuw. In India and the Philippines, HLB disease is also transmitted by psyllid, *Diaphorina citri* (Plate 38a) (Capoor *et al.*, 1967, 1974; Martinez and Wallace, 1967; Salibe and Cortez, 1966; Celino *et al.*, 1966). Indian citrus decline (HLB) is also transmitted by *Trioza erytreae,* the vector of South African greening (Massonie *et al.*, 1976) and African form of HLB is transmitted by *Diaphorina citri,* vector of Asian greening (Lallemand *et al.*, 1986). *D. citri* nymphs acquired Ca. L. asiaticus (LAS) and Ca. L. americanus (LAM) more efficiently than adults. No single insect acquired both bacterial species. LAS concentration was highest when acquired by 4th and 5th instar nymphs of *D. citri* (do Nascimento *et al.*, 2010). These bacteria are persistent and propagative in the vector. American form of the HLB is also transmitted by *D. citri*. *T. erytreae* transmit HLB pathogen in South Africa, Madagascar Island and Yemen. *D. citri* transmit HLB pathogen in Asia, Southeast Asia, North, Central, South America, Western Saudi Arabia, and Iran. In Brazil, *D. citri* transmits both Asian and American liberibacters (Yamamato, 2006). In Reunion and Mauritius Islands and region between Yemen and Southwestern Saudi Arabia both types of HLB and the two vector species are found and transmit both types of HLBs (Asian and African). HLB pathogen multiplies in psyllid vector but transovarial transmission is not known. Acquisition access period of *D. citri* is 30 min, latent period is 1–25 days and inoculation feeding is 1.5 h (Xu *et al.*, 1985). For *T. erytreae,* acquision access period is 24 h and latent period is also 24 h. Psyllids acquire HLB pathogen during nymphal stage and probably transmit it throughout its life span. *D. citri* is found in Asia, Southeast Asia, the Indian subcontinent, Saudi Arabia, Reunion, Mauritius, South, central and North America. Population build up of the vector is mainly related to the flushing rhythms of the plants. *Diaphorina citri* in Florida was detected in 1998 (Halbert *et al.*, 2003). It was distributed throughout the state primarily through migration, hitchhiking on its host plants, and retail trade in

Murraya paniculata sold as ornamental plant (Halbert *et al.*, 2003; Halbert *et al.*, 2008). *Diaphorina citri* has also been recorded in California in 2008 along the border in Mexico threatening the citrus production in that region (Lugue-Williams and Polek, 2010). Preferred hosts of *D. citri* are: *Murraya paniculata, M. exotica, C. aurantifolia, Bergera koinigii, Clausena* spp. and for *T. erytreae,* lemon and most citrus species and *Vepris undulata, Clausena anisata, Zanthoxylum capense, Toddalia lanceolata* are preferred hosts.

Citrus psyllids are very common on citrus species in almost all parts of India, becoming a serious pest of citrus during spring and early summer in western and central India and some parts of the North india. Population decline with the onset of the hot season and the monsoon, but reappear during autumn flushes. Populations are low during winter months in northern India (Raychaudhuri *et al.*, 1977). Maximum population of *D. citri* is found on Mosambi sweet orange and citropsis (*Citropsis schweinfurthii*) during January and February in Maharashtra State (Chavan and Summanwar, 1993).

Both Psyllid species occur together at four places in the world: Reunion and Mauritius in the Indian Ocean, Sta Helena in the Atlantic Ocean, along the West border between Saudi Arabia and Yemen in the Arabian Peninsula (Garnier and Bove, 1993). Both African and Asian HLB forms are transmitted by any of the two *Diaphorina* species (Massonie *et al.*, 1976). *Diaphorina citri* in India can survive and breed to several hosts such as *Murraya koenigii, M. paniculata* (Singh and Nimbalkar, 1977), *Clausena excavata, Citrus histrix, Limonia acidissima (wood apple), Atalantia* sp., *Severina buxifolia,* Trifoliate, som-pan and Som-keo-wan mandarins. Vector of South African HLB, *Trioza erytreae* also breeds on several hosts other than citrus. These hosts are *Clausena anisata, Vepris lanceolata* and *Toddalia lanceolata.* Preferential host for breeding both the psyllid species is Indian Key lime or Mexican lime. All forms of *D. citri* are not efficient vector of HLB pathogen but different biotypes are known to be present in *D. citri* which differ in their vectoral capacity of HLB transmission (Ahlawat *et al.*, 1995). *Murraya* species are also known to be a choicest host of *D. citri.* It is widely grown in southern Florida and in several Southeast Asian countries as ornamental hedge and psyllid can readily colonized on it.

HOST RANGE

HLB pathogen infects many citrus cultivars, species, hybrids and citrus relatives. Cultivars of following *Citrus* spp. irrespective of the root-scion combination are susceptible: Citrons, Grapefruits, Kumquats, Lemons, Mandarins, Pummelos, Rough lemon, Rangpur lime, Sweet lime, Sweet oranges, Tangelos and Limes (Kagzi, Key, Mexican, West Indian lime).

Mandarins, Sweet oranges, grapefruits and Tangelos are highly susceptible to HLB pathogens followed by lemon, Rough lemon and sour orange. Citranges (Rusk and Troyer), Indian key lime, trifoliate orange and its hybrids are tolerant to HLB infection and occasionally develop leaf mottle symptoms. Severity of symptoms increases if a tree is infected by both HLB bacterium and tristeza virus especially in mandarins and sweet oranges. Citrus relatives are also hosts of HLB. Among these *Murraya paniculata* (Lopes *et al.*, 2005, 2006; Zhou *et al.*, 2007), *Clausena lansium*, (Ding *et al.*, 2006), *Severinia buxifolia* (Hung *et al.*, 2000) and *Calodendron capense* (Garnier *et al.*, 2000) are important hosts of HLB pathogen. HLB pathogen infects periwinkle and tobacco experimentally through dodder (Garnier and Bove, 1993). Tomato has also been infected with HLB pathogen through dodder (Duan *et al.*, 2008a). All the three Liberibacters are transmitted to periwinkle through dodder, *Cuscuta campestris*. Dodder, *Cuscuta indecora* is also host of *Ca Liberobacter* spp. and these pathogens multiply well in *Cuscuta indecora* (Hartung *et al.*, 2010). Multiplication of the pathogen in dodder can be used to study host–pathogen interactions. The liberibacter titers in periwinkle plants are higher than in citrus. For instance, *Ca*. L. americanus reaches titers of 3.4×10^8 liberibacters per gram of midribs (Teixeira *et al.*, 2008a). *D. citri* transmits Asian form of HLB (*Candidus* Liberibacter asiaticus) into *M. paniculata* and *M. exotica* but not in *Bergera* (*Murraya*) *koenigii*. Disease symptoms rarely develop in *Murraya* plants but pathogen is determined by conventional and real-time polymerase chain reaction (PCR). Back-inoculations of '*Ca*. L. asiaticus' from *M. paniculata* to Madam Vinous sweet orange resulted in disease development (Damsteegt *et al.*, 2010). At the time when HLB was becoming established in Florida, *Murraya paniculata* was one of the more popular landscape plants produced in Florida. *Murraya paniculata* was found to be a host of Florida HLB bacterium (Zhou *et al.*, 2007). *Clausena lansium* has also been shown to be susceptible to HLB bacterium (Deng *et al.*, 2007).

INDICATOR

Sweet orange and tangelo are used as indicator hosts of HLB bacterium. Graft inoculated Sweet orange seedling develop symptoms as Leaf yellowing and mottling with zinc deficiency-like patterns which may develop in several months. Asian form develops better symptoms at warm temperature but S. African form develops symptoms at cool temperature.

THE PATHOGEN

Earlier studies such as successful transmission of HLB disease indicated that the associated pathogen may be a virus. Later a Rickettsia- like organism has been reported in citrus yellow shoot disease from China (Ke *et al.*, 1980). Results of electron microscopy, however, showed that a microorganism similar to Mycoplasma-like organisms (MLOs) was present in phloem sieve tubes of HLB affected trees (Lafleche and Bove, 1970a, b). Yellow shoot samples in China when observed in electron microscopy also showed MLOs (Xu *et al.*, 1985). Nature of Likubin associated pathogen and strains have been described from Philippines (Su and Li, 1990). However, HLB organism has 25 nm thick cell envelop which is double in thickness of the unit membrane envelop of MLOs (now known as phytoplasmas) (Saglio *et al.*, 1971; Garnier *et al.*, 1984a) hence MLO etiology was ruled out. Finally, based on the peptiglycan in the cell envelop of HLB organism, it was established that HLB disease is associated with a bacterium of gram negative type (Garnier and Bove, 1993; Garnier *et al.*, 1984b, 1993). Indian isolates of HLB also showed bacteria-like organisms in electron microscopy (Naidu and Govindu, 1981). Characterization of the HLB pathogen was done through sequencing of its 16S ribosomal DNA and ribosomal protein gene. The term '*Candidus*' has been proposed for organisms in "waiting position' for putative taxa of bacteria like organisms causing Huanglongbing (HLB) disease that could not be described in enough detail to warrant establishment of novel taxa (Murray and Stackebrandt, 1995). The term *Candidus* was finally approved by International Committee on Systematic Bacteriology. HLB bacterium is now a new genus in the alphaproteobacteria division (gram negative bacteria) and given the name of new genus as "*Candidus* Liberibacter". Initially, it was named "Liberobacter" from the Latin liber = bark and bacter = bacterium (Jagoueix *et al.*, 1994). Liberobacter was later replaced by liberibacter (Garnier *et al.*, 2000). Penicillin treatment of HLB affected trees showed remission of symptoms further suggesting the bacterial nature of the pathogen. In Citrus, three species of *Candidus* Liberibacter have been identified to be associated with HLB agents of Asian, South African and American. These species for Asian, African and American HLBs are: "*Candidus* Liberibacter asiaticus" (Ca.L.a), "*Candidus* Liberibacter africanus" (Ca.L.af.) and "*candidus* Liberibacter americanus" (Ca.L.am.) respectively (Jagouiex *et al.*, 1996; Coletta-Filho *et al.*, 2004; Teixeira *et al.*, 2005; Halbert, 2005). Ca.L.am. is found only in Brazil but it was found in one sample from Hunan in China (Lou *et al.*, 2008). If this result is confirmed, it will be the first time that *Ca.* L. americanus has been detected out of Brazil. A new subspecies of Ca.L.africanus named "capensis" was also isolated from *Calodendron capense,* an ornamental rutaceous tree in Western Cape Province of South Africa (Garnier *et al.*, 2000; Phahladira, 2008). These Liberibacter species are restricted in sievetubes of the infected host. To determine the taxonomy of the causal bacterium,16S rDNA of HLB bacterium of Indian isolates on sweet orange from pune has been amplified using PCR, cloned and sequenced (Garnier *et al.*, 1993). These genes have also been sequenced from HLB infected Nagpur mandarin (Das, 2009). Based on these studies, HLB bacterium has been shown to be á class of Protobacteria. Organisms of this class have the ability to grow and survive in eukaryotic cells and also in an arthropod vector. HLB-bacterium has this capacity as it grows in phloem sieve tubes of eukaryotic plant host, and it is transmitted by two arthropod psyllid vectors, *Trioza erytreae* and *Diaphorina citri,* in which it circulates and replicates. HLB bacterium has been classified as under:

Kingdom	:	Procaryotae
Phylum	:	Gracilicutes
Class	:	á-Proteobacteria
Genus	:	*Liberibacter*
Species	:	*asiaticus, africanus, americanus*

Once HLB pathogen was identified as MLO, it's culturing on synthetic media has been initiated and reported as cultured from India and South Africa (Garnett, 1984; Ghosh *et al.*, 1971, 1975; Nariani *et al.*, 1975; Sodhi *et al.*, 1973). However, the results of these workers could not be repeated and confirmed. Davis *et al.* (2008) reported co-cultivation of "*Candidatus* Liberibacter asiaticus with actinobacteria from citrus with huanglongbing disease. Recently, Sechler *et al.* (2009) reported successful culturing of *Candidus* Liberibacter asiaticus and Ca. L. americanus in a specialized growth medium. Colonies on the medium were shown to be positive for Ca. L. asiaticus by a 16s-based rDNA real time PCR reaction. Cells of Ca. L. asiaticus and Ca. L. americanus in medium observed by scanning electron microscope were ovoid to rod-shaped measuring 0.3–0.4 by 0.5 to 2.0 μm. Cultured organisms was reported to be pathogenic to citrus and could be reisolated from inoculated plants. Results of this work are quite interesting and convincing but need confirmation.

GENOME

A complete circular '*Candidus* Liberibacter asiaticus' (LAS) genome has been obtained by metagenomics, using the DNA extracted from a single LAS infected psyllid (Duan *et al.*, 2008b, 2009). The 1.23 Mb genome has an average 36.5% GC content. Annotation revealed a high percentage of genes involved in both cell motility (4.5%) and active transport in general (8.0%), which may contribute to its virulence. '*Ca.* L. asiaticus' appears to have a limited ability for aerobic respiration and is likely auxotrophic for at least five amino acids. Multiprotein phylogenetic analysis confirmed '*Ca.* L. asiaticus' as an early-branching and highly divergent member of the family Rhizobiaceae (Duan *et al.*, 2010). This is the first genome sequence of an α-proteobacteria that is both an intracellular plant pathogen and insect symbiont. (Duan *et al.*, 2009). Zinc is an essential cofactor in many proteases and also forms the core of the zinc-finger domain. Two putative operons involved in zinc uptake were identified in LAS. Complementation studies indicates that out of the two operons only one may be functional suggesting that disruption of zinc system would either decrease the severity of symptoms or possibly kill the bacterium (Duan *et al.*, 2010). These results confirm the classification of HLB bacterium as shown above

Ca.L.asiaticus and *Ca*.L.africanus are differentiated on their temperature sensitivity, nucleic acid (DNA) hybridization and genomic properties (sequence studies of 16S/23S rRNA). *Ca*. L. americanus is a distinct species based on 16S/23S rRNA sequences. Draft genome of liberibacter contained at least 1.161,087 base pairs with an average of 36% GC contents. The genome contains three ribosomal operons, 34 tRNAs and 1077 open reading frames of which 814 had a functional assignment (Duan *et al.*, 2008b). Ca.L. americanus genome is close to 1.31 Mbp and appears circular (Wulff *et al.*, 2008).

Recently, a new species of *Candidus* Liberibacter known as *Candidus* Liberibacter solanacearum (Ca. L. solanacearum) has been reported from New Zealand (Liefting *et al.*, 2008). Ca.L. solanacearum infects tomato, capsicum, chilli, potato tamarillo (*Solanum betaceum*), and cape gooseberry (*Physalis peruviana*) (Liefting *et al.*, 2009). Ca. L. solanacearum is also affecting carrots in Europe (Munyaneza *et al.*, 2010). Ca L. solanacearum has 95 to 97% 16SrDNA sequence identity with citrus liberibacters. The vector of Ca. L. solanacearum is potato/tomato psyllid, *Bactericera cockerelli* and found in USA, Canada, Mexico, Guatemala and Honduras. In Texas, both *D. citri* and *Bactericera cockerelli* are present. Interestingly, a Liberibacter species, *Candidatus* Liberibacter psyllaurous has also been reported in potato (Hansen *et al.*, 2008; Cavarrublas *et al.*, 2010). A zebra chip (ZC) disease of potato also caused by *Ca*. L. solanocearum has been reported from USA, Mexico, Central America and New Zealand (Lin *et al.*, 2009; Secor *et al.*, 2009). ZC can cause general cell death of the tubers. This cell death was confirmed using high-resolution, field-emission scanning electron microscopy (FE-SEM) of fresh-cut tuber tissue (Godfrey *et al.*, 2010). Recently, a phytoplasma related to '*Candidatus* Phytoplasma asteris' of group 16SrI has been detected in citrus plants showing huanglongbing (yellow shoot diseases) symptoms in Guangdong, P. R. China (Chen *et al.*, 2008) and a phytoplasma of group 16SrIX closely related to the pigeon pea witches' broom phytoplasma was also observed with citrus huanglongbing symptoms in the State of sao Paulo, Brazil (Teixeira *et al.*, 2008b) and called as HLB phytoplasma. A leafhopper, *Scaphytopius marginelineatus* (Cicadellidae: Deltocephalinae) naturally collected in Brazil was found to be positive of HLB phytoplasma in PCR but transmission results are being evaluated to confirm this leafhopper as vector of HLB phytoplasma (Marques *et al.*, 2010). However, relationship of *candidus* Liberibacter species with HLB phytoplasmas needs further studies.

DIAGNOSIS AND MANAGEMENT

Huanglongbing can be easily detected in hosts like sweet oranges just by observing mottling of leaves–a characteristic symptom of HLB disease. Sparse foliage and 'open' type growth in affected trees, lopsided fruits with colour inversion and aborted brown black seeds are other diagnostic symptoms in the field. Since HLB remains symptomless in some hosts, it is essential to index field samples in glasshouse on indicator seedlings of sweet orange (Madam Vinous, Pineapple, Hamlin, mosambi and tangelo (Orlando, Seminole) by graft or vector inoculation. HLB inoculated plants will show typical symptoms of yellowing and mottling after several months at higher temperature of Asian HLB (25°C for 8 hours in the dark, and 30 to 32°C for 16 hours in the light) and at cool temperature for African and American HLBs (20°C for 8 hours in the dark, and 24 to 27°C for 16 hours in the light). Best way of biological diagnosis of HLB disease is by transmitting the disease to periwinkle through dodder. In periwinkle, symptoms of smalling of leaves, chlorosis and witches' broom are developed in much shorter time than in *Citrus* spp. (Plate 38b). Electron microscopy is a good technique for detection of liberibacter species. *Ca.* Liberibacter species are observed in sieve tubes of phloem tissues when thin sections of infected tissues are observed in electron microscopy (Garnier and Bove, 1996). Pleomorphic bodies are seen in sieve tubes of phloem tissue of affected plants (Fig. 25). Bacterium associated with HLB disease is pleomorphic in nature, hence difficult to purify for development of polyclonal antibodies. Therefore, monoclonal antibodies (MAbs) have been developed against Ca. Liberibacter species. Thirteen MAbs have been

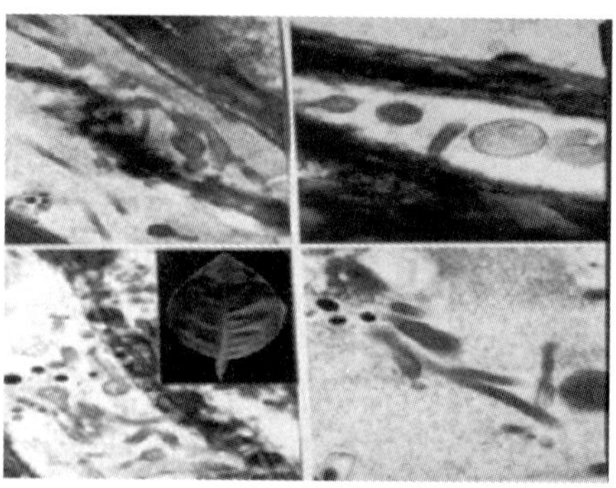

Fig. 25: Electron micrograph of thin secions showing pleomorphic HLB bacteria in sievetubes of infected tissue.

developed against several isolates from Southeast Asia and South Africa (Garnier *et al.*, 1987). These antibodies are highly specific and react only to isolate against which it has been prepared (Table 9) (Garnier *et al.*, 1991; Gao *et al.*, 1993). MAb 10A6 made against pune strain of HLB pathogen was purified and the antigenic protein was used to prepare another MAb, 1A5 which recognized most of the Asian strains except those from China (Garnier and Bove, 1993). Enzyme-linked immunosorbent assay (ELISA) with monoclonal antibodies (MA) is also used for detection of HLB bacterium. Biological and serological methods are helpful to identify HLB even in a mixed infection with citrus tristeza virus (Ghosh *et al.*, 2003). Results of Table 9 suggest that serotypes do exist in HLB pathogens. Citrus HLB pathogens are easily detected by immunofluorescence (IF) both in sections and extracts using greening specific MAbs (Plate 39a, b, c) (Garnier and Bove, 1993; Ahlawat, 1993; Ahlawat, 2010; Ahlawat and Pant, 2010) and this is one of the best serological tests to detect Ca. L. spp. HLB bacterium is also detected by immunoelectron microscopy (Ke *et al.*, 1993).

Table 9: Detection of Greening organism in immunofluorescence with MAbs

Greening isolates	*MAbs*							
	India			*China*			*S. Africa*	
	$2D_{12}$	$10A_6$	$1A_5$	$10F_4$	$5H_4$	$11H_6$	$2H_9$	$14A_1$
India								
Pune strain	+	+	+	–	–	–	–	–
Bangalore strain	+	+	+	–	–	–	–	–
China	–	–	–	+	+	+	+	–
Malaysia	–	+	+	–	–	–	–	–
Thailand	–	+	+	–	–	–	–	–
Taiwan	–	–	–	–	+	–	–	–
Indonesia	–	+	+	–	–	–	–	–
The Philippines	–	–	+	–	–	–	–	–
South Africa	–	–	–	–	–	–	–	+

The *rpl*KAJL-*rpo*BC gene cluster codes for ribosomal proteins K, A, J, L, and RNA polymerase proteins B and C. The gene clusters from *Ca*. L. asiaticus and *Ca*. L. africanus were obtained and used, respectively, as hybridization probes In-2.6 for specific detection of the Asian liberibacter, and probe As-1.7 in the case of the African liberibacter (Villechanoux *et al.*, 1992, 1993; Planet *et al.*, 1995). A fragment of bacterial genome (In 2.6) from Pune strain of Ca. L. asiaticus has been used as a probe for the detection of HLB bacterium in DNA-DNA hybridization in field trees and in individual *Diaphorina citri* collected on infected trees in Delhi and Bangalore in India (Figs. 26, 27) (Bove *et al.*, 1993). The technique of DNA-

DNA hybridization detected the pathogen in less than 1% psyllids from India as against psyllids collected from Malaysia in May (5%) and September (39%) indicating a variation in the percentage of contaminated *D. citri* insects during the year. DNA-DNA hybridization is an important technique to study epidemiology of HLB pathogen.

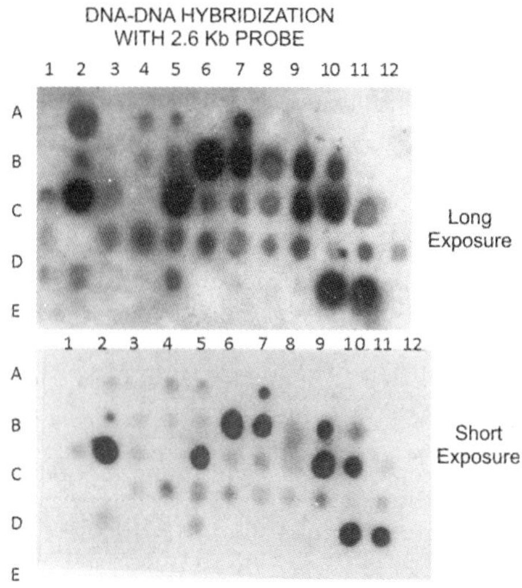

Fig. 26: Detection of HLB bacteria in DNA-DNA hybridization using In 2.6kb probe. Upper one is with long exposure and lower one is with short exposure

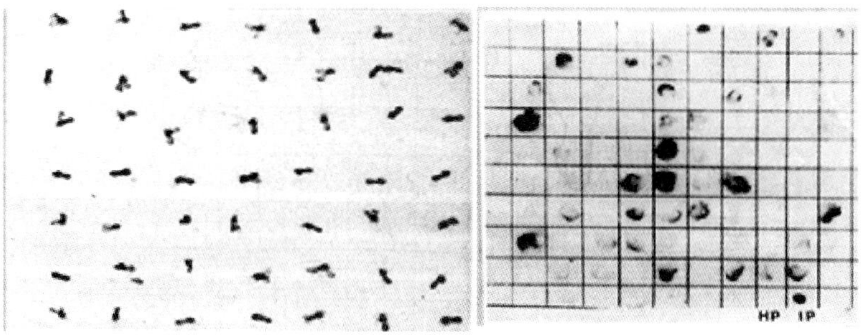

Fig. 27: Detection of HLB bacteria in DNA-DNA hybridization using in 2.6 kb probe. Upper one is with long exposure and lower one is with short exposure

For detection of Liberibacters, two PCR systems have been used. First system is based on the amplification of a 1160 bp fragment of liberibacter 16S rDNA with primer pair f-OA1/r-OI2c for *Ca.* L. africanus and f-OI1/r-OI2c for *Ca.* L. asiaticus (Jagoueix *et al.*, 1996). With primers f-GB1/r-GB3 for *Ca.* L. americanus, a 1027 bp amplicon was obtained (Teixeira *et al.*, 2005b). The second PCR system is based on the *rpl*KAJL-*rpo*BC gene cluster.

With primer pair f-*rpl*A2/r-*rpl*J5, a 701 bp amplicon was obtained with the Asian liberibacter, and a 667 bp amplicon with the African Liberibacter (Hocquellet *et al.*, 1999). A "multiplex" PCR system for routine detection of all three liberibacters in one test tube within a single reaction mixture containing the two sets of primers described above has been developed. Nested PCR, Real time PCR and Quantative real time PCR (qPCR) have been applied to the detection and quantification of liberibacters in plants and insect vectors (Li *et al.*, 2006, 2007; Wang *et al.*, 2006; Teixeira *et al.*, 2008b; Cordero *et al.*, 2010). PCR and Real time PCR were also used to detect *Candidatus* Liberibacter asiaticus in *Diaphorina citri* (Hung *et al.*, 2004; Manjunath *et al.*, 2008; Das *et al.*, 2007). Using PCR, distribution of HLB pathogen in infected plants was also detected (Tatineni *et al.*, 2008). Detection of HLB bacterium through PCR, real time PCR and nested PCR is now commonly used in India and elsewhere (Jagouiex *et al.*, 1996; Ahlawat *et al.*, 2003; Baranwal *et al.*, 2003, 2004; Gouda *et al.*, 2006; Das, 2004; Cordero *et al.*, 2010). It has been observed in India that better detection of greening pathogen by PCR is obtained during winter season. DNA extraction by CTAB method was suitably done from midrib and veins including bark of sweet orange for DNA extraction by SS Tris-EDTA method (Gopal *et al.*, 2007). A membrane based DNA template preparation protocol has been recently developed for simultaneous detection of *Candidus* Liberibacter asiaticus and citrus yellow mosaic badnavirus (Fig. 18) (Baranwal and Arya, 2009; Baranwal *et al.*, 2005). Using metagenic approach, it has been established that in HLB affected trees only *Ca.* L. asiaticus was present in phloem tissues and 1.7 '*Ca.* Liberibacter' cells were present per phloem cell (Heather *et al.*, 2009). A technology known as laser induced fluorescence spectroscopy (LIFS) has also been used to detect HLB infection earlier than the symptoms appear in inoculated plants (Cardinali *et al.*, 2010).

Effective control of HLB disease is difficult because diagnostic symptoms are variable and propagation is done inadvertently from infected budwood. Immediately after the discovery that HLB pathogen is a gram negative bacterium, tetracycline injection to the trunk of infected trees were given in several countries. Tetracycline therapy provided remission of disease symptoms but the symptoms reappeared as soon as the therapy was discontinued. Regular treatment with tetracycline resulted in phytotoxicity hence it is not used now. However, penicillin was shown to give remission of HLB symptoms (Aubert and Bové, 1980; Bové *et al.*, 1980) but it is also not a viable option. A new chemotherapeutant, B.P.-101, produced by the Hindustan Antibiotics Ltd., Pimpri, Pune, was effective to control HLB disease in lime, sweet orange, and mandarin trees. BP 101 was effective when put under bark or injected into affected tree but was not effective by foliar spray (Capoor *et al.*, 1973; Capoor and Thirumalachar, 1973). Plants sprayed with terramycin, ledermycin, or streptocycline at 1000 pprn or BP-101 at 500 pprn recovered completely from HLB symptoms. Plants sprayed with terramycin at 500 ppm, ledermycin at 100 or 500 ppm, and BP-101 at 100 pprn showed recovery of disease symptoms in majority of cases (Nariani and Bhagabati, 1979). A chemical known as CPAN No.MJN 1891 was found effective to make symptomatic Mosambi sweet orange trees free from symptoms in two months after application the paste in to trunk (Ahlawat *et al.*, 1989). However, chemical treatment is not a practical option to manage HLB disease. Once HLB is established in a region, it is difficult to eradicate it. The Florida citrus industry producers are using wettable powder micronutrients, phosphate and methyl salicylate (a systemic acquired resistance product) to prevent infection of trees. However, these nutritional supplements have no effect on tree health, yield or fruit quality (Gottwald *et al.*, 2010). Preventing citrus trees from becoming infected with HLB-associated agents is the only way to manage HLB most effectively. Following preventive measures are suggested:

1. Symptomatic trees must be removed to eliminate liberibacter inoculum.
2. To reduce psyllid vector population by insecticide treatments or by biological control approaches.
3. Production of healthy citrus planting material and to maintain insect-proof nurseries for developing new orchards and to replace infected trees.

To achieve the first measure (removal of symptomatic trees) it is important to identify HLB infected trees through regular inspections. Four inspections per year are normally taken but if required more number of surveys can be planned. The trees with symptoms and escapes (symptomless) are identified during inspection and removed. In large citrus groves of São Paulo state and Florida, inspections of adult trees are conducted with "platforms" on which two or four inspectors can be accommodated. The platform moves in between rows at a speed of 3 to 4 km/hour allowing inspection of ~5000 trees/day. Even for young orchards, platforms with two inspectors are now used. After the removal of infected tree, sprouts, stumps and roots of infected tree must be carefully removed.

Psyllids are not only the vector of greening disease but they are also serious pest of citrus in Southeast Asia. Control of this vector is difficult but can be minimized through insecticidal sprays (Monocrotophos, organophosphates and pyrethroids) in spring when the population starts building up. Mineral oils may be used as an alternative being environment friendly and less destructive (Davis *et al.*, 2005). It has been commonly observed that more trees at the border are infected than in the centre of a grove. It is therefore advisable that either a suitable barrier crop/trees should be identified at regional basis or trees at the border should be sprayed more frequently. Psylla-free areas may be identified to grow certified nurseries and mother trees (Ahlawat and Sardar, 1976). In Reunion Island, both psyllid vectors (*D. citri* and *T. erytreae*) and both Asian and African form of HLB were present. Asian form of HLB and its vector, *Diaphorina citri* were present below 400–500 m, and African form of HLB and its vector, *Trioza erytreae*, above 400–500 m. High populations of these insects were observed at each flushing period. This is the only region where biological control of both psyllids was successfully achieved (Etienne and Aubert, 1980; Aubert and Quilici, 1984). In Reunion Island, *Tamarixia* (ex-*Tetrastichus*) *dryi*, an ectoparasite of *T. erytreae* was imported from Transvaal, South Africa, in December 1974. *Tamarixia (ex-tetrastichus) radiata*, an ectoparasite of *D. citri*, was imported in April 1978 from Punjab, India with a view to control psyllids by these parasites (Aubert and Quilici, 1984). These two imported psyllid parasites were absent in this Island and obviously parasites of these imported parasites were not present there. Absence of *Tamirixia* parasites (or Psyllid hyperparasites) made it possible to eliminate psyllid vectors from this Island and explain the success of biological control of psyllids in Reunion Island. Interestingly, *Tamarixia dryi* not only destroyed *Trioza erytreae*, but also destroyed an additional psyllid, *Trioza eastopi*, living on a very common shrub, *Litsea chinensis* in Reunion Island. Biological control of psyllids in other countries is not successful because psylla parasites have their own parasites in these countries which was lacking in Reunion island resulting successful eradication of psyllids through wide spread release of these parasites. After the control of psyllids in Reunion Island there is significant improvement in citrus plantations as compared prior to introduction of parasites (Aubert *et al.*, 1984). RNA interference (RNAi) is a fundamental component of gene regulation and antiviral defence within cells of eukaryotic organisms. RNAi has been suggested as a strategy to develop "insect-proof plants" and terms such as "genetic insecticide" have been used. However, practical applications of RNAi for insect vector(s) control of *Xylella fastidiosa*, *Candidus* Liberibacter psyllaurous and huanglongbing remain to be implemented (Rosa *et al.*, 2010).

The farmers in Mekong Delta of Vietnam observed an interesting interaction between citrus and guava (*Psidium guajava*) that commercial life of citrus plantings was tremendously increased when interplanted with white guava cultivar xaly as compared to monoculture citrus. Vietnamese/Japanese/Australian collaborative research projects have demonstrated that psyllid (*Diaphorina citri*) vector populations was reduced to near non-detectable levels in interplantings compared to citrus monoculture. These findings were reported at the *International Workshop for Prevention of Citrus Greening Disease in Severely Infected Areas*, held in Ishigaki Japan, in December 2006, sponsored by Japanese International Research Center for Agricultural Sciences (JIRCAS). At this meeting, Dr.

Nguyen Van Hoa (Student of the author), from the Vietnamese Southern Fruit Research Institute, described collaborative research projects between researchers at JIRCAS, SOFRI, and the University of Western Sydney (UWS), New South Wales, Australia. Greenhouse studies in Florida revealed high mortality rates of psyllids (95% mortality) within 6–9 days when confined to guava in no-choice situation. The effect may be due to volatile compounds produced by guava that are deleterious to psyllids (Gottwald *et al.*, 2010). It has recently been shown that the volatiles of guava repel *D. citri*, and this explains the guava-tree effect (Noronha *et al.*, 2008). The effect of various volatile extracts on psyllid behavior and survival are under investigation. In addition, large-scale replicated field plots have been established in multiple locations in Florida to examine the dynamics of HLB epidemics in citrus/guava interplantings versus citrus monocultures (Gottwald *et al.*, 2010). However, more experiments are required to establish whether this strategy is viable. Nutritional treatment has been suggested to manage HLB but studies revealed that nutritional treatment does not sustain tree health, yield or fruit quality (Gottwald *et al.*, 2010).

In addition, eradication of sources of inoculum, control of psyllid vectors and growing disease-free plants are important factors to manage HLB. HLB-free healthy citrus plants can now be obtained by shoot-tip grafting (Vijyakumari *et al.*, 2006). Use of tolerant rootstocks is another option to manage the disease. *Citrus macroptera* and *C. indica* were found free from greening infection in Northeastern region of India and may be used as sources of resistance (Bhagabati *et al.*, 1989). More details to obtain pathogen-free planting material is given in Chapter 8.

Citrus Variegated Chlorosis

Citrus variegated chlorosis disease (CVC) caused by a bacterium, *Xylella fastidiosa* is widely distributed in Brazil. It is known to occur in the States of Sao Paulo and minas Gerais since 1987 (Rossetti *et al.*, 1990; Lee *et al.*, 1991). This disease is also reported from Paraguay and Argentina (Brlansky *et al.*, 1991). In Argentina, this disease is locally called as 'pecosita' (Brlansky *et al.*, 1991, 1993). In Brazil, variegated chlorosis causes approximately 20% loss to the growers and is now estimated to exceed several million dollars loss per year in Brazil (Rossetti, 1993, 2000).

SYMPTOMS

Symptoms of variegated chlorosis are similar to zinc deficiency or huanglongbing disease of citrus. Chlorosis begins on younger leaves and by the time of maturity of leaves, intensity of chlorosis is greatly increased which is visible only on upper surface of leaves (Plate 40a). In young infected trees, severe leaf chlorosis develops within leaf veins. Gummy Lesions develop underside of the leaves corresponding yellow areas on the upper surface of leaves. These lesions appear as dark brown, necrotic and with slightly raised surface due to gum deposit (Plate 40b). The lesions resemble boron toxicity or molybdenum deficiency (Rossetti, 1993). Fruits on affected trees remain small, hard and have higher contents of sugar. Fruits ripe earlier on affected trees. More fruits are set on trees following first year of infection but in subsequent years following infection, the fruits are decreased in number (Lee *et al.*, 1991). Initially, symptoms appear on one or two branches but subsequently whole tree is infected. If the affected limb is pruned out, the remaining part of the canopy may remain symptomless for some time. Symptoms of CVC are quite common on trees of 1–8 years but nursery plants can also be affected. Two years after infection, infected trees become uneconomical. Symptoms of this disease are more severe under conditions of high temperature and drought (Beretta *et al.*, 1993). Affected trees are stunted with protruding branches. Young trees are more susceptible than older trees.

TRANSMISSION

Variegated chlorosis is transmitted by grafting from field infected trees to healthy citrus plants. CVC is also transmitted by natural root grafting (He *et al.*, 2000). Bacterium, *Xylella fastidiousa* associated with variegated chlorosis disease is transmitted through xylem feeding leafhoppers (Plate

40c) and sharpshooters from even non-symptomatic alternate hosts to citrus. At least 11 species of sharpshooters are known to vector CVC in Brazil. Main vectors are, *Dilobopterua costalimai, Acrogonia terminalis, Oncometopia facialis, Bucephalogonia xantopis* and *Plesiommata corniculata* (Grevena *et al.*, 1998). In Argentina, sharpshooters, *Bucephalogonia xanthophis; Scopogonalia subolivacea; Hortensia similis; Rotigonalia limbatula; Sonesimia grossa; Chlorotettix latocintus and Curtara samera* from cicadellidae family and two species of embracidae family; *Ceresa ustulata and Entylia carinata* are main vector of *Xylella fastidious* (de Coll *et al.*, 2000). The bacterium is also seed transmitted (Rodrigues Neto *et al.*, 1992).

HOST RANGE

Most of the sweet orange cultivars are susceptible to variegated chlorosis such as Pera, Natal, Valensia, Hamlin and Bahia-Navel on rootstocks, Rangpur lime, *Cleopatra mandarin*, and *Volkamer lemon*. Murcott mandarin on Rangpur lime is symptomless carrier of CVC pathogen (Beretta *et al.*, 1993). Tangerins and Tahiti lime are also symptomless carriers of the pathogen. Severity of symptoms may be variable in lemons, lime, mandarin and its hybrids, kumquats, trifoliate orange and grapefruits. Rangpur lime, citron and Pummelo are tolerant to CVC. Plant species in at least 28 families of monocotyledonous or dicotyledonous are natural hosts of the bacterium associated with variegated chlorosis disease and could be the natural source of pathogen survival. Some of the important weed hosts of *X. fastidiosa* are: *Conium maculatum; Chloris halophila; Digitaria* sp.; *Paspalum regnelli; Axonopus compresus; Blainvillea biaristata; Commelina erecta; Medicago polymorpha; Phalaris angusta; Lepidium aetes; Stachys arvensis; Talinum paniculata* and *Sida rhombifolia* (de Coll *et al.*, 2000). CVC does not kill trees, but trees become less productive within a few years following infection.

THE PATHOGEN

A gram-negative xylem limited bacterium, *Xylella fastidiosa* is associated with Citrus variegated chlorosis (Beretta *et al.*, 1992; Chagas *et al.*, 1992). Similar strains of *X. fastidiosa* cause Pierce's disease of grapevine (Li *et al.*, 2002; Kitajima, 2010), phony peach, alfalfa dwarf; leaf scorch of almond, coffee, elm, sycamore, oak, plum, mulberry, maple, and oleander; and periwinkle wilt (Purcell and Hopkins, 1996). The bacterium, *X. fastidiosa* is found in leaf midribs, fruit columella and petioles (Rossetti, 1993). CVC bacterium has rippled cell wall which composed of three layers and bacterial cells are 0.3–0.5 μm in diameter and 1 to 3.5 μm in length (Plate 40d) (Chagas *et al.*, 1992). Causal bacterium of variegated chlorosis has been cultured on a modified PW medium. Koch's postulate has been fulfilled by inoculating cultured bacterium to sweet orange and re-isolated the same bacterium from inoculated plants (Chang *et al.*, 1993). Coffee leaf scorch is also caused by a strain of *Xylella fastidiosa* from citrus (Li *et al.*, 2001) suggesting that the CVC strain is apparently closely related to the bacterium causing coffee leaf scorch. Inoculation of coffee with the CVC bacterium causes coffee leaf scorch symptoms suggesting that the two diseases are caused by the same bacterium. Studies on genetic diversity of *Xylella fastidiosa* isolated from diseased citrus and coffee will help to characterize relationships among different *X. fastidiosa* strains (Qin *et al.*, 2001). The genome sequence of the CVC strain of *X. fastidiosa*, clone 9a5c, was recently determined, and the nature of genes that were identified by annotation suggests a number of potential pathogenicity mechanisms, such as cell-wall hydrolysis, adhesion, intervessel migration, and toxicity (Simpson *et al.*, 2000). However, gene function must be determined experimentally. It appears that *X. fastidiosa* causing pierce's disease of grapevine and citrus infectious variegation are also closely related. These two organisms have identical metabolic functions and are likely to use a common set of genes in plant colonization and pathogenesis, permitting convergence of functional genomic strategies (Van Sluys *et al.*, 2003).

DIAGNOSIS AND MANAGEMENT

Variegated chlorosis bacterium is detected by light and electron microscopy. Bacterial cells are seen in xylem tissue. Polyclonal and monoclonal antibodies have been prepared against cultured CVC bacterium and these antibodies are being used for bacterium detection in DAS-ELISA and by immunoprinting (Garnier *et al.*, 1993). Dot-immunobinding assay (DIBA) (Lee *et al.*, 1993), Western blot assay and PCR are also used for bacterium detection (Beretta *et al.*, 1993, 1997). *Xylella fastidiosa* is also detected in ultrathin sections of affected citrus by immunogold labeling. Molecular assays based on DNA hybridization or PCR diagnostics are also available for CVC diagnosis (Pooler and Hartung, 1995). Primers were designed based on a unique region of the CVC 9a5c genome, not found in other *Xylella fastidiosa (Zf)* infecting grapes, almond and oleander. This set of primer produces a diagnostic amplicon of 131 bp that can be used for detection by standard PCR and is also suitable for quantitative (Q)PCR using a Taq man probe labeled with texas Red. A second set of pairs was designed from the same region for QPCR with Sybr® green. This protocol can detect the pathogen upto 0.1 pg of Zf CVC-DNA and can be used for routine diagnosis of CVC (Francis *et al.*, 2010).

Remission of disease symptoms is possible by treatment of infected trees with oxytetracycline. In Brazil, if 30% of trees in a grove are infected, the recommendation is to remove the entire grove. In Brazil, control strategies to manage CVC include nursery certification, protection of nursery stock from infection, and pruning of symptomatic parts of trees to remove inoculum sources. A new strategy has been suggested to manage CVC. According to this strategy, entire canopy of a condomned tree (heavily infected tree) is removed at soil level below the graft line. Two-three months later, one healthy bud is grafted on each of the 2–3 new selected shoots (comming out from cut portion) and the shoot portion above graft is removed after about 6 weeks of grafting. The new scion is allowed to grow forming a new canopy. The new canopies grew faster and produce fruits earlier than the new young trees planted in the same orchard. It is probably due to high xylem sap influx coming from the already established root system. This practice may help to reuse the rootstock of even severely affected trees with CVC and also to bring the fruiting on such trees much earlier than new planting. This demonstrates the possibility of reusing the rootstock of a tree severely affected by CVC to generate a healthy and productive sweet orange scion (Lopes *et al.*, 2010). Control of sharpshooters is very difficult. Long term approaches involve the identification and development of CVC resistant sweet orange cultivars (Chung and Brlansky, 2009). High level of resistance has been reported in Rangpur lime against *X. fastidiosa* (Lopes *et al.*, 2010). For effective management of the disease, following practices are recommended:

1. Infected trees up to three years of age should be removed.
2. Effected branches of trees 3–5 years of age should be removed.
3. Infected trees older than 5 years should be pruned 5–10 cm below any symptomatic leaf (Beretta *et al.*, 1996).
4. Propagation material must be obtained from registered nurseries (Rossetti, 2000).

REFERENCES

Ahlawat, Y.S. 1993. Newer techniques for quick detection of greening disease of citrus. *Indian Phytopath.* 46: 276.

Ahlawat, Y.S. 1997. Virus, greening bacterium and viroids associated with citrus (*Citrus* species) decline in India. *Indian J. Agric. Sci.*, 67: 51–57.

Ahlawat, Y.S. 2007. Citrus decline: Problems and Preventions. *Indian Phytopath.* 60: 1–12.

Ahlawat, Y.S. and Sardar, K.K. 1976. Distribution of greening and tristeza diseases in various blocks of Darjeeling district. *Sci. & Cult.*, 42: 275–277.

Ahlawat, Y. S., Payak, M.M. and Thirumalachar, M.J. 1989. Control of citrus greening disease in India. *Curr. Sci.* 58: 921–922.

Ahlawat, Y.S., Varma, A., Chakraborty, N.K. and Srivastava, K.P. 1995. Biotypes and vectoral capacity of *Diaphorina citri* in transmitting citrus greening BLO. pp. 222–232 *In*: Proc. 2nd Agric. Sci. Cong. (*Eds.*) Swaminathan, M.S., Randhawa, N.S., Ananthakrishanan, T.N. and Narain, P. Nat. Acad. Agric. Sci., IARI, New Delhi-110012.

Ahlawat, Y.S., Baranwal, V.K., Thinlay, Doe Doe, and Majumdar, S. 2003. First report of citrus greening disease and associated bacterium Candidatus Liberobacter asiaticus from Bhutan. *Plant Disease* 87: 448.

Ahlawat, Y.S. 2010. Diagnosis of plant viruses and allied pathogens. Studium Press (India) Pvt. Ltd., New Delhi.

Ahlawat, Y.S. and Pant, R.P. 2010. pp. 71–94. *In*: IPM system in Agriculture. Virus and virus-like diseases of plants Vol. 10. (*Eds.*). Upadhyay, R.K., Ahlawat, Y.S., Dubey, O.P., Baranwal, V.K. and Srivastava, K.P. Aditya Books Pvt. Ltd. New Delhi. p. 381.

Arif, M., Ahmad, A., Ibrahim, M. and Hassan, S. 2005. Occurrence and distribution of virus and virus like diseases of citrus in North-West Frontier province of Pakistan. *Pak. J. Bot.* 37: 407–421.

Asana, R.D. 1958. The citrus dieback problem in relation to cultivation of citrus fruits in India. *Indian Journal of Horticulture* 15: 283–286.

Aubert, B. and Bové, J.M. 1980. Effect of penicillin or tetracycline injections of citrus trees affected by greening disease under field conditions in Reunion island. pp. 103–108. *In*: Proc. 8th Conf. IOCV, IOCV, Riverside, CA.

Aubert, B. and Quilici, S. 1984. Biological control of the African and asian citrus psyllids (Homoptera: Psylloidea), through eulophid and Encyrtid parasites (Hymenoptera: Chalcidoidea) in Reunion Island. pp. 100–108 *In*: Proc. 9th Conf. IOCV, IOCV, Riverside, CA.

Aubert, B., Sabine, A. and Picard, P. 1984. Epidemiology of the greening disease in Reunion Island before and after the biological control of the African and Asian citrus psyllas. *Proc. Inernl. Soc. Citriculture* 1: 440–442.

Baranwal,V.K., Majumder, S., Ahlawat, Y.S. and Singh, R.P. 2003. Sodium sulphite yields improved DNA of higher stability for PCR detection of Citrus Yellow Mosaic Virus from citrus leaves in *Journal of Virological Method* 112: 153–156.

Baranwal, V.K., Mazumder, S., Singh, Jitendra, Suryanarayana, V., Ghosh, D.K. and Ahlawat, Y.S. 2004. PCR detection of Candidatus Liberibacter asiaticus, the agent of Huanglongbing disease in citrus. *Indian Phytopath* 57: 164–168.

Baranwal, V.K., Mazumder, S., Ahlawat, Y.S. and Singh, R.P. 2005. A novel approach for simultaneous detection of citrus yellow mosaic virus and citrus greening bacterium by multiplex polymerase chain reaction. *Indian Journal of Biochemistry* 4: 528–533.

Baranwal, V.K. and Arya, M. 2009. Novel approaches of nucleic acid template preparation for the detection of RNA and DNA viruses and greening bacterium in citrus by RT-PCR and PCR. P. 47 *In*: 5th Housing Internl. Conf. on Plant Path. *In*: Globalized era. Nov. 10–13, 2009, New Delhi, India.

Beretta, M.J.G., Lee, R.F., Derrick, K.S., Davis, C.L. and Barthe, G.A. 1992. Culture and serology of a *Xylella fastidiosa* associated with citrus variegated chlorosis in Brazil. *In*: Proc. 7th Intl. citrus Cong. Italy.

Beretta, M.J.G., Lee, R.F., Barthe, G.A., Thome Neto, J., Derrick, K.S. and Davis, C.L. 1993. Citrus variegated chlorosis: detection of *Xylella fastidiosa* in symptomless trees. pp. 306–311 *In*: Proc. 12th Conf. IOCV, IOCV, Riverside, CA.

Beretta, M.J.G., Garcia, A. Jr., Lee, R.F., Derrick, K.S., Barthe, G. and Thome Neto, J. 1993. Observation on citrus variegated chlorosis in Brazil. pp. 291–293. *In*: Proc. 12th Conf. IOCV, IOCV, Riverside, CA.

Beretta, M.J.G., Rodas, V., Garcia Jr. A., and Derrick, K.S. 1996. Control of citrus Variegated Chorosis by pruning. pp. 378–379 *In*: 13th Conf. IOCV, IOCV, Riverside, CA.

Beretta, M.J.G., Barthe, G.A., Ceccardi, T.L., Lee, R.F. and Derrick, K.S. 1997. A survey for strains of *Xylella fastidiosa* in citrus affected by citrus variegated chlorosis and citrus blight in Brazil. *Plant Dis.* 81: 1196–1198.

Bhagabati, K.N., Ahlawat, Y.S., Chakraborty, N.K. and Borthakur, B.C. 1989. Distribution of greening, tristeza and mosaic diseases of citrus in Northeastern States of India. *Indian Phytopath.* 42: 552–555.

Bové, J.M. 2006. Huanglongbing: a destructive, newly-emerging, century-old disease of citrus. *J. Plant Pathology* 88: 7-37.

Bové, J.M. and Saglio, P. 1974. Stubborn and greening: a review, 1969–1972. pp. 1–11 *In*: Proc. 6th Conf. IOCV, IOCV, Riverside, CA

Bové, J.M., Bonnet P., Garnier M. and Aubert B. 1980. Penicillin and tetracycline treatments of greening disease-affected citrus plants in the glasshouse, and the bacterial nature of the procaryote associated with greening. pp. 91–102. *In*: Proc. 8th Conf. IOCV, IOCV, Univ. Calif., Riverside, CA.

Bove, J.M., Garnier, M., Ahlawat, Y.S., Chakraborty, N.K. and Varma, A. 1993. Detection of Asian strains of the greening BLO by DNA-DNA hybridization in Indian orchard trees and Malaysian *Diaphorina citri* psyllids. pp. 258–263 *In*: Proc. 12th Conf. IOCV, IOCV, Riverside, CA.

Brlansky, R.H., Davis, C.L., Timmer, L.W., Howd, D.S., and Contreras, J. 1991. Xylem limited bacteria in citrus from Argentina with symptoms of citrus variegated chlorosis. *Phytopathology* 81: 1210 (Abst.).

Brlansky, R.H., Davis, C.L., Lee, R.F. and Timmer, L.W. 1993. Immunogold localization of xylem-inhabiting bacteria affecting citrus in Argentina and Brazil. pp. 311–319 *In*: Proc. 12th Conf. IOCV, IOCV, Riverside, CA.

Capoor, S.P. 1963. Decline of citrus trees in India. Symp. on plant and animal viruses. *Nat. Inst. Sci. India Bull.* 24: 48–64.

Capoor, S.P., Rao, D.G. and Viswanath, S.M. 1967. *Diaphorina citri* Kuway., a vector of greening disease of citrus in India. *Indian J. Agric. Sci.* 37: 572–576.

Capoor, S.P. and Thirumalachar, M.J. 1973.Cure of greening-affected citrus plants by chemotherapeutic agents. *Plant Dis. Reptr.* 57: 160–163.

Capoor, S.P., Thirumalachar, M.J., Pandey, P.K. and Chakraborty, N.K. 1973. Control of the Greening Disease of Citrus by B.P.-101:A New Chemotherapeutant. pp. 50–52 *In*: Proc. 6th IOCV, IOCV, Riverside, CA.

Capoor, S.P., Rao, D.G. and Viswanath, S.M. 1974. Greening disease of citrus in the Deccan Trap country and its relationship with the vector, *Diaphorina citri* Kuwayama. pp. 43–49. *In*: Proc. 6th Conf: IOCV, IOCV, Riverside, CA.

Cardinali, M.C.B., Venancio, A.L., Kubota, T.M.K., Debora, M., Milori, B.P., Ferreira, F.J., Martins, P.K. and Freitas-Astua 2010. Evaulation of early diagnosis of citrus greening using laser-induced fluorescence spectroscopy. p. 130. *In*: Abs. 18th Conf. IOCV Campinas, Brazil, 7–12 November, 2010.

Cavarrubias, R., Oswaldo A., Almeyda-Leon, I.H., Lobato-Sanchez, R. and Cadena-Hinojosa, M.A. 2010. Relationship among bactericera cockerlli Sulc population, abnormal tuber sprouting, internal tuber discoration and the presence of *Candidatus* Liberibacter psyllaurous in potato fields grown in Toluca, Mexico. *American Journal of Potato Research* 87: 140–141.

Celino, C.S., Salibe, A.A. and Cortez, R.E. 1966. *Diaphorina citri* Kuway., the insect vector for the leaf mottle virus of citrus in the Philippines. Manila, Bureau of Plant Industry.

Chavan, V.M. and Summanwar, A.S. 1993. Population dynamics and aspects of the biology of citrus psylla, *Diaphorina citri* Kuw. In Maharashtra, pp. 286–290 *In*: Proc. 12th Conf. IOCV, IOCV, Riverside, CA.

Chang, C.J., Garnier, M., Zreik, L., Rossetti, V. and Bove, J.M. 1993. Culture and serological detection of the xylem limited bacterium causing citrus variegated chlorosis and its identification as a strain of *Xylella fastidiosa*. *Curr. Microbiol.* 27: 137–142.

Chagas, C.M., Rossetti, V. and Beretta, M.J.G. 1992. Electron microscopy studies of xylem limited bacterium in sweet orange affected with citrus variegated chlorosis disease in Brazil. *J. Phytopathology* 134: 306–312.

Cheema, G.S. and Bhat, S.C. 1929. The dieback disease of citrus trees and its relation to the soils of western india. *Bull. Bombay Dept. Agric.* 155: 1–48.

Chung, K.R. and Brlansky, R.H. 2009. Citrus diseases exotic to Florida: Citrus variegated chlorosis. Fact Sheet p. 223, one of a series of the Plant Pathology Department, Florida Cooperative Extension Service, Institute of Food and Agricultural Sciences, University of Florida.

Chen, J., Pu, X., Deng, X., Liu, S., Li, H. and Civerolo, E. 2008. A phytoplasma related to `*Candidatus* Phytoplasma asteris' detected in citrus showing huanglongbing (yellow shoot diseases) symptoms in Guangdong, P.R. China. *Phytopathology* 99: 236–242.

Cochran, L.C. 1976. The occurrence of greening disease in Pakistan. p. 21. *In*: Proc. 7th Conf. IOCV, IOCV. Riverside, CA.

Coletta-Filho, H.D., Targon, M.L.P.N., Takita, M.A., De Negri Jr, J.D., Pompeu, J., Machado, M.A. 2004. First report of the Causal Agent of Huanglongbing ("*Candidatus* Liberibacter asiaticus") in Brazil. *Plant Disease* 88: 1382.

Cordero, C., Luis, M., Llauger,R., Teixeira, D.C., Eveillard, S., Batista, L., Pena, I., Lopez, D., Perez, D., Cueto, J.R. and Bove, J.M. 2010. Huanglongbing in Cuba, diagnostics, characterization and management. p. 30. *In*: Abstract 18th Conf. IOCV held in Campinas/SP, Brazil, Nov. 7–12, 2010.

da Graça, J.V. 1991. Citrus greening disease. *Ann. Rev. Phytopathol.* 29: 109–36.

da Graça, J.V., and Korsten, L. 2004. Citrus huanglongbing: review, present status and future strategies. *In*: Naqvi, S.A.M.H. (*Ed*.). Diseases of Fruits and Vegetables, pp. 229–245. Kluwer Academic Publishers, Dordrecht, The Netherlands.

Das, A.K. 2004. Rapid detection of *Candidatus* Liberibacter asiaticus, the bacterium associated with citrus huanglongbing (greening) disease using PCR. *Curr. Sci.* 87: 1183–1185.

Das, A.K., Rao, C.N. and Singh, S. 2007. Presence of citrus greening (Huanglongbing) disease and its vector in the Northeastern region of India confirmed by PCR technique. *Curr. Sci.* 92: 1759–1763.

Das, A.K. 2009. Molecular identification and characterization of citrus greening bacterium, *Candidatus* Liberibacter asiaticus associated with decline of Nagpur mandarin orange in Vidarbha region, Maharashtra. *Curr. Sci.* 96: 890–892.

Damsteegt, V.D., Postnikova, E.N., Stone, A.L., Kuhlmann, M., Wilson, C., Sechler, A., Schaad, N.W., Brlansky, R.H., and Schneider, W.L. 2010. *Murraya paniculata* and related species as potential hosts and inoculum reservoirs of '*Candidatus* Liberibacter asiaticus', causal agent of Huanglongbing. *Plant Dis.* 94: 528–533.

Davis, R., Gunua, T., Bellis, G. and Grimshaw, J. 2005. Huanglongbing disease of citrus trees. Pest advisory leaflet no. 45. Plant Protection Service. Secretariat of the Pacific Community. ISSN 1017–6276 1017–6276

Davis, M.J., Mondal, S.N., Chen, H., Rogers, M.E. and Brlansky, R.H. 2008. Co-cultivation of "*Candidatus* Liberibacter asiaticus? with Actinobacteria from citrus with huanglongbing. *Plant Dis.* 92: 1547–1550.

de Coll, O.R., Remes Lenicov, A.M.M., Agostini, J.P. and Paradell, S. 2000. Detection of *Xylella fastidiosa* in weeds and sharpshooters in orange groves affected with citrus variegated chlorosis in Misiones, Argentina. pp. 216–222. *In*: Proc. 12th Conf. IOCV, IOCV, Riverside, CA.

Deng, X., Zhou, G., Li, H., Chen, J. and Civerrolo, E.L. 2007 Detection of *Candidatus* Liberibacter asiaticus from Wampee (Clausena lansium Skeels) by nested PCR. Online Plant Healthy Progress doi:10.1094/PHP-2007-0419-01-BR.

do Nascimento, F.E., Coletta-filho, H.D., Tomaseto, A.A., de Olinda, R.A. and Lopes, J.R.S. 2010. Effect of developmental stage of *Diaphorina citri* on acquisition of *Candidatus* Liberibacter asiaticus and Ca. L. Americanus. p. 39. *In*: Abstract 18th Conf. IOCV held in Campinas/SP, Brazil, November 7–12, 2010.

Ding, F., Wang, G., Yi, G., Zhong, Y., Zeng, J. and Zhou, B. 2006. Infection of Wampee and lemon by the citrus huanglongbing pathogen (*Candidatus* Liberibacter asiaticus) in *China. J. Plant Pathol.* 87: 207–212.

Duan, Y.P., Gottwald, T., Zhou, L.J. and Gabriel, D.W. 2008a. First report of dodder transmission of *Candidatus* Liberibacter asiaticus to tomato (*Lycopersicon esculentum*). *Plant Disease* 92: 831.

Duan, Y.P., Zhou, L.J., Hall, D., Li, W.B., Liu, L. and Gottwald, T. 2008b. Genome sequencing of *Candidatus* Liberibacter asiaticus, the causal agent of citrus huanglongbing (greening) in Florida. Program and Abstracts Book, Int. Congress Citrus, Wuhan, China, October 2008, abstract 87: 52.

Duan, Y.P., Zhou, L., Hall, D.G., Li, W., Doddapaneni, H., Lin, H., Liu, L., Gabriel, D.W., Vahling, C.M., Williams, K., Dickerman, A. and Gottwald, T. 2009. Complete genome sequence of citrus huanglongbing bacterium, *Candidatus* Liberibacter asiaticus? obtained through metagenomics. *Mol. Plant-Microbe Interact.* 22: 1011–1020.

Duan, Y.P.,Vahling, C.M. and Zhou, L.J. 2010. Genomics and functional genomics of citrus huanglongbing bacterium '*Candidatus* Liberibacter asiaticus. p. 18. *In*: Abs. 18th Conf. IOCV held in Campinas/S.P., brazil, Nov. 7–12, 2010.

Etienne, J. and Aubert, B. 1980. Biological control of psyllid vectors of greening disease on Reunion Island. pp. 118–121 *In*: Proc. 8th Conf. IOCV, IOCV, Riverside, CA.

Fawcett, H.S. and Klotz, L.J. 1948. Diseases and their control. pp. 495–496 *In*: Batchelor, L.D. and Webber, H.J. (*Eds.*), The Citrus Industry, Vol. 2, Univ. California Press, Berkeley.

Francis, M., Lin, H., Pena, A., Graham, J.H. and Civerolo, E.L. 2010. SYBR® green for specific detection and quantification of citrus variegated chlorosis strains of *Xylella fastidiosa*. p. 116. *In*: Abs. 18th Conf. IOCV held in Campinas/S.P., Brazil, Nov. 7–12, 2010.

Fraser, L.R. and Singh, D. 1966. Greening virus – a new threat to citrus Industry. *Indian Hort.* 10: 21–22.

Fraser, L.R., Singh, D., Capoor, S.P. and Nariani, T.K. 1966. Greening virus, the likely cause of citrus dieback in India. *FAO Plant Prot. Bull.* 14: 127–130.

Gao, S., Garnier, M. and Bové, J.M. 1993. Production of monoclonal antibodies recognizing most strains of the greening BLO by *in vitro* immunization with an antigenic protein purified from the BLO. pp. 244–249 *In*: Proc. 12th Conf. IOCV, IOCV, Univ. Calif., Riverside, CA.

Garnier, M. and Bové, J.M. 1983. Transmission of the organism associated with citrus greening disease from sweet orange to periwinkle by dodder. *Phytopathology* 73: 1358–1363.

Garnier, M., Danel, N. and Bové, J.M. 1984a. Aetiology of citrus greening disease. *Ann. Microbiology.* (Inst. Pasteur) 135A: 169–179.

Garnier, M., Danel, N. and Bové, J.M. 1984b. The greening organism is a Gram-negative bacterium. pp. 115–125. *In*: Proc. 9th Conf. IOCV, IOCV, Riverside, Univ. CA.

Garnier, M., Martin-Gros, G. and Bove, J.M. 1987. Monoclonal antibodies against the bacterial-like organism associated with citrus greening disease. *Annales del Institut Pasteur/Microbiologie* 138: 639–659.

Garnier, M., Gao, S.J., He, Y., Villechanoux, S., Gandar, J. and Bové, J.M. 1991. Study of the greening organism (GO) with monoclonal antibodies: serological identification, morphology serotypes and purification of the GO. pp. 428–435. *In*: Proc. 11th Conf. IOCV, IOCV, Riverside, CA.

Garnier, M., Villechanoux, S., Jagoueix, S. and Bove, J.M. 1993. Nature of greening bacterium–like organism (BLO): taxonomic characterization by use of cloned DNA fragments. pp. 250–257. *In*: Proc. 12th Conf. IOCV, IOCV, Riversised, CA.

Garnier, M. and Bove, J.M. 1993. Citrus greening disease and the greening bacterium. pp. 212–219 *In*: Proc. 12th Conf. IOCV, IOCV, Riverside, CA.

Garnier, M., Chang, C.J., Zreik, L., Rossetti, V. and Bove, J.M. 1993. Citrus variegated chlorosis:serological detection of *Xylella fastidiosa*, the bacterium associated with the disease. pp. 301–305. *In*: Proc. 12th Conf. IOCV, IOCV, Riverside, CA.

Garnier, M. and Bové, J.M. 1996. Distribution of the huanglongbing (greening) liberobacter species in fifteen African and Asian countries. pp. 388–391. *In*: Proc. 13th Conf. IOCV, IOCV, Riverside, CA.

Garnier, M., Jagoueix-Eveillard, S., Cronje, P., Le Roux, H. and Bové, J.M. 2000. Genomic characterisation of a liberibacter present in an ornamental Rutaceous tree, *Calodendrum capense*, in the Western Cape province of South Africa. Proposal for a "*Candidatus* Liberibacter africanus subsp. capensis". *Int. J. Sys. Evol. Microbiol.* 50: 2119–2125.

Garnett, H.M. 1984. Isolation of the greening organism. *Citrus Subtrop. Fruit J.* 611: 4–5.

Grevena, S.J., Lopes, R.S., Paiva, P.E.B., Yamamoto, P.T. and Roberto, S.R. 1998. The *Xylella fastidiosa* vectors. *In*: Citrus variegated chlorosis, 36–53, Bebedouro-SP.

Ghosh, S.K., Raychaudhuri, S.P., Varma, A. and Nariani, T.K. 1971. Isolation and culture of mycoplasma associated with citrus greening disease. *Curr. Sci.* 40: 299–300.

Ghosh, S.K., Raychaudhuri, S.P., Chenulu, V.V. and Varma, A. 1975. Isolation, cultivation and characterization of mycoplasma-like organisms from plants. *Proc. Indian Nat. Sci. Acad. Ser. B.* 41: 334–339.

Ghosh, D.K., Das, A.K. and Singh, S. 2003. Individual and mixed infection of *Citrus tristeza* virus and greening Liberobacter and their association with sweet orange decline in the state of Maharashtra. *J. Mycol. Pl. Pathol.* 33: 69–72.

Godfrey, P., Miles, Marcus, Samuel, A., Chen, Jianchi, Civerolo, Edwin L. and Munyaneza, Joseph E. 2010. Evidence that cell death is associated with zebra chip disease in potato tubers. *Am. J. Pot Res* 87: 337–349.

Golmohammadi, M., Aliabadi, A.A., Bertolini, E. Llop, P. and Lopez, MM. 2010. Current situation and impact of huaglongbing (HLB) caused by *Candidatus* Liberibacter asiaticus in the South of Iran. pp. 75 *In*: Abs. 18th Conf. IOCV held in Campinas/SP, Brazil, Nov. 7–12, 2010.

Gopal, K., Pradeepthi, E.R., Gopi, V., Khayum Ahammed, S., Sreenivasulu, Y., Reddy, M.K., Baranwal, V.K. and Purushotham, K. 2007. Occurrence, molecular diagnosis and suitable time of detection of citrus greening disease in sweet orange. *Acta Phytopathologica et Entomologica Hungarica* 42: 49–58.

Gottwald, T.R., da Graça, J.V. and Bassanezi, R.B. 2007. Citrus huanglongbing: The pathogen, its epidemiology, and impact. Plant Healthy Progress doi:10.1094/PHP-2007-0906-01-RV.

Gottwald, T.R., Hall, D.G., Beattie, G.A.C., Ichinose, K., Nguyen, M.C., Le, Q.D., Bar-Joseph, M., Lapointe, S., Stover, E., Parker, P.E., McCollum, G. and Hilf, M.E. 2010. Investigations of the effect of guava as a possible tool in the control/ management of huanglongbing. Proc. 17th Conf. IOCV, IOCV, Riverside, CA.

Gottwald, T.R., Irey, M., Gast, T., Parnell, S., Taylor, E., Hilf, M.E. 2010. Spatio-temporal analysis of an HLB epidemic in Florida and implications for future spread. *In*: Proc. 17th Conf. IOCV, IOCV, Riverside, CA.

Gottwald, T., Irey, M., Graham, J. and Wood, B. 2010. Inconsequential effect of nutritional treatments to control huanglongbing. pp. 33. *In*: Abstract 18th Conf. IOCV held in Campinas/SP, Nov. 7–12, 2010.

Gouda, K.A., Baranwal, V.K. and Ahlawat, Y.S. 2006. Simplified DNA extraction and improved PCR based detection of Citrus greening bacterium. *J. Plant Biochemistry and Biotechnology* 15: 117–121.

Gupta, O.P., Nauriyal, J.P. and Knorr, L.C. 1972. Citrus greening in the Indian Punjab. pp. 14–18. *In*: Proc. 5th Conf: IOCV, IOCV, Riverside, CA.

Halbert, S.E. 2005. The discovery of huanglongbing in Florida. Proceedings of 2nd International Citrus Canker and Huanglongbing Research Workshop, Florida Citrus Mutual, Orlando 2005, H-3.

Halbert, S.E., Niblett, C.L., Manjunath, K.L., Lee, R.F. and Brown, L.G. 2003. Establishment of two new vectors of citrus pathogens in Florida. *Proc. Interl. Soc. Citriculture* 9: 1016–1017.

Halbert, S.E. and Manjunath, K.L. 2004. Asian citrus psyllids (Sternorrhycha: Psyllidae) and greening disease of citrus: a literature review and assessmentof risk in Florida. *Florida Entomologist* 87: 330–53.

Halbert S.E., Manjunath, K.L. and Brodie, M.W. 2008. Large-scale distribution of *Diaphorina citri* Kuwayama and citrus Huanglongbing in Florida. IRCHLB Proceedings Dec. 2008: www.plantmanagementnetwork.org

Hansen, A.K., Trumble, J.T., Stouthamer, R. and Paine, T.D. 2008. A new huanglongbing species, "*Candidatus* Liberibacter psyllaurous", found to infect tomato and potato, is vectored by the psyllid Bactericera cockerelli (Sulc.). *Applied and Environmental Microbiology* 74: 5862–5865.

Hartung, J.S., Paul, C., Achor, D. and Brlansky, R.H. 2010. Colonization of dodder, *Cuscuta indecora*, by '*Candidatus* Liberibacter asiaticus' and 'Ca. L. americanus'. *Phytopathology* 100: 756–762.

Hartung, J.S., Halbert, S.E., Pelz-Stelinski, K., Brlansky, R.H., Chen, C. and Gmitter, F.G. 2010. Lack of evidence for transmission of '*Candidatus* Liberibacter asiaticus' through citrus seed taken from affected fruit. *Plant Dis.* 94: 1200–1205.

Heather, L. Tyler, F.W. Roesch, Luiz, Gowda, Siddarame, Dawson, William, O. and Triplett, Eric W. 2009. Confirmation of the sequence of '*Candidatus* Liberibacter asiaticus' and assessment of microbial diversity in huanglongbing-infected citrus phloem using a metagenomic approach. *Molecular Plant Microbe Interactions* 22: 1624–1634.

He, C.X., Li, W.B., Ayres, A.J., Hartung, J.S., Miranda, V.S. and Teixeira, D.C. 2000. Distribution of *Xylella fastidiosa* in citrus rootstocks and transmission of citrus variegated chlorosis between sweet orange plants through natural root grafts. *Plant Dis*. 84: 622–626.

Hocquellet, A., Toorawa, P., Bové, J.M. and Garnier, M. 1999. Detection and identification of the two "*Candidatus* Liberobacter sp." Associated with citrus huanglongbing by PCR amplification of ribosomal protein genes of the beta operon. *Mol. Cell. Probes* 13: 373–379.

Hung, T.H., Wu, M.L. and Su, H.J. 2000. Identification of alternative hosts of the fastidious bacterium causing citrus greening disease. *Journal of Phytopathology* 148: 231.

Hung, T.H., Hung, S.C., Chen, C.N., Hsu, M.H. and Su, H.J. 2004. Detection by PCR of *Candidatus* Liberibacter asiaticus, the bacterium causing citrus greening in vector psyllids: application to the study of vector-pathogen relationship. *Plant Pathology* 53: 96–102.

Jagoueix, S., Bové, J.M. and Garnier, M. 1994. The phloem-limited bacterium of greening disease of citrus is a member of a subdivision of the Proteobacteria. *International Journal of Systematic Bacteriology* 44: 397–86.

Jagouiex, S., Bove, J.M. and Garnier, M. 1996. PCR detection of two 'Candidus' Liberobacter species associated with greening disease of citrus. *Molecular and Cellular Probes* 10: 43–50.

Ke, C., Chen, H. and Lin, S.C. 1980. Study on transmission of rickettsia-like organism associated with citrus yellow shoot by psylla. *Fulian Agric. Sci. Tech*., 10: 10–11.

Ke, C., Ke, S., Wu, R.J., Yang, H. and Hsu, P.T. 1993. Purification and serology of the organism associated with citus huanlungbin. pp. 220–223 *In*: Proc. 12th Conf. IOCV, IOCV, Riversised, CA.

Kitajima, E.W. 2010. Citrus viruses and bud transmitted bacteria in Brazil: History and perspectives. pp. 2. *In*: Abs. 18th Conf. IOCV held at Campinas/SP, Brazil 7–12 Nov. 2010.

Laflèche, D. and Bové, J.M. 1970a. Structures de type mycoplasme dans les feuilles d'orangers atteints de la maladie du greening. *C.R. Acad. Sci., Ser. D*, 270: 1915–1917.

Lafleche, D. and Bove, J.M. 1970b. Mycoplasmes dans les argumes attaints de "greening"des stubborn, ou des maladies similaries. *Fruits* 25: 455–465.

Lallemand, J., Fos, A. and Bové, J.M. 1986. Transmission de la bactèrie associée à la forme africaine de la maladie du "Greening" par le psylle asiatique *Diaphorina citri* Kuwayama. *Fruits*, 41: 341–343.

Lee, R.F., Derrick, K.S., Beretta, M.J.G., Chagas, C.M. and Rossetti, V. 1991. Citrus variegated chlorosis: A new destructive disease of citrus in Brazil. *Citrus Ind*. 72(10): 12–13.

Lee, R.F., Beretta, M.J.G., Hartung, J.S., Hooker, M.E. and Derrick, K.S. 1993. Citrus variegated chlorosis: Confirmation of a *Xylella fastidiosa* as the causal agent. *Summa Phytopathol*. 19: 123–125.

Li, W.B., Pria, W.D., Jr., Teixera, D.C., Miranda, V.S., Ayres, A.J., Franco, C.F., Costa, M.G., He, C.X., Costa, P.I. and Hartung, J.S. 2001. Coffee leaf scorch caused by a strain of *Xylella fastidiosa* from citrus. *Plant Disease* 85(5): 501–505.

Li, W., Zhou, C., Pria, Jr., W.D., Teixera, D.C., Miranda, V.S., Pereira, A.J., He, C.-X., Costa, P.I. and Hartung, J.S. 2002. Citrus and coffee strains of *Xylella fastidiosa* induce Pierce's disease of grapevine. *Plant Disease* 86: 1206–1210.

Li, W., Hartung, J.S. and Levy, L. 2006. Quantitative real time PCR for detection and identification of *Candidatus* Liberibacter species associated with citrus huanglongbing. *J. Microbiol. Methods* 66: 104–115.

Li, W., Hartung, J.S. and Levy, L. 2007. Evaluation of DNA amplification methods for improved detection of "*Candidatus* Liberibacter species" associated with citrus huanglongbing. *Plant Disease* 91: 51–58.

Liefting, L.W., Weir, B.S., Pennycook, S.R. and Clover, G.R.G. 2009. '*Candidatus* Liberibacter solanacearum', a liberibacter associated with plants in the family Solanaceae. *Int. J. Syst. Evol. Microbiol.* 59: 2274–2276.

Liefting, L.W., Perez-Egusquiza, Z.C., Clover, G.R.G. and Anderson, J.A.D. 2008. A New *Candidatus* Liberibacter' species in *Solanum tuberosum* in New Zealand. *Plant Dis.* 92: 1474–1474.

Lin, K.H. 1944. A report of field observations on citrus diseases in Fujian and Jiangxi provinces. *Bull. Citrus Res. Lingnan Univ.* 7: 1–11.

Lin, K.H. 1956. Observation on yellow shoot of citrus. Etiological study of yellow shoot of citrus. *Acta Phytopathol Sinica* 2: 1–42.

Lin, Kung Hsiang.1956 a. Yellow shoot of citrus (in Chinese). *Acta Phytopathologica Sinica* 2: 1–12.

Lin, Kung Hsiang 1956b. Etiological studies of yellow shoot of citrus (in Chinese). 42. *Acta Phytopathologica Sinica* 2: 12–42.

Lin, Hong, Doddapaneni, H., Munyaneza, J., Civerolo, E., Venkatesan, S., Buchman, J. and Stenger, D. 2009. Molecular characterization and phylogenetic analysis of 16S rRNA from "*Candidatus* Liberibacter" strains associated with Zebra chip disease of potato (*Solanum tuberosum* L.) and the potato psyllid (*Bactericera cockerelli* Sulc). *J. Plant Path.* 91: 215–219.

Lopes, S.A., Martins, E.C. and Frare, G.F. 2005. Deteção de *Candidatus* Liberibacter americanus em *Murraya paniculata. Summa Phytopathologica* 31: 48–49.

Lopes, S.A., Martins, E.C. and Frare, G.F. 2006. Deteção de *Candidatus* Liberibacter asiaticus em *Murraya paniculata. Fitopatologia Brasileira* 31: 303.

Lopes, A.S., Frare, G.F., Sausa, M.C., Fernandes, N.G. and Ayres, A.J. 2010. A new strategy to control citrus variegated chlorosis disease. p. 13. *In*: Abs. 18th Conf. IOCV held in Campinas/S.P., Brazil, Nov. 7-12, 2010.

Lou, B.H., Zhou, C.Y., Zhao, X.Y., Li, Z.G., Xu, M., Liu, J.X., Zhou, Y. and Tang, K.S. 2008. Primary study on species and intraspecific differentiations of HLB pathogens in eight provinces of China. Program and Abstracts book, 11th international citrus congress, p. 232, abstract p. 333.

Lugue-Williams, M.J. and Polek, M. 2010. The state of California action plan for huanglongbing and its vectors: current status of the ACP detection and eradiction program. p. 36. *In*: Abstract 18th Conf. IOCV, held in Campinas/SP, Brazil, Nov. 7–12, 2010.

Luis Pantoja, M., Collazo Cordero, C., Llauger Riveron, R., Peña Barzaga, L., Lopez Hernandez, D., Blanco, E., Casi, J.C., Batista Le Riverend, L., Tanaka, F.A.O., Salaroli, R.B., Martins, E., Teixeira, D., Kitajima, E., Ayres, J., Bové, J.M., Perez, J.L. and Cueto, J.R. 2008. Huanglongbing in Cuban Citriculture. Book of Program and Abstracts, 11th International Citrus Congress, Wuhan, China, 26–30 Oct. 2008, Abstract # 84, pp. 50–51.

Manjunath, K.L., Halbert, S., Ramadugu, C., Webb, S. and Lee, R.F. 2008. Detection of *Candidatus* Liberibacter asiaticus in *Diaphorina citri* and its importance in the management of citrus huanglongbing disease. *Phytopathology* 98: 387–396.

Martinez, A.L. and Wallace, J.M. 1967. Citrus leaf-mottle yellows disease in the Philippines and transmission of the causal virus by a psyllid: *Diaphorina citri*. *Plant Dis. Rep.*, 51: 692–695.

Marques, R.N., Teixeira, D.C., Martins, E.C., Leite, A.P.R., Yamamoto, P.T., Sanches, A.L. and Lopes, J.R.S. 2010. Detection of the Phytoplasma associated with citrus huanglongbing symptoms in the leafhopper, *Scaphytopius marginelineatus*. p. 40. *In*: Asb. 18th Conf. IOCV held in Campinas/SP, November 7–12, 2010.

Massonie, G., Garnier, M. and Bové, J.M. 1976. Transmission of India citrus decline by *Trioza erytreae* (Del Guercio), the vector of South Africa greening. pp. 18–20. *In*: Proc. 7th Conf. IOCV, IOCV, Riverside, CA.

McClean, A.P.D. and Oberholzer, P.C.J. 1965a. Greening disease of sweet orange: evidence that it is caused by a transmissible virus. *South Africa Journal of Agricultural Science* 8: 253–276.

McLean, A.P.D. and Oberholzer, P.C.J. 1965b. Citrus psylla, a vector of the greening disease of sweet orange. *S. African J. Agric. Sci.* 8: 297–298.

Munyaneza, J.E., Fisher, T.W., Sengoda, V.G., Garcznyski, S.F., Nissinen , A. and Lemmetty, A. 2010. First report of "*Candidatus* Liberibacter solanacearum" associated with psyllid-affected carrots in Europe. *Plant Disease* 94: 639.

Murray, R.G.E. and Stackebrandt, E. 1995. Taxonomic note: Implementation of the provisional status Candidus for incompletely described Prokaryotes. *Intl. J. Syst. Bacteriol.* 47: 240–241.

Naidu, R. and Govindu, H.C. 1981. Association of bacterium-like organism with citrus greening in India. *Curr. Sci.*, 50: 242–244.

Nariani, T.K., Raychaudhuri, S.P. and Bhalla, R.B. 1967. Greening virus of citrus in India. *Indian Phytopathol.* 20: 146–150.

Nariani, T.K. and Raychaudhuri, S.P. 1968. Occurrence of tristeza and greening viruses in Bihar, West Bengal and Sikkim. *Indian Phytopathol.* 21: 343–344.

Nariani, T.K., Viswanath, S.M. and Menon, M.R. 1970. Occurrence of tristeza and greening diseases of citrus in Kerala. *Agric. Res. J. Kerala*, 8: 123–124.

Nariani, T.K., Ghosh, S.K., Kumar, D., Raychaudhuri, S.P. and Viswanath, S.M. 1975. Detection and possibilities of therapeutic control of the greening disease of citrus caused by mycoplasma. *Proc. Indian Nat. Sci. Acad. Ser. B.* 41: 362–366.

Nariani, T.K. and Bhagabati, K.N. 1979. Studies on the therapeutic and insect vector control of the greening disease of citrus in India. Proc. 8th Conf. IOCV, IOCV, Riverside.

Noronha Jr, N.C., Bento, J.M.S. and Parra, J.R.P. 2008. Guava and citrus plant volatiles. Fundecitrus workshop on HLB-resistant citrus. Araraquara, February 2008.

Phahladira, M.N.B. 2008. Identification of alternative hosts to citrus of "*Candidatus* Liberibacter africanus" amongst indigenous Rutaceae of South Africa. MSc. study., University of Pretoria, Pretoria, South Africa.

Planet, P., Jagoueix, S., Bové, J.M. and Garnier, M. 1995. Detection and characterization of the African citrus greening Liberobacter by amplification, cloning and sequencing of the rplKAJL-rpoBC operon. *Current Microbiology* 30: 137–141.

Pooler, M.R. and Hartung, J.S. 1995. Specific PCR detection and identification of *Xylella fastidiosa* strains causing citrus variegated chlorosis. *Curr. Microbiol.* 31: 377–381.

Purcell, A.H. and Hopkins, D.L. 1996. Fastidious xylem-limited bacterial plant pathogens. *Annu. Rev. Phytopathol.* 34: 131–151.

Qin, X., Miranda,V.S., Machado, M. Lemos and Eand Hartung, J.S. 2001. An evaluation of the genetic diversity of *Xylella fastidiosa* isolated from diseased citrus and coffee. *Phytopathology* 91: 599–605.

Raychaudburi, S.P., Nariani, T.K., Lele, V.C. and Singh, G.R. 1972. Greening and citrus Decline in India. p. 35-37. *In*: Proc. 5th Conf: IOCV, IOCV, Riverside, CA.

Raychaudhuri, S.P., Nariani, T.K. and Ahlawat, Y.S. 1977. Dieback of citrus in India. *Proc. Interl. Soc. Citriculture* 3: 914–918.

Reinking, O.A. 1919. Diseases of economic plants in south China. *Philippine Agr*. 8: 109–135.

Rodrigues Neto, J., Berian, L.O.S., Baptista, C.R. and Oliveira, A.R. 1992. Seed transmission of an organism associated to CVC-V. RAIB-23a 27/11/92 Instituto Biologico, Arq., do Instituro Biologico (Suplemento).

Rosa, C., Wuriyanghan, Ha and Falk, B.C. 2010. RNAi approaches for helping to control insect vectors of plant pathogens. P. 7. *In*: 18th Conf. IOCV held at Campinas/SP, Brazil 7–12, Nov. 2010

Rossetti, V., Garnier, M., Bove, J.M., Beretta, M.J.G., Teixeira, A.R.R., Quaggio, J.M. and De negri, J.D. 1990. Presence de bacteries dan le xyleme d'orangers attaints de chlorose variegee, une nouvelle maladie des agrumes au Brazil. C.R. Acad. Sci. Paris 310, serie III: 345–349.

Rossetti, V. 1993. Citrus variegated chlorosis, a new severe disease in Brazil: A review. pp. 449–452 *In*: Proc. 12th Conf. IOCV, IOCV, Riverside, CA.

Rossetti, V. 2000. Citrus variegated chlorosis (CVC) in Brazil – an overview. pp. 213–215 *In*: Proc. 14th Conf. IOCV, IOCV, Riverside, CA.

Saglio, P., Lafleche, D., Bonnisol, C. and Bove, J.M. 1971. Isolement, culture et observations au microscope electronique des structures de type mycoplasme associes a la maladie du stubborn des agrumes et leur comparison avec les structures observes dens le cas de la maladie du greening des agrumes. *Physiol. Veg*. 9: 569–582.

Salibe, A.A. and Cortez, R.E. 1966. Studies on the leaf mottling disease of citrus in the Philippines. *FAO Plant Prot. Bull*. 14: 141–144.

Schwarz, R.E. 1968. Mechanical transmission of the greening virus to cucumber. pp. 264–266 *In*: Proc. 4th Conf. IOCV, IOCV, Riverside, CA.

Sechler, A., Schuenzel, E.L., Cooke, P., Donnua, S., Thaveechai, N., Postnikova, E., Stone, A.L., Schneider, W.L., Damsteegt, V.D. and Schaad, N.W. 2009. Cultivation of "*Candidatus* Liberibacter asiaticus", Ca. L. americanus associated with Huanglongbing. *Phytopathology* 99: 480–486.

Secor, G.A., Rivera, V.V., Abad, J.A., Lee, I.-M., Glover, G.R.G., Liefting, L.W., Li, X. and De Boer, S.H. 2009. Association of '*Candidatus* solanacearum' with zebra chip disease of potato established by graft and psyllid transmission, electron microscopy and PCR. *Plant Disease* 93: 574–583.

Simpson, A.J.G., Reinach, F.C., Arruda, P. *et al*. 2000. The genome sequence of the plant pathogen *Xylella fastidiosa*. *Nature* 406: 151–157.

Singh, A.B. and Nimbalkar, M.R. 1977. *Murraya koenigii* and *M. paniculata*-preferable hosts of *Diaphorina citri* Kuway. *Sci. Cult*. 43: 97–98.

Sodhi, S.S., Dhillon, S.S., Jeyarajan, R. and Cheem, S.S. 1973. Mycoplasma-like organisms associated with citrus greening. *Proc. Indian Nat. Sci. Acad. Ser. B*. 41: 386–389.

Su, H.J. and Li, H.A. 1990. The nature of Likubin organism, life cycle, morphology and possible strains. pp. 106–110. (*Eds*.) Aubert, B., Tontyaporin, S. and

Buangsuwon, D. *In*: Proc. of the Asia Pacific International Conf. on Citriculture, Chiang Mai, Thailand 4–10 Feb. 1990.

Tatineni, S., Sagaram, U.S., Gowda, S., Robertson, C.J., Dawson, W.O., Iwanami, T. and Wang, N. 2008. In plant distribution of "*Candidatus* Liberibacter asiaticus" as revealed by polymerase chain reaction (PCR) and real-time PCR. *Phytopathology* 98: 592–599.

Texeira, D.C., Ayres, A.J., Kitajima, E.W., Tanaka, F.A.O., Danet, J.L., Jagoueix-Eveillard, S., Saillard, C. and Bové, J.M. 2005a. First Report of a Huanglongbing-like Disease of Citrus in Sao Paulo State, Brazil, and Association of a New Liberibacter Species, "*Candidatus* Liberibacter americanus", with the Disease. *Plant Disease* 89: 107.

Teixeira, D.C., Danet, J.L., Eveillard, S., Martins, E.C., de Jesus Junior, W.C., Yamamoto, P.T., Lopes, S.A., Bassanezi, R.B., Ayres, A.J., Saillard, C. and Bové, J.M. 2005b. Citrus huanglongbing in São Paulo State, Brazil: PCR detection of the "*Candidatus* Liberibacter species associated with the disease. *Molecular and Cellular Probes* 19: 173–179.

Teixeira, D.C., Lopes, S.A., Yamamoto, P.T., Eveillard, S., Martins, E.C., de Jesus Junior, W.C., Bassanezi, R.B., Ayres, A.J., Danet, J.L., Saillard, C. and Bové, J.M. 2005c. PCR detection of the two Liberibacter species associated with citrus huanglongbing (HLB) in São Paulo State, Brazil. pp. 432–438 *In*: Proc. 16th conf. IOCV, IOCV, Riverside, CA.

Teixeira, D.C., Saillard, C., Eveillard, S., Danet, J.L., Ayres, A.J. and Bové, J.M. 2005d. *Candidatus* Liberibacter americanus sp. nov., associated with citrus huanglongbing (greening disease) in São Paulo State, Brazil. *Intern. J. Syst. Evol. Microbiol.* 55: 1857–1862.

Teixeira, D.C., Saillard, C., Eveillard, S., Danet, J.L., Ayres, A.J. and Bové, J.M. 2005e. A new liberibacter species, *Candidatus* Liberibacter americanus sp. nov., is associated with citrus huanglongbing (greening disease) in São Paulo state, Brazil. pp. 325–340. *In*: Proc. 16th conf. IOCV, IOCV, Riverside, CA.

Teixeira, D.C., Saillard, C., Couture, C., Martins, E.C., Wulff, N.A., Eveillard-Jagoueix, S., Yamamoto, P.T., Ayres, A.J. and Bové, J.M. 2008a. Distribution and quantification of *Candidatus* Liberibacter americanus, agent of huanglongbing disease of citrus in São Paulo state, Brazil, in leaves of an affected sweet orange tree as determined by PCR. *Mol. Cell. Probes* 22: 139–150.

Teixeira, D.C., Wulff, N.A., Marins, E.C., Kitajima, E.W., Bassanezi, R., Ayres, A.J., Eveillard, S., Saillard, C. and Bove, J.M. 2008b. A phytoplasma closely related to the pigeonpea witches' broom phytoplasma (16Sr IX) is associated with citrus huanglonging symptoms in the State of Sao Paulo, Brazil. *Phytopathology* 93: 977– 984.

Tirtawidjaja, S. 1980. Citrus virus research in Indonesia. *In*: Proc. 8th Conf. IOCV, IOCV, Riverside, CA.

Van Vuuren, S.P., Cook, G. and Pieterson G. 2010. No evidence for seed transmission of *Candidus* Liberibacter africanus in citrus. p. 79. In: Abs. Proc. 18th Conf. IOCV held in Campinas, Brazil, 7–12 Nov. 2010.

Van de Merwe, A.J. and Andersen, F.G. 1937. Chromium and manganese toxicity. Is it important in Transvaal citrus greening? *Farming S. Africa* 12: 439–440.

Van Sluys, M.A., de Oliveira, M.C., Monteiro-Vitorello, C.B., Miyaki, C.Y., Furlan, L.R., Camargo, L.E.A., da Silva, A.C.R., Moon, D.H., Takita, M.A., Lemos, E.G.M., Machado, M.A., Ferro, M.I.T., da Silva, F.R., Goldman, M.H.S., Goldman, G.H., Lemos, M.V.F., El-Dorry, H., Tsai, S.M., Carrer, H., Carraro, D.M., de Oliveira, R.C., Nunes, L.R., Siqueira, W.J., Coutinho, L.L., Kimura, E.T., Ferro, E.S., Harakava, R., Kuramae, E.E., Marino, C.L., Giglioti, E., Abreu, I.L., Alves, L.M.C., do Amaral, A.M., Baia, G.S., Blanco, S.R., Brito, M.S., Cannavan, F.S., Celestino, A.V., da Cunha, A.F., Fenille, R.C., Ferro, J.A., Formighieri, E.F., Kishi, L.T., Leoni, S.G., Oliveira, A.R., Rosa Jr, V.E., Sassaki, F.T., Sena, J.A.D., de Souza, A.A., Truffi, D., Tsukumo, F., Yanai, G.M., Zaros, L.G., Civerolo, E.L., Simpson, A.J.G., Almeida Jr., N.F., Setubal, J.C. and Kitajima, J.P. 2003. Comparative analyses of the complete genome sequences of pierce's disease and citrus variegated chlorosis strains of *Xylella fastidiosa*. *Journal of Bacteriology* 185: 1018–1026.

Varma, A., Ahlawat, Y.S., Chakarborty, N.K., Garnier, M. and Bove, J.M. 1993. Detection of greening BLO by electron microscopy, DNA hybridization in citrus leaves with and without mottle from various regions in India. pp. 280–285, *In*: Proc. 12th Conf. IOCV, IOCV, Riverside, CA.

Vijyakumari, N., Ghosh, D.K., Das, A.K. and Singh, S. 2006. Elimination of CTV and greening pathogens through *in vitro* soot grafting from citrus exotic germplasm. *Indian J. Agric Sci.* 76: 209–210.

Villechanoux, S., Garnier, M., Renaudin, J. and Bové, J.M. 1992. Detection of several strains of the bacterium-like organism of citrus greening disease by DNA probes. *Current Microbiology* 24: 89–95.

Villechanoux, S., Garnier, M., Laigret, F., Renaudin, J. and Bové, J.M. 1993. The genome of the non-cultured, bacterial-like organism associated with citrus greening disease contains the nusG-rplKAJL-rpoBC gene cluster and the gene for a bacteriophage type DNA polymerase. *Current Microbiology* 26: 161–166.

Wang, Z.,Yin, Y., Hu, H., Yuan, Q., Peng, G. and Xia, Y. 2006. Development and application of molecular-based diagnosis for '*Candidatus* Liberibacter asiaticus', the causal pathogen of citrus huanglongbing. *Plant Pathology* 55: 630–638.

Wulff, N.A., Eveillard, S., Foissac, X., Ayres, A.J. and Bové, J.M. 2008. Ribosomal RNA operons and genome size of '*Candidatus* Liberibacter americanus', a bacterium associated with citrus huanglongbing in Brazil. Proceedings of the 2008 International Research Conference on Huanglongbing, Orlando, FL, U.S.A.

Xu, C.F., Li, K.B. and Liao, J.Z. 1985. On the transmission of citrus yellow shoot by psylla and observation with electron microscopy. *Acta Phytopathologica Sinica*, 15(4): 241–245.

Yamamoto, P.T., Felippe, M.R. and Garbim, L.F. 2006. *Diaphorina citri* (Kuwayama) (Hemiptera: Psyllidae): vector of the bacterium *Candidatus* Liberibacter americanus. p. 96 *In*: Proc. of the Huanglongbing-Greening Intl. Workshop, Ribeirão Preto, SP, Brazil.

Zhou, L.J., Gabriel, D.W., Duan, Y.P., Halbert, S.E. and Dixon, W.N. 2007. First report of dodder transmission of huanglongbing from naturally infected *Murraya paniculata* to citrus. *Plant Disease* 91: 227.

Zhou, L., Duan, Y., Gabriel, D. and Gottwald, T.R. 2008. Seed transmission of *Candidatus* Liberibacter asiaticus in periwinkle and dodder resulted in low bacterial titer and very mild disease in periwinkle. *Phytopathology* 98: S181.

6

CITRUS DISEASES CAUSED BY MOLLICUTES

Following three diseases are caused by the pathogens belonging to class Mollicutes of Bacteria kingdom

1. **Citrus Stubborn Disease**
2. **Witches' Broom Disease of Lime**
3. **Citrus Rubbery Wood Disease**

The word "Mollicute" is derived from the Latin *mollis* (meaning "soft" or "pliable"), and *cutis* (meaning "skin"). Mollicute comes in a class of bacteria which are distinguished by the absence of a cell wall. Mollicutes are parasites of various animals and plants, living on or in the host's cells. These organisms are very small, measuring only 0.2–0.3 μm in size and have a very small genome size. Cells of these organisms are pleomorphic in nature and require sterols that make the cell membrane somewhat more rigid. Many are able to move about through gliding, but members of the genus *Spiroplasma* are helical and move by twisting. The best-known genus in Mollicutes is *Mycoplasma*. Mollicutes cause diseases in humans, attaching to cells in the respiratory or urogenital tracts, particularly species of *Mycoplasma* and *Ureaplasma*. Phytoplasma and *Spiroplasma* are plant pathogens but some may survive in their insect vectors. Classification of Mollicutes is given below:

Kingdom	:	Bacteria
Phylum	:	Firmicutes or tenericutes
Class	:	Mollicutes
Orders	:	Acholeplasmatales
		Anaeroplasmatales
		Entomoplasmatales
		Haloplasmatales
		Mycoplasmatales

Citrus Stubborn Disease

Stubborn disease was present even in about 1915 in Washington Navel orange trees near Redlands in California State of USA (Fawcett *et al.*, 1944). Citrus diseases earlier known as 'pink nose' and 'acorn fruit' are now known to be similar to stubborn disease. Stubborn disease was shown to be transmitted by grafting and was considered to be caused by a virus (Fawcett *et al.*, 1944; Haas *et al.*, 1944; Fawcett, 1946). Most of the earlier studies on stubborn disease of citrus and its pathogen have been carried out by the University of California and the United States Department of Agriculture (USDA) in 1954 (Calavan, 1969). Stubborn disease is present in North Africa, the eastern Mediterranean Basin, and the Middle East. In Israel, this disease is called as little leaf and in Egypt as Safargali. Stubborn disease is also present in warmer areas like Corsica (France). Stubborn diseases is now known to occur in the following countries: Algeria, Cyprus, Egypt, France (Corsica only), Greece, Iran, Iraq, Israel, Italy (few records; at least Sardinia, Sicily), Jordan, Lebanon, Libya, Mexico, Morocco, Pakistan, Saudi Arabia, Spain, Syria, Tunisia, Turkey, USA (Arizona, California, Illinois, Maryland), Yemen, and possibly in some countries of South Africa (unconfirmed reports) and New Zealand. It is economically important disease as the yield and quality of fruits are greatly reduced. Severe form of stubborn reduces yield in young Frost Lisbon lemon (*Citrus limon*) upto 59%, in Frost Washington navel upto 92%, Tangelo (*C. paradisi* × *C. reticulata*) upto 41–95%, and in Tangerines (*C. reticulata*) upto 81–98% (Calavan and Christiansen, 1966). Stubborn disease is now known to be caused by a Mollicute, *Spiroplasma citri.* Stubborn disease has been reviewed by Bové and Saglio (1974).

SYMPTOMS

Leaves on stubborn affected trees are small, abnormally upright, stiff and pinched at the tip or cup or heart-shaped. In huanglongbing diasease, foliage is sparse but in stubborn foliage becomes bunchy or rosette-like with numerous short leafy branches. Growth of leafy branches from multiple axillary buds is common. Leaves become chlorotic resembling zinc, iron or manganese deficiency. Premature leaf and fruit drop is common. Twigs show dieback symptoms and affected trees remain stunted and bushy in growth (Plate 41a). There is normal but unseasonal flowering resulting fruits of various ages on the same tree. Fruits are acorn-shaped, malformed, lop sided with curved columella (Plate 41b). Fruits may colour abnormally with abnormal retention of green colour especially on the styler end. There is excessive abortion of seeds in fruits on affected trees and fruits have insipid sour or bitter flavor. Fruits are smaller with blue, dense or cheesy albedo having prominent vascular network (Carpenter *et al.*, 1965; Calavan, 1969). Leaf mottling symptoms similar to huanglongbing disease develop in greenhouse grown sweet orange and grapefruit seedling test plants when grafted with tissue from citrus trees showing symptoms of stubborn disease (Olson, 1969). Best symptoms of stubborn in citrus develop under hot conditions (28–32°C) and may not give rise to conspicuous symptoms at lower temperatures (Oldfield, 1988). Stubborn symptoms develop faster when inoculated plants are maintained at temperatures of 35°C day and 27–30°C night (Olson and Rogers, 1969; Bove *et al.*, 1974).

TRANSMISSION

The pathogen associated with stubborn disease (*Spiroplasma citri*) is transmitted by tissue grafting (Fawcett *et al.*, 1944). Since associated pathogen is irregularly distributed in infected field trees, graft transmission from infected trees to greenhouse grown indicator plants is erratic. Though the disease is not seed transmitted but stubborn pathogen is known to be present in the coats of aborted seeds from fruits of affected trees (Calavan *et al.*, 1974). Vectors of stubborn disease are several species of leafhoppers (Kaloostian *et al.*, 1975; Rana *et al.*, 1975). *S. citri* is naturally transmitted by leafhoppers: *Circulifer* (*Neoaliturus*) *tenellus*, *Scaphytopius nitridus* (Plate 41c-1, c-2), and *S. acutus* in California (USA) (Oldfield *et al.*, 1976; Oldfield, 1988), *Circulifer* (*Neoaliturus*) *haematoceps* (Bové, 1986; Fos *et al.*, 1986; Klein *et al.*, 1988) in the Mediterranean area. *S. citri* is transmitted in a persistent and circulative manner and multiplies in its vector. Vector, *C. tenellus* requires acquisition access period of 2 hours or more, transmission access period of at least 6 h and it starts transmitting the pathogen 16–18 days after transmission feeding and continued throughout its life (Liu *et al.*, 1983a, b). Both male and female *S. nitridus* can transmit but no transovarial transmission was obtained (Oldfield *et al.*, 1984). Successful transmission of *S. citri* by the leafhopper depends on the ability of the *Spiroplasmas* to pass through the insect gut cells, to multiply in various tissues, and to finally cross the salivary gland cells. The crossing of these different barriers requires protein interactions between the *Spiroplasma* cells and cells of the insect host. It is solely from the salivary gland that *Spiroplasmas* are inoculated into the sieve tubes during insect feeding, thus initiating a new infection (Labroussaa *et al.*, 2010). *S. citri* bodies have been observed in salivary glands of innocuous leafhopper vector, *circulifer tenellus* (Fig. 29). None of these vectors have a particular preference for citrus as a host; they may therefore acquire *S. citri* from other hosts and probably transmit to citrus species. Other Homopteran insects may acquire *S. citri*, but not able to transmit it (Rana *et al.*, 1975; Bové *et al.*, 1979). It is possible that other species may act as vectors in the Mediterranean area (Klein *et al.*, 1982).

Circulifer tenellus belongs to family Cicadellidae and is found in Algeria, Egypt, France (Corsica) (Giustina, 1989), Israel, Italy (Sicily), Libya, Morocco, Spain, Tunisia, Turkey, India, Kyrgyzstan, Tajikistan, Turkey, Turkmenistan, Uzbekistan, Namibia, South Africa, Sudan, Tunisia. Mexico, USA, Jamaica, and Puerto Rico. *Neoaliturus haematoceps* is found in France (including Corsica), Israel, Morocco, Spain, Syria, Turkey, Iran and Morocco.

HOSTRANGE

Spiroplasma citri infects plant species in 18 dicotyledonous families and one in monocotyledonous family and can infect both citrus and non-citrus hosts. Following are susceptible hosts of stubborn pathogen: Calamondin, Carrizo, Rusk, Troyer Citranges, Grapefruit, Kumquat, Lemons, Mandarin, limes, Pummelo (Chandler), Rough lemon, Satsuma, Sour orange, Sweet lime, sweet orange, and Limes (Kagzi, Key, Mexican, West Indian). Sweet oranges, grapefruit, mandarin and mandarin hybrids are highly susceptible to *S. citri*. Acid lime, lemons, trifoliate orange and its hybrids are infected experimentally through graft inoculation but develop milder symptoms. Tangelos (Orlando).Citranges (Cunningham, Troyer) and *P. trifoliata* are symptomless or tolerant hosts. In addition to citrus, *Spiroplasma citri* also infects several other hosts and cause diseases in them (Gumpf, 1988). For example, it is the causal agent of brittle root disease of horseradish in Illinois (Sullivan *et al.*, 1987) and sesame yellows in Turkey and Iran. White clover, chrysanthemum and periwinkle (Plate 42a) are hosts of stubborn pathogen and can be infected by direct inoculation by leafhopper vectors (Reuther *et al.*, 1978). London rocket, many *Brassica* spp., wild radish and marigold (Plate 42b) are some of the natural/experimental hosts of *S. citri*. *Catharanthus roseus, Crepis echiodes, Echium* sp. and *Sesamum indicum* are found to be naturally infected in eastern Mediterranean region of Turkey (Kersting *et al.*, 1993). Carrot purple leaf disease is also known to be associated with *Spiroplasma citri*. Recent studies confirmed that carrot is a host of *S. citri* which got infected by beet leafhopper, *circulifer tenellus* (Mello *et al.*, 2009). There is no indication at present of special races or strains attacking citrus.

INDICATOR

Madam Vinous or Hamlin sweet orange seedlings and tangelo are used as indicator hosts in *Citrus* species. Symptoms in inoculated plants appear as leaf tip chlorosis, leaf mottle, leaf cupping, stunting and short internodes. These symptoms are developed within 4–12 weeks at warm temperature.

THE PATHOGEN

Initially, stubborn disease was established as a virus disease (Fawcett, 1946). Further investigations revealed that safargali (stubborn) in Egypt is caused by a culturable chytrid-like organism (Nour-Eldin, 1967). Subsequently, the stubborn pathogen was established as a Mycoplasma-like organism based on electron microscopic studies of infected tissues and tetracycline therapy of infected plants (Igwegbe and Calavan, 1970; Lafleche and Bove, 1970). Finally in 1971, Saglio *et al.* from France and Fudl-Allah *et al.* from California successfully cultured a pleomorphic organism from stubborn infected plants on liquid and on agar media (Fig. 30). *Spiroplasma citri* was also isolated from bodies of field collected *C. tenellus* (Lee *et al.*, 1973). *S. citri* cells are wall-less, pleomorphic with characteristic spiral morphology. The minimum viable length of a helix is 2.0 × 0.1–0.2 μm. The helices are motile by flexing or rotation. Some strains are non-motile and non-helical. *S. citri* is one of the very few plant pathogenic mollicutes to have been cultured. Based on helical morphology of this cultured mollicute, name "Spiroplasma" was coined to the causal organism of stubborn disease (Saglio *et al.*, 1973). Taxonomic position of *S. citri* is given below:

Kingdom	:	Bacteria
Phylum	:	Firmicutes or tenericutes
Class	:	Mollicutes
Order	:	Mycoplasmatales
Family	:	Spiroplasmataceae
Genus	:	*Spiroplasma*
Species	:	*citri*

DIAGNOSIS AND MANAGEMENT

For diagnosis stubborn disease, graft inoculations are made on sensitive indicator plants such as Madam Vinous sweet orange. Inoculated plants should be maintained under warm temperature 32°C during day and 27°C at night (Bové, 1988). Other indicators are grapefruit cv. Marsh and tangelo cv. Sexton. The best inoculum source for transmission is young leaf patches including midrib. *Spiroplasma citri* is easily detected by culturing on synthetic media (Fig. 30) (Lee and Davis, 1984) and by electron microscopy of infected plants and vector (Fig. 28, 29) (Fudl-Allah *et al.*, 1971; Lafleche and Bove, 1970; Bove *et al.*, 1984). It is also detected in ELISA (Saillard and Bové, 1983; Clark *et al.*, 1989) and by immunodiffusion test (Fig. 31). The latex agglutination technique is also suitable for rapid detection of *S. citri* (Fletcher and Slack, 1986). DAPI staining technique can also detect *S. citri* (Dale, 1988). To optimize the detection of *S. citri*, its culturing in liquid

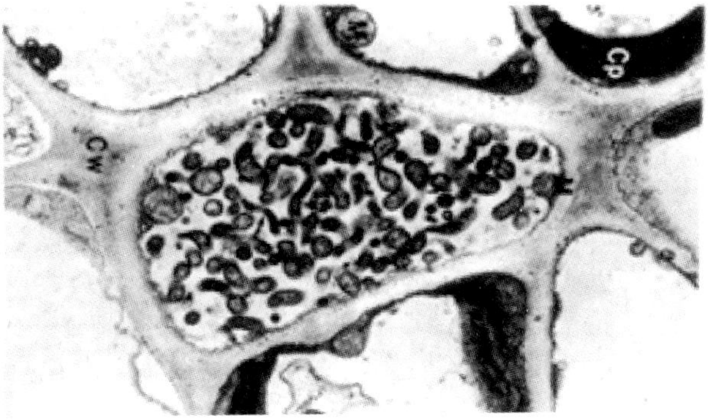

Fig. 28: Thin section electron micrograph showing pleomorphic bodies of stubborn pathogen (Courtesy D.J. Gumpf)

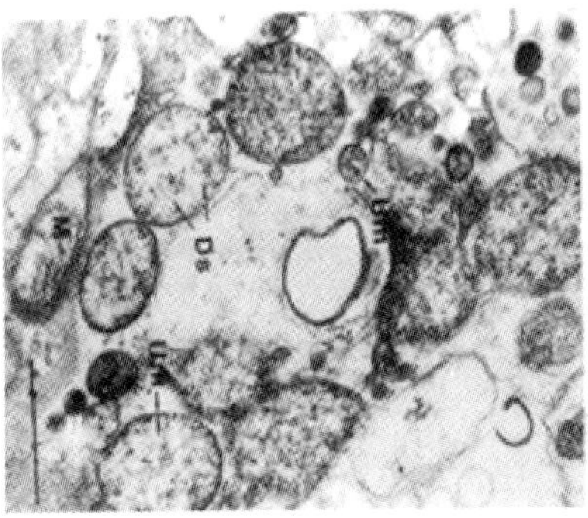

Fig. 29: *Spiroplasma citri* in slavery glands of stubborn vector, *C. tenellus* (Courtesy G.N. Oldfield)

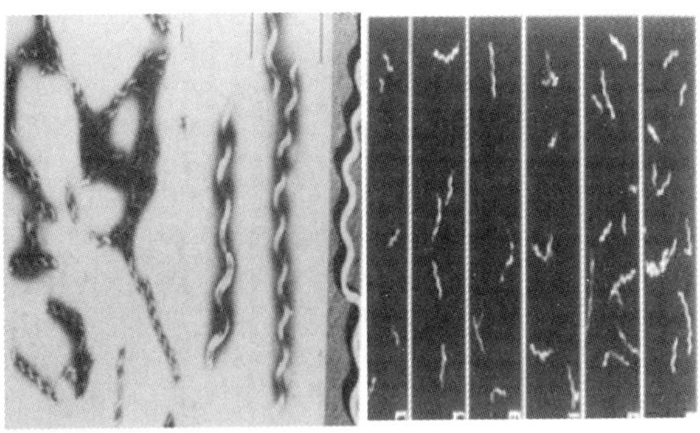

Fig. 30: Spiral form of *S. citri* in culture after different intervals (collected during UCR visit)

media will increase the number of spiroplasmas in the culture. DNA extracted from organisms in liquid medium facilitated assay for *S. citri* by PCR. This method has proven to be more rapid and sensitive than the

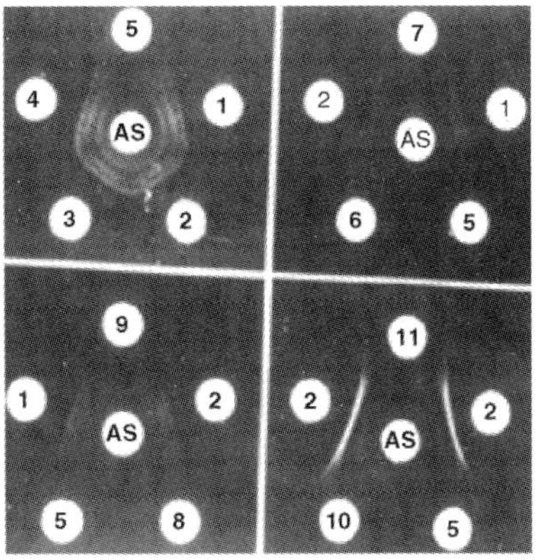

Fig. 31: Immunodiffusion test for the detection of *S. citri* (Courtesy D.J. Gumpf)

traditional culturing method (Rangel *et al.*, 2005). *S. citri* is also detected by IC-PCR (Saillard *et al.*, 1996). A real-time polymerase chain reaction (PCR) assay has been used to detect *S. citri* in infected citrus trees. This technique can also helps to determine incidence of stubborn disease. Best tissue for detection of *S. citri* in real time PCR is fruit columella because pathogen titer remains highest in tissues of columella (Yokomi *et al.*, 2010).

Stubborn disease is difficult to manage because the pathogen, *S. citri* has several natural weed hosts and is transmitted by several leafhopper vectors. Production of healthy budwood is therefore a practical means of control. Trees showing symptoms should be rogued and replaced. Healthy trees can be obtained by eliminating the pathogen through shoot tip grafting. Growing citrus nurseries and disease-free mother trees in vector-free areas is one of the possible options. Control of weed hosts and vectors are important for effective management of stubborn disease. Tetracycline therapy is although effective but not a practical option. It is suggested that *S. citri* has to be covered by virus-free certification schemes for citrus, such as the one now under development by EPPO. It has been suggested that trap plants should be planted in the vicinity of orchards to manage vectors (Gumpf, 1988).

Witches' Broom Disease of Lime

Witch's broom disease of lime (WBDL) was first observed in coastal plain of the Sultanate of Oman near the border of United Arab Emirates around 1970 (Bove, 1986). It was estimated that over 98% of limes currently grown in Oman are infected with WBDL. WBDL kills lime trees in less than 5 years and has become a major limiting factor for lime production in Oman. This disease has now spread to UAE, Muscat and is spreading fast in Iran (Bove, 1995; Salehi *et al.*, 2001). Witches' broom disease is highly destructive to lime trees in these countries (Garnier *et al.*, 1991). In India, Witches' broom disease has also been found on acid lime (*Citrus aurantifolia*) trees in some citrus growing areas of Maharashtra State with an incidence of 5% (Ghosh *et al.*, 1999a). The disease has also been recorded with variable incidence in some major acid lime growing states of Andhra Pradesh, Tamilnadu and Karnataka. Blastomania is a somewhat similar to witches' broom disease reported on *C. aurantifolia* and *C. limonia* in India (Mali *et al.*, 1975). Blatomania has been shown to be associated with phytoplasmas (Sharma and Singh, 1988).

SYMPTOMS

Symptoms of witches' broom disease (WBDL) in infected trees appear as small pale leaves, highly proliferating shoots, short internodes and witches' broom (Plate 43). Leaves drop prematurely and infected branches have distorted twigs characteristic of witches' broom symptoms. In advanced stages, infected branches show die-back symptoms. Witches' broom symptoms can be easily marked on a tree due to its contrast with normal foliage of the tree. During early stages of infection only one branch shows symptoms but subsequently several branches of infected tree will develop witches' broom symptoms. Leaves on older witches' brooms die but remain attached to them (Plate 43). Fruits on healthy part of the tree are normal but on affected branches, fruit size is reduced and such fruits often dried. In final stages of infection, most infected shoots and twigs die and only a few witches' brooms are left. At this stage, there is no yield and growers have no option except to remove such trees (Bove *et al.*, 1988). Trunk and branches of tree having witches' broom symptoms are excessively ramified.

TRANSMISSION

WBDL is transmitted by grafting witches' broom affected twigs to healthy lime (Plate 44a) and Troyer Citrange seedlings (Garnier *et al.*, 1991). WBDL is also transmitted by dodder to periwinkle (Plate 44b) (Garnier *et al.*, 1991; Bove, 1986; Bove *et al.*, 1988; Ghosh *et al.*, 1999a). A leafhopper, *Hishimonas phycitis* is widely present and reproduces on infected lime orchard trees in Oman and UAE and is suspected to be the vector of the disease. *H. phycitis* is vector of the brinjal (egg plant) little leaf phytoplasma disease in India (Bindra and Singh, 1969; Mitra, 1988). *H. phycitis* collected from infected lime tees were positive to witches' broom in ELISA tests and by DNA probe in hybridization tests suggesting that this leafhopper may be a putative vector of witches' broom disease (Bove *et al.*, 1993). These results need confirmation through experimental transmission as even the non-vector insect species can also acquire the pathogen which can be detected in the insect by serological or molecular diagnostic tests. However, no experimental transmission was obtained with *H. phycitis* with Indian isolate of witches' broom disease of lime (Ghosh *et al.*, 1999a). A recent report indicates that WBDL phytoplasma can be seed-transmitted based on PCR tests, but needs further examination.

HOSTRANGE

Witch's broom disease agent naturally infects lime (kagzi, Key, Mexican, acid), citron, Palestine sweet lime and sweet limetta. Experimentally, this disease is graft transmissible to lemon, Rough lemon, Trifoliate orange, Troyer citrange, *Citrus excelsa, C. macrophylla*, Etrog citron, *C. ichangensis, C. karna*, Rangpur lime, Meyer lemon and *C. hystrix*. No symptoms have been seen in sweet orange, Clementine, pummelo, sour orange, grapefruit or mandarins and attempts to experimentally infect these plants have failed.Persian lime (*C. latifolia*) was not infected in experimental trials (Bove *et al.*, 1996).

Causal agent of WBDL is transmitted to eggplant, ornamental eggplant, two species of tobacco, Jimsonweed, black nightshade and tomato by grafting and dodder. Disease symptoms on these hosts are yellowing, smalling of leaves, short internodes, virescence, phyllody, witches' broom, stunting, wilting and death of infected plants. Phytoplasmal infection to these hosts was confirmed by PCR and back inoculation (Salehi *et al.*, 2001).

INDICATOR

Acid lime or Mexican lime seedling are used as indicator host in which extremely reduced leaves and witches' brooms symptoms develop in several months after grafting (Plate 44a). Symptoms develop within six months when inoculated plants are kept at warm temperature. The best indicator host is periwinkle. Dodder inoculated periwinkle plants show yellowing of mid vein, smalling of leaves and witches' broom symptoms (Plate 44b).

THE PATHOGEN

Witches' broom disease is caused by a phytoplasma both in Oman and in India. The phytoplasma bodies have been observed in sieve tubes of phloem tissues of diseased plants (Plate 44c) (Bove *et al.*, 1988; Ghosh *et al.*, 1999b). WBDL phytoplasma was characterized by studying its genome size, the sequences of its 16S ribosomal DNA and the 16S–23S ribosomal DNA spacer region, and hybridization profiles obtained by using WBDL phytoplasma-specific probes. The size of the WBDL phytoplasma genome is 720 kbp. Genomic similarities with the phytoplasmas of sunhemp, sesame, and alfalfa phyllodies were demonstrated, and it was established that the WBDL phytoplasma belongs to the sunhemp phyllody phylogenetic subgroup (Zreik *et al.*, 1995). Phytoplasma associated with WBD has been classified as *Candidus* Phytoplasma aurantifolia (Zreik *et al.*, 1995).

DIAGNOSIS AND MANAGEMENT

Witches' broom disease can be easily identified in field trees by its specific symptoms. WBDL is easily detected by inoculating indicator plants by grafting and periwinkle plants by dodder. Typical symptoms of the disease will appear on inoculated lime seedlings within six months. Witches' broom phytoplasma is detected by thin section electron microscopy. Pleomorphic phytoplasma bodies can be viewed in sieve tubes of phloem tissues in sections of midrib of diseased periwinkle or lime leaves (Bove *et al.*, 1988; Ghosh *et al.*, 1999b). Presence of *Candidus* phytoplasma aurantifolia (Ca.P. a) is also detected by DAPI test in which a bright yellow green fluorescence is observed in the phloem sieve elements of diseased tissue when stain with DAPI (4′,6-diamidino-2-phenylindole) a DNA binding fluorochrome (Plate 44d) (Ghosh *et al.* 1999b). Eighteen Monoclonal antibodies have been prepared against witches' broom disease of lime by Professor Bove and his group and are being used in DAS-ELISA or immunofluorescence to detect *Ca.*P.a. in plants and insect, *H. phycitis* (Garnier *et al.*, 1991). DNA probes specific to Ca.P.a. have been prepared and used to detect witches' broom Phytoplasma in DNA-DNA hybridization tests from plants as well as from insects (Bove *et al.*, 1993). PCR amplification of "*Candidus* Phytoplasma aurantifolia" 16S rDNA was obtained with universal primer

P1 (forward) and specific primer WB3 (reverse) (Zreik *et al.*, 1995; Bove *et al.*, 2000).

Strict regulations are required for movement of propagation material from countries or regions where witches' broom is present to avoid the introduction of this disease in places where it is not found. In endemic areas, eradication of infected plants should be done as quickly as possible. Resistant lines of lime, if available, should be planted. Attemps are being made to regenerate somatic hybrid plants of *Citrus sinensis* (sweet orange) and *Citrus micrantha,* a progenitor species of lime (Khan and Grosser, 2004). Healthy planting material can be obtained by shoot tip grafting.

Citrus Rubbery Wood Disease

Rubbery wood (RB) disease of citrus is known to occur in India for a long time (Ahlawat and Chenulu, 1985a, b). RB is quite prevalent in Northeastern region of India, Kota region of Rajasthan and Nagpur region of Maharashtra state in India (Chenulu and Ahlawat, 1993). In some orchards of lime and lemon, incidence of RW disease was recorded upto 30% in Northeastern India. In Akola and Tirupati regions of India, a disease of Nagpur mandarin locally known as 'Valia' appears similar to rubbery wood disease. RW is a disease of minor importance in India. A disease with similar symptoms also occurs in Brazil which is now considered to be a reaction of exocortis viroid.

Rubberywood affected trees are characterized with development of thin and linear shoots which bend towards ground owing to unusual flexibility or elasticity (Plate 45a). Such trees are stunted and bushy, and become totally unproductive. Leaves are slightly chlorotic and smaller in size. Glasshouse inoculated plants of kagzi lime (*Citrus aurantifolia*), lemon and mandarin develop willowing habit and their shoots bend to an extent that they can touch the ground. In subsequent years of infection, numerous thin branches develop at right angle which first grow very fast till they attain the height of 15–23 cm and then their growth is restricted. Usually there is no leaf fall of lemon and mandarin seedlings in the glasshouse but RW infected plants show heavy leaf fall. Severely affected glasshouse inoculated plants show dieback and die following 4–5 years of infection. After the death of main shoot, numerous thin branches come out at the base of the plant. These branches have short internodes and smaller leaves giving the plant a bushy appearance. Swelling of buds are also noted in Eureka lemon plants (Ahlawat and Chenulu, 1985a)

Rubberywood disease is transmitted by grafting when branches showing rubberywood are grafted to test plants (Plate 45b). RW is also transmitted by dodder, *Cuscuta reflexa* from lemon to lemon. In Electron microscope, Phytoplasma like bodies were seen in sieve tubes of phloem tissues of ultrathin sections of diseased plant but not in tissues of healthy plant (Ahlawat, 1987, 1989; Ahlawat *et al.*, 1986). Rubbery wood phytoplasma did not show any reaction in immunofluorescence tests with monoclonal antibodies of Indian, Chinese and South African isolates of greening bacterium and monoclonal antibodies of *Candidus* Phytoplasma aurantifolia, the causal pathogen of witches' broom disease of lime. RW phytoplasma is detected either by electron microscopy or through biological indexing on lemon seedlings (Ahlawat and Pant, 2003). Propagation from infected trees should be avoided for better management of RW phytoplasma disease.

REFERENCES

Ahlawat, Y.S. 1987. Association of mycoplasma-like organisms with rubbery wood disease of citrus. p. 12. *In*: 3rd Reg. Workshop on Plant Mycoplasma, INSA, New Delhi.

Ahlawat, Y.S. 1989. Association of mycoplasma-like organisms with rubbery wood disease of citrus. p. 105–108 *In*: Plant Diseases caused by Fastidious prokaryotes (*Eds.*) Raychaudhuri, S.P. and Varma, A. Today and Tomorrow Printers and Publishers, New Delhi.

Ahlawat, Y.S. and Chenulu, V.V. 1985a. Rubberywood - a hetherto unrecorded disease of citrus. *Curr. Sci.* 54: 580–581.

Ahlawat, Y.S. and Chenulu, V.V. 1985b. Rubberywood and multiple sprouting-two newly discovered diseases of citrus. *Indian Phytopath.* 38: 625.

Ahlawat, Y.S. and Pant, R.P. 2003. Major virus and virus-like diseases of citrus in India, their diagnosis and management. *Ann. Rev. Plant Path.* 2: 447–474.

Ahlawat, Y.S., Chenulu, V.V., Gumpf, D.J. and Pandey, P.K. 1986. Some newly discovered virus and mycoplasma diseases of citrus in India. *In*: Proc. 10th Conf: IOCV, Riverside, Univ. Calif. (abstract).

Bindra, O.S. and Singh, B. 1969. Biology and bionomics of *Hishimonus phycitis* (Distant), a jassid vector of little-leaf disease of brinjal (*Solanum melongena*). *Indian Journal of Agricultural Sciences* 39: 912–919.

Bove, J.M. 1986. Witches' broom disease of lime. *FAO Plant Prot. Bull.* 34: 217–218.

Bové, J.M. 1986. Stubborn and its natural transmission in the Mediterranean area and in The Near East. *FAO Plant Protection Bulletin* 34: 15–23.

Bové, J.M. 1988. *Spiroplasma citri*. *In*: European handbook of plant diseases (*Ed.* by Smith, I.M., Dunez, J., Lelliott, R.A., Phillips, D.H. and Archer, S.A.), pp. 129–131. Blackwell Scientific Publications, Oxford, UK.

Bové, J.M. 1995. Oman-Witches' broom disease of lime. *In*: Virus and virus-like diseases of citrus in the Near East region, 217–229. FAO, Rome.

Bové, J.M. and Saglio, P. 1974. Stubborn and greening: a review, 1969–1972. pp. 1–11 *In*: Proc. 6th Conf. IOCV, IOCV, Riverside, CA.

Bové, J.M., Calavan, E.C., Capoor, S.P., Cortez, R.E. and Schwarz, R.E. 1974. Influence of temperature on symptoms of California stubborn, South Africa greening, India citrus decline and Philippine leaf mottling disease. pp. 12–15. *In*: Proc. 6th Conf. IOCV, IOCV, Riverside, CA.

Bové, J.M., Moutous, G., Saillard, C., Fos, A., Bonfils, J., Vignault, J.-C., Nhami, A., Abassi, M., Kabbage, K., Hafidi, B., Mouches, C. and Viennot-Bourgin, G. 1979. Mise en évidence de *Spiroplasma citri*, l'agent causal de la maladie du 'stubborn' des agrumes dans 7 cicadelles du Maroc. Comptes Rendus Hebdomadaires des Séances de l'Académie des Sciences D No. 288: 335–338.

Bove, J.M., Saillard, C., Vignault, J.C. and Fos, A. 1984. Citrus stubborn disease in Iraq and Syria: correlation between symptom expression and detection of *Spiroplasma citri* by culture and ELISA. pp. 145–152 *In*: Proc. 9th Conf. IOCV, IOCV, Riverside CA.

Bove, J.M., Garnier, M., Mjeni, A.M. and Khayrallah 1988. Witches' broom disease of small fruited acid lime treesin Oman: First MLO disease of citrus. pp. 307–309 *In*: Proc. 10th Conf. IOCV, IOCV, Riverside, CA.

Bove, J.M., Zreik, L., Danet, J.L., Bonfils, J., Majeni, A.M.M. and Garnier, M. 1993. Witches' broom disease of lime trees: monoclonal antibody and DNA probes for the detection of the associated MLO and the identification of a possible vector. pp. 342–348 *In*: Proc. 12th Conf. IOCV, IOCV, Riverside, CA.

Bove, J.M., Navarro, L., Bonnet, P., Zreik, L. and Garnier, M. 1996. Reaction of citrus cultivars to graft inoculation of *Phytopasma aurantifolia*-infected lime shoots. pp. 249–251. *In*: Proc. 13th Conf. IOCV, IOCV, Riverside, CA.

Bove, J.M., Danet, J.L., Bananej, K., Hassanzadeh, N., Taghizadeh, M., Salehi, M. and Garnier, M. 2000. Witches' broom disease of lime (WBDL) in Iran. pp. 207–212. *In*: Proc. 14th Conf. IOCV, IOCV, Riverside, CA.

Calavan, E.C. 1969. Investigations of stubborn disease in California: indexing, effects on growth and production, and evidence for virus strains. Proc. First Internl. *Citrus Symp.* 3: 1403–1412.

Calavan, E.C. and Christiansen, D.W. 1966. Effects of stubborn disease on various varieties of citrus trees. *Israel J. Bot.* 15: 121–132. .

Calavan, E.C., Harjung, M.K., Fudl-Allah, A.E.-S.A. and Bowyer, J.W. 1974. Natural incidence of stubborn in field grown citrus seedlings and budlings. pp. 16–19 *In*: Proc. 6th Conf. IOCV, IOCV, Riverside, CA.

Carpenter, J.B., Calavan, E.C. and Christiansen, D.W. 1965. Occurrence of excessive seed abortion in citrus fruits affected with stubborn disease. *Plant Dis. Reptr.* 49: 668–672.

Chenulu, V.V. and Ahlawat, Y.S. 1993. p. 34 *In*: Virus and mycoplasma diseases of fruit crops in India (*Eds.*) Chenulu, V.V. and Ahlawat, Y.S.. ICAR, New Delhi, India.

Clark, M.F., Davies, D.L., Buss, S.L. and Morton, A. 1989. Serological discrimination among mycoplasma-like organisms using polyclonal and monoclonal antibodies. *Acta Horticulturae* 235: 107–113.

Fawcett, H.S. 1946. Stubborn disease of citrus, a virosis. *Phytopathology* 36: 675–677.

Fawcett, H.S., Perry, J.C. and Johnston, J.C. 1944. The stubborn disease of citrus. *Calif. Citrograph* 29: 146–147.

Fletcher, J. and Slack, S.A. 1986. Latex agglutination as a rapid detection assay for *Spiroplasma citri*. *Plant Disease* 70: 754–756.

Fos, A., Bové, J.M., Lallemand, J., Saillard, C., Vignault, J.C., Ali, Y., Brun, P. and Vogel, R. 1986. The leafhopper *Neoaliturus haematoceps* is a vector of *Spiroplasma citri* in the Mediterranean area. *Annales de l'Institut Pasteur Microbiologie* 37A: 97–101.

Fudl-Allah, A.E.C., Calavan, E.C. and Igwegbe, E.C.K. 1971. Culture of a mycoplasma-like organism associated with stubborn disease of citrus. *Abstr. Phytopathology* 61: 1321.

Garnier, M., Zreik, L. and Bove, J.M. 1991. Witches' broom, a lethal mycoplasmal disease of lime trees in the Sultanate of Oman and the United Arab Emirates. *Plant Dis.* 75: 546–551.

Garnier, M., Zreik, L. and Bové, J.M. 1991. Witches' broom disease of lime trees in Oman: transmission of a mycoplasma-like organism (MLO) to periwinkle and citrus and the production of monoclonal antibodies against the MLO. pp. 448–453. *In*: Proc. 11th Conf. IOCV, IOCV Riverside, CA.

Ghosh, D.K., Das, A.K., Singh, S., Singh, S.J. and Ahlawat, Y.S. 1999a. Occurrence of witches' broom-a new phytoplasmal disease of acid lime (*Citrus aurantifolia*) in India. *Plant Dis.* 83: 302.

Ghosh, D.K., Das, A.K., Singh, S., Singh, S.J. and Ahlawat, Y.S. 1999b. Association of a phytoplasma with witches' broom, a new disease of acid lime (*Citrus aurantifolia*). *Curr. Sci.* 77: 174–177.

Giustina, W. Della 1989. Homoptéres Cicadellidae. *Faune de France* 73: 350.

Gumpf, D.J. 1988. Stubborn diseases of citrus caused by *Spiroplasma citri*. *In*: Mycoplasma diseases of crops. Basic and applied aspects (*Ed*. by Maramorosch, K. and Raychaudhuri, S.P.), pp. 327–342. Springer-Verlag, New York, USA.

Haas, A.R.C., Klotz, L.J. and Johnston, J.C. 1944. A corn disease in oranges. *Calif. Citrograph* 29: 168–169.

Igwegbe, E.C.K. and Calavan, E.C. 1970. Occurrence of mycoplasma-like bodies in phloem of stubborn infected citrus seedlings. *Phytopathology* 60: 1525–1526.

Kaloostian, G.H., Oldfield, G.N., Pierce, H.D. and Calavan, E.C., Granett, A.L., Rana, G.L. and Gumpf, D.J. 1975. Leafhopper-natural vector of citrus stubborn disease? *Calif. Agr*. 29: 14–15.

Kersting, U., Baspinar, H., Cinar, A., Sengonca, C. and Uygun, N. 1993. New findings on the epidemiology of *Spiroplasma citri* in the eastern Mediterranean region of Turkey. pp. 336–341. *In*: Proc. 12th Conf. IOCV, IOCV, Riverside, CA.

Khan, I.A. and Grosser, J.W. 2004. Regeneration and characterization of somatic hybrid plants of *Citrus sinensis* (sweet orange) and *Citrus micrantha*, a progenitor species of lime. *Euphytica* 137: 271–278.

Klein, M., Raccah, B. and Oman, P.W. 1982. The occurrence of a member of the *Circulifer tenellus* species complex (Homoptera: Cicadellidae: Euscelini) in Israel. *Phytoparasitica* 10: 237–240.

Klein, M., Rasooly, P. and Raccah, B. 1988. New findings on the transmission of *Spiroplasma citri*, the citrus stubborn disease agent in Israel, by a beet leafhopper from the Jordan valley. *Hassadeh* 68: 1736–1737

Labroussaa, F., Arricau-Bouvery, N., Dubrana, M.B.M, Duret, S., Bove, J.M., Renaudin, J. and Saillard, C. 2010. *Spiroplasma citri*-insect vector interactions. p. 8. *In*: Abs. 18th Conf. IOCV held in Campinas/SP, Nov. 7–12, 2010.

Lee, I.M. and Davis, R.E. 1984. New media for rapid growth of *Spiroplasma citri* and corn stunt spiroplasma. *Phytopathology* 74: 84–89.

Lee, I.M., Cartia, G., Calavan, E.C. and Kaloostian, G.H. 1973. Citrus stubborn disease organism cultured from beet leafhopper. *Calif. Agric*. 27(11): 14–15.

Lafleche, D. and Bove, J.M. 1970. Mycoplasmes dans les agrumes attients de "greening" de "stubborn" ou de maladiessimilaires. *Fruits* 25: 455–465.

Liu, H.Y., Gumpf, D.J., Oldfield, G.N. and Calavan, E.C. 1983a. Transmission of *Spiroplasma citri* by *Circulifer tenellus*. *Phytopathology* 73: 582–585.

Liu, H.Y., Gumpf, D.J., Oldfield, G.N. and Calavan, E.C. 1983b. The relationship of *Spiroplasma citri* and *Circulifer tenellus*. *Phytopathology* 73: 585–590.

Mali, V.R., Choudhari, K.G. and Rane, S.D. 1975. Blastomania - a new bud-transmissible disorder of citrus. *Current Science* 44: 627–628.

Mello, A.F., Wayadande, A., Yokomi, R.K. and Fletcher, J. 2009. Transmission of different strains of *Spiroplasma citri* to carrot and citrus by *Circulifer tenellus* Baker (Hemiptera: Cicadellidae). *Journal of Economic Entomology* 102(4): 1417–1422.

Mitra, D.K. 1988. Little leaf disease of eggplant. *In*: Mycoplasma diseases of crops. Basic and applied aspects (*Ed*. by Maramorosch, K. and Raychaudhuri, S.P.), pp. 343–348. Springer Verlag, New York, USA.

Nour-Eldin, F. 1967. A tumor inducing agent associated with citrus trees infected with safargali (stubborn) disease in the United Arab Republic. *Phytopathology* 57: 108–113.

OEPP/EPPO 1995. Certification schemes No. 12. Pathogen-tested citrus trees and rootstocks. Bulletin OEPP/EPPO Bulletin 25.

Olson, E.O. 1969. Mottled leaf symptom on indexed plants graft-inoculated from citrus trees showing various symptoms of stubborn disease. *Proc. First Internl. Citrus Symp.* 3: 1413–1420.

Olson, E.O. and Rogers, B. 1969. Effect of temperature on expression and transmission of stubborn disease of citrus. *Plant Dis. Reptr.* 53: 45–49.

Oldfield, G.N. 1988. Ecological associations of *Spiroplasma citri* with insects, plants and other plant mycoplasmas in the western United States. *In*: Mycoplasma diseases of crops. Basic and applied aspects (*Ed.* by Maramorosch, K. and Raychaudhuri, S.P.), pp. 175–191. Springer-Verlag, New York, USA.

Oldfield, G.N., Kaloostian, G.H., Pierce, H.D., Calavan, E.C., Granett, A.L. and Blue, R.L. 1976 Beet leafhopper transmits Citrus stubborn disease. *California Agriculture* 30: 6- 15.

Oldfield, G.N., Sullivan, D.A. and Calavan, E.C. 1984. Aspects of transmission of the citrus stubborn pathogen by *Scaphytopius nitrides* (DeLong). pp. 131–136 *In*: Proc. 9th Conf. IOCV, IOCV, Riverside CA.

Rana, G.L. Kaloostian, G.H., Oldfield, G.N., Granett, A.L., Calavan, E.C., Pierce, H.D., Lee, I.M. and Gumpf, D.J. 1975. Acquisition of *Spiroplasma citri* through membrane by homopterous insects. *Phytopathology* 65: 1143–1145.

Rangel, B., Krueger, R.R. and Lee, R.F. 2005. Current research on *Spiroplasma citri* in California. pp. 439–442. *In*: Proc. 16th Conf. IOCV, IOCV, Riverside, CA.

Reuther, W., Calavan, E.C. and Carman, G.E. 1978. The citrus Industry. Vol. IV (revised), University of California, Division of Agric. Sciences.

Saglio, P.D., Lafleche, D., Bonissol, C. and Bove, J.M. 1971a. Isolement et culture in vitro des mycoplasmes associes au 'stubborn' des agrumes et leur observation au microscope electronique. *Compy. Rend. Acad. Sci. Paris*, Series D 272: 1387–1390.

Saglio, P., Laflèche, D., Bonissol, C. and Bové, J.M. 1971b. Isolement, culture et observation au microscope électronique des structures de type mycoplasme associées à la maladie du stubborn des agrumes et leur comparaison avec des structures observées dans le cas de la maladie du greening des agrumes. *Physiol. Veg.*, 9: 569–582.

Saglio, P., Hospital, M.L., Lafleche, D., Dupont, G., Bove, J.M., Tully, J.G. and Freundt, E.A. 1973. *Spiroplasma citri* gen. and sp. N.: A mycoplasma-like organism associated with "stubborn" disease of citrus. *Intern. J. Syst. Bacteriology* 23: 191–204.

Saillard, C. and Bové, J.M. 1983. Application of ELISA to spiroplasma detection and classification. *In*: Methods in mycoplasmology (*Ed.* by Razin, S. and Tully, J.G.), Vol. 1, pp. 471–476. Academic Press, New York, USA.

Saillard, C., Nhami, A., Moreno, P., garnier, M. and Bove, J.M. 1996. *Spiroplasma citri* detection by immunocapture PCR. pp. 413. *In*: Proc. 13th Conf. IOCV, IOCV, Riverside, CA. (Abst).

Salehi, M., Izadpanah, K. and Taghizadeh, M. 2001. Witches broom disease on lime in Iran: new distribution areas, herbaceous hosts and transmission trial. pp. 293–296. *In*: Proc. 15th Conf. IOCV, IOCV, Riverside, CA.

Sharma, S.R. and Singh, S.J. 1988. Association of mycoplasma-like organisms with citrus blastomania. *Indian Journal of Plant Pathology* 6: 86–92.

Sullivan, D.A., Oldfield, G.N., Eastman, C.E., Fletcher, J. and Gumpf, D.J. 1987. Transmission of a citrus-infecting strain of *Spiroplasma citri* to horseradish. *Plant Disease* 71: 469.

Yokomi, R.K., Mello, A.F., Fletcher, J. and Saponari, M. 2010. Estimation of citrus stubborn disease incidence in citrus groves by real-time PCR. 17th Conf. IOCV, IOCV, Riverside (Available: http://www.ivia.es/iocv/)

Zreik, L., Carle, P., Bove, J.M. and Garnier, M. 1995. Characterization of mycoplasma-like organism associated with witches' broom disease of lime and proposition of a *Candidus taxon* for the organism "*Candidatus Phytopasma aurantifolia*". *Intern. J. Syst. Bacteriol.* 45: 449–453.

7

CITRUS DISEASES CAUSED BY VIROIDS

Following diseases of viriod nature are known to affect citrus plantations

1. **Citrus Exocortis**
2. **Cachexia (Xyloporosis)**
3. **Citrus Yellow Corky Vein**
4. **Citrus Gummy Bark**
5. **Gum Pocket and Gummy Pitting**
6. **Shell Bark**
7. **Dwarfing Citrus Trees by Viroids**
8. **Citrus Bent Leaf**

Viroids are small, naked, single stranded RNA molecules that infect plant cells and cause diseases. Viroids are not enclosed in a protein coat like viruses. They generally consist of less than 400 nucleotides and do not contain any genes. The circular RNA strand undergoes extensive base pairing within itself, forming a double-stranded structure that mimics DNA and is apparently replicated by the host cell's enzymes. Viroids include many commercially important disease agents, such as coconut cadang-cadang, citrus exocortis, and potato spindle tuber viroid. Citrus exocortis viroid (CEVd) was first identified as a group of small, pathogenic RNA molecules (Semancik and Weathers, 1972). CEVd remained the only well described viroid of citrus up to 1985, although a viroid aetiology was proposed for cachexia disease (Roistacher *et al.*, 1983). Subsequently more numbers of viroid-like RNAs have been identified in citrus. A proposal has been made for the ordering of this collection of known citrus viroids into five groups using the parameters of molecular size, nucleotide sequence homology and host reaction (Duran-Vila *et al.*, 1988).

Citrus species are natural hosts of five viroid species: Citrus exocortis viroid (CEVd), Citrus bent leaf viroid (CBLVd), Hop stunt viroid (HSVd), Citrus viroid III (CVd-III), and Citrus viroid IV (CVd-IV). CEVd and specific sequence variants of HSVd are the causal agents of the well-known diseases of citrus exocortis and cachexia. Other viroids have been found to induce different degrees of stunting. Citrus bent leaf viroid has little or no effect in growth and yield (Vernière *et al.*, 2004) and has 318–329 nucleotides. CBLVd belongs to family: Pospiviroidae and Genus: Apscaviroid. Based on the symptoms expression in *Etrog citron* (the indicator host for most citrus viroids), five citrus viroids have been identified: CEVd (citrus exocortis), citrus viroid-1 (CVd-1), CVd-11, CVd-111, CVd-1V (Duran-Vila *et al.*, 1986; Semancik and Duran-Vila, 1991). These viroids range from 275 nucleotides in CVd-1V to 371 nucleotides in CEVd (Duran-Vila *et al.*, 1988). The classification of viroids has been suggested and is described in Table 10.

Table 10: Classification of viroids

Citrus viroids (CV)	*Bases/nucleotide*	*Genus*	*Viroid subfamily*
Citrus exocortis	368–463	Pospiviroid	Pospoviroidae
CVd-1a	315–327	Apscaviroid	Apscaviroidae
CVd-1b	318	–	–
CVd-11a (non-cachexia)	302	Hostuviroid	Pospiviroidae
CVd-11b (Cachexia)	293–297	–	–
CVd-111a	294	Apscaviroid	Pospiviroidae
CVd-111b	294	–	–
CVd-111c	294	–	–
CVd-111d	–	–	–
CVd-1V	284	Cocadviroid	Pospiviroidae

Viroids in each group differ with only a few nucleotides. CVd-1 and CVd-111 are species of genus *Apscaviroid* with a central conserved region like the type member apple scar skin viroid. CVd-1V is a species of genus *Cocadviroid* with the conserved region like the type member cadang cadang viroid of coconut. This chapter will describe citrus exocortis, cachexia, yellow corky vein and other citrus viroid diseases of minor importance.

The molecules of citrus viroids consist of ribonucleotides ranging in size from 284 to 371, with the exception of one CEVd isolate variant of 463 nucleotides. Most of viroid species have been found to exist as a number of sequence variants. For example, CEVd has over 40 sequence variants. In the viroid molecule five structural domains have been defined. These domains are designated as Left terminal ($T_{L'}$) Pathogenicity (P), Central conserved region (C), Variable (V) and the Right terminal (T_R). Although the (P) domain is considered the main domain for the symptom modulation, other domains may also contribute. The RNA is a circular molecule, consisting of double-stranded and single stranded regions. In electron microscopical analysis, the particle appears as a rod-like structure. However, on heating (over 70°C) the rod-like structures become single-stranded circles. This property of viroid molecule has been used in the diagnosis of viroids from various plant species using Reverse or sequential polyacridmide gel electrophoresis.

Citrus Exocortis Disease

Exocortis disease of citrus is also known as "scaly butt" in Australia. Both these diseases were initially shown to be caused by a virus (Benton et al.1950, Bitters, 1952). Exocortis disease is mainly a disease of trifoliate orange rootstock and resulting in failure of several scion varieties on trifoliate orange. Another disease known as "Rangpur lime disease" causing a bark disorder in Rangpur lime rootstock is also similar to exocortis disease (Moreira, 1955). Exocortis disease is found wherever susceptible rootstocks like Trifoliate orange, Rangpur lime, Citranges, Sweet lime and Citron are used. However, in Japan, no exocortis disease is known to occur on trifoliate orange rootstock although it is a widely used rootstock for Satsuma mandarin (Tanaka and Yamada, 1969). Citrus exocortis viroid is one of the most ancient viroids among pospiviroids. The yield upto 55% is reduced in exocortis infected Valencia sweet orange trees on trifoliate rootstock. Exocotis affected trees remain stunted and this dwarfing factor of exocortis had been practically exploited to develop dwarfing trees for density planting in Israel and some other countries (Cohen, 1968; Ashkenazi and Oren, 1988). Natural variants of citrus dwarfing viroid have now been characterized (Murcia *et al.*, 2009). In India, exocortis disease is commonly present in Nagpur mandarin and Mosambi sweet orange on Rangpur lime and rough lemon rootstocks (Nariani *et al.*, 1968; Patil and Warke, 1968; Ghosh *et al.*, 2002; Kapur *et al.*, 1974). Hop stunt viroid has also been isolated from Nagpur mandarin and Mosambi sweet orange on rough lemon and Rangpur lime rootsocks in India (Ramachandran *et al.*, 2005).

SYMPTOMS

Canopy of exocortis affected tree is normally symptomless. Trees raised on susceptible rootstocks develop symptoms below the bud union on the rootstock at or below ground level. Characteristic symptoms of exocortis disease on susceptible rootstocks are shelling, scaling or splitting of rootstock bark above the ground level (Plate 46a). Affected trees remain stunted but degree of stunting on trees on trifoliate orange rootstock varies with the source of infection. In orchard trees, first scaling symptom develops either at or below the ground line or at the bud union. Small elongated areas of dead bark may have slight amount of gum beneath the bark flakes. Bark-scaling may extend below ground up to the main root. Bark-scaling takes longer time to develop ranging from 4–8 years after infection (Bitters,

1952). The inner bark remains alive, and new layer of scales are developed when outer ones slough off.

TRANSMISSION

Citrus exocortis disease is transmitted by grafting and natural spread may take place through contaminated budwood. In nurseries and orchards, exocortis disease spreads through contaminated operational tools (budding knives and pruning shears) or even by propagating tools and worker's contaminated hands (Garnsey and Jones, 1967; Garnsey, 1968). Citrus exocortis is also transmitted mechanically to some herbaceous hosts. Exocortis pathogen is not transmitted by seed in citrus species but it is seed-borne in seeds of *Impatiens walleriana* to the extent of 66% and in *Verbena* species upto 28% (Singh *et al.*, 2009). Exocortis is also transmitted by dodder, *Cuscuta subinclusa* from citrus to citrus and Petunia (Weathers 1965a, 1965b). No insect vector is known for exocortis disease.

HOST-RANGE

Natural infection of citrus exocortis viroid (CEVd) has been reported in grapevine (Flores *et al.*, 1985), tomato, eggplant, carrot, turnip and broad bean. CEVd can infect large number of citrus species, its hybrids and citrus relatives. Exocortis affects large number of rootstock varieties including the ones commonly used as root-stocks like Trifoliate orange, Citron, Lemons and Rangpur lime. Citranges, Etrog citron, Harvey lemon, Mandarin, limes, *Poncirus trifoliata*, Rangpur lime, Sweet lime and Tahiti lime are susceptible hosts to exocortis. Among these susceptible hosts, Citranges and *P. trifoliata* are highly susceptible. Clementine, Grapefruits, Lemons (Eureka, Lisbon), Mandarins, Pummelos, Rough lemon, Sour orange, Sweet oranges, Tangelos, Tangor, West Indian lime are symptomless hosts of exocortis. Tangelo (Orlando), sweet orange, Tangors, Sour orange, Rough lemon, Pummelos, mandarins, Lime (Kagzi, Key, Mexican, West Indian), lemons, Grapefruit, *Cucumis sativus* are tolerant hosts of exocortis. Reaction of exocortis viroid varies on different Citrus species/cultivars and there is considerable range in time of symptom appearance, degree of symptom severity and the amount of stunting. Trifoliate orange and Rangpur lime show bark splitting but Citron may show leaf epinasty and veinal necrosis. In India, exocortis has been isolated from mandarin and sweet orange budded on *Citrus karna* and Rangpur lime (Nariani *et al.*, 1968; Raychaudhuri *et al.*, 1977). Sweet orange, grapefruit and mandarin trees are symptomless but if they are on sensitive rootstock, the rootstock will show bark-splitting symptoms and the affected tree will become stunted. Herbaceous hosts, *Gynura aurantiaca* (Flores and Rodriguez, 1981), tomato, *Chrysanthemum morifolium, Datura stramonium* (Weathers *et al.*, 1967; Duran-Vila *et al.*, 1988; Schwinghamer and Broadbent, 1986), petunia (Weather, 1965b), potato and various species of *Scopolia* (Singh and Clark, 1973) are also susceptible to exocortis.

INDICATOR HOSTS

Arizona 861-S1 Etrog citron: The symptoms on this indicator host develop as leaf epinasty and/or stunting, midvein necrosis, petiole necrosis, leaf tip browning within weeks to months after inoculation at warm temperature (Plate 46b). Cracking of mid vein, browning of underside of veins, blotching and cracking of stem, wrinkling and browning of petioles and leaf tips are other symptoms of exocortis viroid observed on this indicator host. Best symptoms of exocortis disease develop at a temperature of 28–32°C.

Symptoms with different citrus viroids develop differently on Etrog citron indicator host (Duran Vila *et al.*, 1993). For example, CVD-1a shows mild epinasty and localized midvein necrosis. Pin holes in the cambial face of the bark above the bud union are developed after one year of inoculation. In CVd-11a symptom develops as necrotic spot in the stem just below insertion of petiole.

THE PATHOGEN

A circular and linear form of infectious RNA molecules is associated with exocortis disease. Complete genome of exocortis viroid has been sequenced and contains 371 to 375 nucleotides which is the largest of citrus viroids (Gross *et al.*, 1982; Viswader *et al.*, 1982). Citrus exocortis viroid (CEVd) is a species in the genus *Pospiviroid* with a central conserved region similar to the the type member potato spindle tuber viroid (Duran-Vila *et al.*, 1988).

Strains in exocortis viroid do exist which may show insertion and deletion up to 27 nucleotides (Semancik, 1988). Mild and severe forms of CEVd are known to exist but no cross-protection has been demonstrated (Garnsey *et al.*, 1993). Mixed infection of viroids in field affected trees appears to be of common occurrence (Duran-Villa *et al.*, 1986). In exocortis affected trees in India, two viroids have been identified, one Indian tomato bunchy top viroid (Mishra *et al.*, 1991) and the other Citrus B (CVd-11a) group of viroids (Ramachandran *et al.*, 1993).

DIAGNOSIS AND MANAGEMENT

Citrus exocortis viroid (CEVd) can be easily identified in field trees if raised on sensitive rootstocks like Trifoliate orange, Citron and Rangpur lime. Symptoms of bark scaling and stunting of affected tree are quite conspicuous and can be easily observed. Indexing on Arizona 861-S1 Etrog citron will detect the presence of exocortis viroid by producing characteristic symptoms as described above. Exocortis viroid is detected by sequential gel electrophoresis (sPAGE) under native and denaturing (dPAGE) condions and silver staining. dPAGE increases the resolution of infectious linear and circular forms of viroids (Baksh *et al.*, 1984; Flores, 1988; Duran-Vila *et al.*, 1993). Reverse PAGE (R-PAGE) is also commonly used to detect viroid infections. CEVd is also detected by molecular hybridization of tissue imprints on a nitrocellulose membrane using radioactive probes (dot blots) and by hybridisation of CEVd-cDNA probes to RNA dot blots and Northern transfers (Gillings *et al.*, 1988; Palacio-Bielsa *et al.*, 2000). In India, exocortis viroid is detected by reverse polyacrylamide gel electrophoresis (R-PAGE) (Rustem and Ahlawat, 1999; Ghosh *et al.*, 2002). CEVd can also be detected by molecular techniques like nucleic acid hybridization and RT-PCR (Bernad *et al.*, 2006; Murcia *et al.*, 2008).

To manage citrus exocortis disease, viroid sensitive rootstocks should not be used for propagation. If there is no alternative available then only exocortis viroid-free budwood must be used for propagation. In Australia, desert lime (*Eremocitrus glauca*) has been suggested to be a good rootstock for citrus in which inoculated viroids did not reach detectable titers (Bani Hashemian *et al.*, 2010). Budwood for propagation should be obtained only from indexed viroid-free trees. CEVd contamination should be avoided by separation of healthy and diseased plants in nurseries. Operational tools like budding knives, shears, clippers etc. must be sterilized. Sterilization of tools can be done by treatment with dilute solutions of sodium hypochlorite or formaldehyde or a mixture of the two.

Cachexia (Xyloporosis)

The term "Cachexia" was coined to a condition of Orlando tangelo showing gumming and browning of phloem tissue, wood pitting and bark cankers (Childs, 1950). Between 1950 and 1956, Childs and co-workers published a series of papers on a disease of tangelos and mandarins (Childs, 1951, 1952, 1958; Childs *et al.*, 1955). They called the disease cachexia (from the Greek: kakos =bad and hexis = condition). Major cachexia disease problem occurs when infected budwood are grafted on sensitive rootstocks or infected trees are top worked with sensitive cultivars. Cachexia is found in most citrus growing regions of the world. A disease with similar symptoms termed as xyloporosis was first reported on citrus trees grafted on Palestine sweet lime in Israel (Reichert and Periberger, 1934). A possible relationship between cachexia (Childs, 1950) and the xyloporosis, a disease found in the Mediterranean region has been speculated (Childs, 1952; Roistacher, 1988). Childs 1951, 1952 concluded that symptoms of the two diseases are not identical. *e.g.* "that phloem discoloration which is characteristically pronounced in cachexia-affected Orlando tangelo trees is not present in xyloporosis-affected trees." Also the wood pitting symptoms reported for xyloporosis on Palestine sweet lime are not present or typical in Orlando tangelo. Similar observations have been made by several workers (Grant *et al.*, 1959; Calavan *et al.*, 1958; Roistacher, 1988). However, it now appears that cachexia and xyloporosis are not synonymous and that the disease in sweet lime (xyloporosis) is probably not caused by the cachexia viroid (Roistacher, 1988).

SYMPTOMS

Cachexia disease induces symptoms of discolouration and gum impregnation of the phloem and cambial pitting of the wood with corresponding pegging of the bark on mandarins, tangelos, tangors, *Citrus macrophylla, C. volkameriana* and some other varieties (Plate 47a) (Calavan and Christiansen, 1985). Bark scaling, severe leaf chlorosis, stunting and dieback are other symptoms of cachexia. It is lethal to citrus cultivars on susceptible rootstocks. Inner bark surface of contaminated trees contains numerous pegs that fit in to corresponding depressions in the wood (Plate 47b). Cachexia disease affects Alemow, Clementine, kumquats, mandarins, mandarin hybrids, Rangpur lime, Satsuma, Sweet lime Tangelos (Nocatee, Orlando, Seminola, Wekiwa) and Ellendale tangor (*Citus reticulata* × *C. sinensis*). Mandarins and Tangelos are highly susceptible hosts to cachexia pathogen. Cachexia is symptomless in grapefruit, sweet orange, lemons, citranges, citrons, citrumelos, trifoliate orange pummelos, rough lemon, sour orange, and West Indian lime. Mild symptoms develop in sweet lime, rough lemon and Rangpur lime. Tolerant hosts of cachexia are: Carrizo, Rusk and Troyer citranges, Citron (Etrog), Citrumelos, Grapefruit, lemons, limes, *Poncirus trifoliata,* Pummelos, Rough lemon, Sour orange, Sweet orange and Tahiti lime. Cachexia also infects cucumber (*Cucumis sativus* cv Suyo) causing mild leaf rugocity in inoculated leaves.

Symptoms of xyloporosis appear as small conoid pits with corresponding small pegs beneath the bud union and brown discoloration of bark and wood of affected trees. Swelling sometimes occurs at the bud union. Leaves on infected trees become small and yellow and fruit set is heavy. Sweet lime is highly susceptible to xyloporosis. Sweet orange, grapefruit and lemons are symptomless carriers. Xyloporosis symptoms of vein chlorosis, tree stunting, cankers, phloem discoloration and stem pitting were consistently present. Phloem discoloration in the tangelo scion is distinct, and diagnostic. Stem pitting consisted of bumps and projections on the bark that fit into depressions in the wood. Mapo tangelo, a hybrid between Avana mandarin and Duncan grapefruit is also susceptible to Xyloporosis-cachexia causing gum-like substances on wood and inner face of bark and the bark scaling. Symptoms of sticky exudation on lower leaf surface and dieback of twigs were also developed (Terranova *et al.,* 1991).

TRANSMISSION

Cachexia disease is transmitted by grafting (Childs, 1952). Seed transmission of cachexia was suggested because symptoms were found in non-inoculated seedlings (Childs, 1958) but not confirmed through experiments. Cachexia is mechanically transmitted when healthy plants were inoculated by knife cuts from a positive cachexia-infected citron. A quick dip of the knife blade in a 1% sodium hypochlorite solution completely prevented transmission.

INDICATOR HOST

Orlando tangelo was first proposed as indicator host of cachexia disease which was used for long time in budwood certification programmes at California and Florida (Childs, 1952). Subsequent experiments showed that Parson's Special mandarin on rough lemon rootstock was as a better indicator host (Calavan and Christiansen, 1985; Roistacher *et al.,* 1973). Parson's special mandarin is now used an indicator host for cachexia. Cachexia induces a strong reaction in Parson's special mandarin showing gum near bud union and point of cut back, pits in wood and gum in bark develop in inoculated indicator plants within one year at warm temperature

(Plate 47a). However, Parson special mandarin on Volkamer lemon rootstock is a superior indicator which induces excellent symptoms of cachexia within six months (Vogel and Bove, 1976). Arizona 861-S 1, has been identified a suitable indicator for indexing exocortis disease. This selection of citron did not produce any symptom when inoculated with cachexia but cachexia pathogen remained symptomless in this host. The reaction of these two viroids on Etrog citron suggested clear differences between exocortis and cachexia viroids (Semancik *et al.*, 1988).

THE PATHOGEN

Based on transmission similarities with exocortis viroid (CEVd), viroid aetiology of cachexia was proposed (Roistacher *et al.*, 1983) which was later confirmed (Semancik *et al.*, 1988). It is now certain that cachexia disease is caused by a viroid. Genome of citrus cachexia viroid (CCaVd) has been sequenced (Semancik *et al.*, 1988; Levy and Hadidi, 1993) which contains 300 nucleotides. Both circular and linear forms of RNAs are present in CCaVd. Cachecia viroid does not have any similarities with CEVd or citron variable viroid and appears to be a new viroid species (Semancik *et al.*, 1988). Higher yield of highly purified CCaVd was obtained from cucumber (*Cucumis sativus* cv. Suyo) through serial elution following CF-11 cellulose chromatography of a 2M-LiCl-soluble nucleic acid preparation. Cachexia viroid cDNA hybridized only to CV-IIa viroid which was identified the causal agent of mild form of citrus exocortis disease. Cachexia and CV-IIa have homology between nucleotide sequences with only a difference of five to ten nucleotides. Therefore, these two RNAs can be considered to be members of common subgroup of citrus viroid (Semancik *et al.*, 1988). Variability in cachexia was confirmed by sequence analysis of selected clones. The variable domain V was extremely conserved in all the cachexia isolates. Five nucleotide differences in both the upper (three nucleotides) and the lower (two nucleotides) strands of the V domain have been identified as the most characteristic feature discriminating cachexia from non-cachexia isolates (Palacio and Duran-Vila, 2000). Nucleotide sequence and structural features of the group III viroids have been described by Rakowski *et al.* (1994).

DIAGNOSIS AND MANAGEMENT

Cachexia viroid can be detected by its specific symptoms it produces in trees on susceptible rootstocks. Indexing is done by graft inoculation on indicator, Parson's special mandarin grafted on a vigorous rootstock or Ellendale tangor on trifoliate orange. Gum deposits near the base of scion appear within 6–12 months after inoculation (Fig. 47a). Numerous tiny pits in the wood and pointed pegs and gum in the bark were detected in Parson`s Special mandarin rootstocks with infected Marsh grapefruit tops 18 months after budding (Calavan and Christiansen, 1985). Cachexia viroid can be detected in direct PAGE (dPAGE) analysis of citron or cucumber tissue 6 to 8 weeks after inoculation (Semancik *et al.*, 1988). Cachexia viroid is also detected by sequential gel electrophoresis (sPAGE) and silver staining (Dura-Vila *et al.*, 1991), dot blot hybridization with labeled nucleic acid probes and RT-PCR (Levy and Hadidi, 1993).

For management of cachexia disease, viroid-free cultivars should be propagated, especially if cachexia viroid-susceptible rootstocks such as Rangpur lime or tangelos are to be used. Healthy cachexia-free planting material can be developed by shoot tip grafting of infected bud lines. However both pathogens are readily eliminated from micro shoot tips by shoot tip grafting in *vitro* (Roistacher *et al.*, 1980). The operational tools like budding knives, shears, clippers etc. must be sterilized to avoid spread of the disease. Sterilization of tools can be done by treatment with sterilant like sodium hypochlorite (1%) or formaldehyde or a mixture of the two.

Citrus Yellow Corky Vein Disease

Citrus yellow corky vein disease is known to occur on *Citrus* species in India and has been reported by several names such as yellow corky vein, yellow mid vein and yellow vein banding (Reddy *et al.*, 1974; Reddy and Naidu, 1989; Sharma and Pandey, 1983; Ahlawat and Chakraborty, 1993). Yellow corky vein disease is now most accepted name of this disease. Yellow corky vein disease causes yield loss from 60.4 to 89.7% in lime in Assam (Azad, 1993). Except graft transmission no other information was available on this disease till it was thoroughly investigated by author and his student, Furdoos Ali Rustem from Yemen between 1996 to 1998 (Rustem and Ahlawat, 1999; Rustem *et al.*, 2000). Recently, a similar disease has been reported on Washington navel orange in Fars province of Iran (Bagherian *et al.*, 2009).

SYMPTOMS

In naturally infected Kagzi lime (*Citrus aurantifolia*), symptoms of yellow corky vein disease develop as yellowing of midrib and lateral veins followed by formation of corky tissues on lower surface of leaf veins. In experimentally inoculated kagzi lime plants, yellowing of leaf veins including midvein were developed 4–6 days after inoculation (Plate 48a). Yellowing and corking may cover all the veins or part of them. Corking is also developed on petioles, both young and old branches of yellow corky vein affected trees (Plate 49b-1). Shrinking of midrib and crinkling of leaves have also been observed in advanced cases of infection (Plate 49a-2). Typical symptoms of yellow corky vein disease are developed at higher temperatures. Dieback symptoms are common in field affected trees. Glasshouse inoculated plants also show dieback symptoms within 3–4 years of inoculation (Rustem and Ahlawat, 1999).

TRANSMISSION

Yellow corky vein disease (CYCVD) is transmitted up to 90% by wedge grafting and 60% by bud grafting. CYCVD is also transmitted by mechanical inoculation from kagzi lime to kagzi lime using glycine buffer. The agent of yellow corky vein is not transmitted through seed and no insect vector is known. Maximum transmission of the disease was obtained at a temperature of 41.7/22.7°C (maximum/minimum). Symptom of yellow corky vein disease appeared in graft inoculated plants in 9 days at this temperature.

HOST RANGE

Most *Citrus* species and cultivars are susceptible to yellow corky vein infection. Susceptible *Citrus* species are: Citranges (Carrizo, Savage, Troyer), Etrog Citron, *Citrus decumana*, Mandarins (Cleopatra, Darjeeling/Khasi, Nagpur, Kinnow), Lemons (Galgal, hill, Volkamer), Limes (Kagzi and kagzi kalan), Nasnaran, Rough lemon (Jatti khatti), Sour orange, Sweet orange (Folkan and Mosambi), and Trifoliate orange. Rough lemon cultivars Karna khatta and Soh Sarkar did not show any symptoms upon inoculation and appear to be resistant to yellow corky vein disease. These hosts are used as rootstocks in India hence the reaction of different scion varieties should be tested on these rootstocks. Symptoms developed on susceptible hosts are given below.

- ***Citranges*** (Carrizo, Savage, Troyer): Yellow vein banding, vein corking.
- ***Etrog Citron*:** Yellow vein banding, vein corking, veinal necrosis and black dots on veins or dark veins (Plate 48b, 49d)
- ***Citrus decumana***: Leaf epinasty.
- ***Mandarins (Cleopatra, Darjeeling/Khasi, Nagpur, Kinnow)*:** Crinkling of leaves, yellow vein banding, vein corking, veinal necrosis, leaf epinasty (Plate 48c, 49a-1, 49a-2).
- ***Lemons (Galgal, hill, Volkamer)*:** Chlorotic spots, yellow vein banding, veinal necrosis, leaf crinkling and rolling and epinasty of leaves.
- ***Lime (Kagzi and kagzi kalan)*:** Chlorotic spots, yellow vein banding, corking of veins, veinal necrosis, and epinasty of leaves. Corking develops on stem and twigs (Plates 48a, 49b-1).
- ***Nasnaran*:** Yellow vein banding and vein corking.
- ***Rough lemon (Jatti khatti)*:** Chlorotic spots, yellow vein banding, vein corcking, veinal Necrosis.
- ***Rangpur lime*:** Partial or complete yellowing of leaf veins (Plate 48d)
- ***Sour orange*:** Leaf crinkling, yellow vein banding, vein corking, veinal necrosis, leaf epinasty.
- ***Sweet orange (Folkan and Mosambi)*:** Crinkling of leaves, partial yellow vein banding, vein corcking, necrosis of veins, rolling and epinasty of leaves.

- ***Trifoliate orange***: partly vein banding and vein corking, corking on shoots and twigs, yellowing and crinkling of leaves (Plate 49b-2, 49c).

INDICATOR HOST

Kagzi lime (*Citrus aurantifolia*) has been established as indicator host of yellow corky vein disease. Typical symptoms of the disease appear within two weeks of grafting and 4–6 weeks after mechanical inoculation at suitable temperature. Kagzi lime is also a good host for propagation of the pathogen.

THE PATHOGEN

Viroid etiology of yellow corky vein disease was established by polyacrlamide gel electrophoresis (PAGE) especially return PAGE (R-PAGE) (Fig. 34) and by nucleic acid hybridization studies. Yellow corky vein pathogen reacted positively with specific probes of exocortis and hop stunt viroids (Fig. 32A, B) suggesting that two viroids, exocortis and hop stunt were associated with yellow corky vein disease (Rustem, 1998; Rustem and Ahlawat, 1999; Ramachandran *et al.*, 2003). Reverse transcription polymerase chain reaction (RT-PCR) followed by cloning and sequencing of PCR products of exocortis component of yellow corky vein revealed that CEVd component consists of 370 nucleotides. This viroid was tentatively designated as yellow corky vein variant of Citrus exocortis viroid (CEVd-ycv). CEVd-ycv showed close phylogenetic relationship with CEVd Gynura variants reported from Australia, but was found to be distantly related to *Citrus exocortis* tomato variant (CEVd-t) previously reported from India (Mishra *et al.*, 1991; Roy and Ramachandran, 2006). Hop stunt component was also subsequently

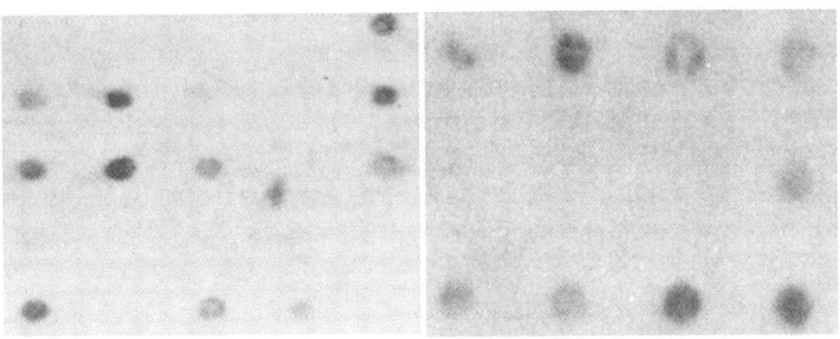

A B

Fig. 32: A. Dot blot hybridization of CYCV with exocortis specific probes; B. Dot blot hybridization of CYCV with hop stunt specific probes

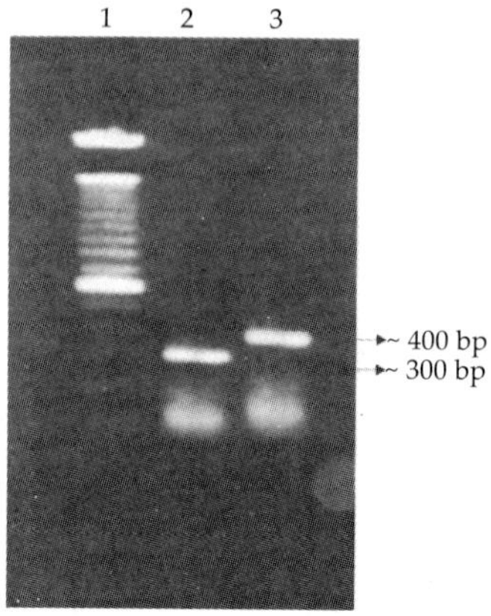

Fig. 33: 1% agarose electrophoresis of RT-PCR amplified product; lane 1:100bp DNA ladder, Lane 2: amplification with HSVD-specific primer-300bp size of amplified product, Lane-3: amplification with CEVd specific primer-400bp size of amplified product (*Source*: P. Ramachandran)

sequenced. It consisted of 295 nucleotides. Hop stunt viroid (HSVd) variants showed nearly 100% sequence identity with six citrus cachexia isolates of HSVd. The viroid was tentatively named as yellow corky vein variant of Hop stunt viroid (HSVd-ycv) (Roy and Ramachandran, 2003). Yellow corky vein disease reported on Washington navel orange in Fars province of Iran was also reported to be associated with hop stunt viroid. This viroid in Iran was found to be a novel variant of hop stunt viroid based on comparison of molecular properties (nucleotide composition, primary and secondary structures, molecular weights, phylogenetic relationships and percent nucleotide similarity and difference) of this viroid variant. It was found that a novel variant of hop stunt viroid (HSVd-sycv) was associated with yellow corky vein disease of Washington navel in Iran. HSVd-sycv was closely related to non-cachexia variant of hop stunt viroid (HSVd-cit) but only with 93.7% homology with HSVd-lycv. It differed in a single nucleotide from HSVd-cit, in the variable domain in the so-called "cachexia expression motif" (Bagherian and Izadpanah, 2010). A variant of hop stunt viroid has also been detected in Nagpur mandarin and Mosambi sweet orange trees on rough lemon and Rangpur lime rootstocks (Ramachandran *et al.*, 2005). However, it could not be experimentally established whether typical symptoms of yellow corky vein disease are caused by hop stunt or exocortis viroids or by synergistic effect of the two viroid species.

DIAGNOSIS AND MANAGEMENT

Yellow corky vein disease is easily detected in field by specific yellow corky vein symptoms. It can be easily detected by graft inoculation on indicator

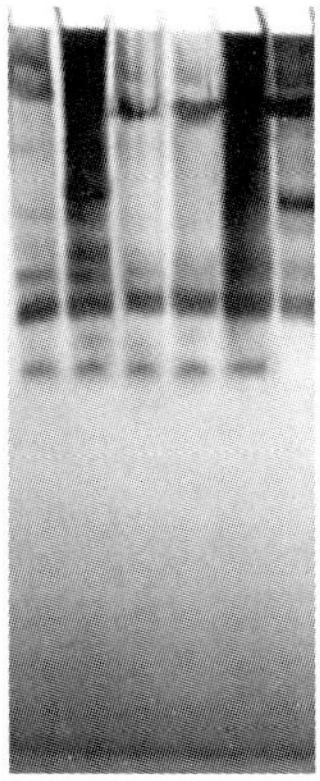

Fig. 34: Detection of CYCV viroid in R-PAGE. Low molecular weight RNAs are shown from CYCV infected sample.

host, Kagzi lime seedlings which will develop typical symptoms of the disease one week after inoculation if kept at a desired temperature. Yellow corcky vein viroid is also detected in R-PAGE and dot-blot hybridization using CEVd and hop stunt viroid probes (Fig. 32A, B). It is also possible to detect both the viroids associated with yellow corky vein disease in RT-PCR (Fig. 33) (Roy and Ramachandran, 2006). Yellow corky vein disease can be managed by use of resistant rootstocks like Karna khatta and Soh Sarkar. These viroids can also be eliminated by shoot tip grafting. Kagzi lime has high degree of polyembryony and hence nucellar plants can be used as planting material.

Citrus Gummy Bark

Gummy bark disease was first reported as phloem discolouration of Baladi and other sweet orange trees grafted on sour orange rootstock in Egypt in 1954 (Nour-Eldin, 1956, 1959). This disease is now known to occur in a number of eastern Mediterranean, North African, near East countries and Middle East countries including Egypt, Greece, Iran, Iraq, Syria, Libya, Sultanate of Oman, Saudi Arabia, Yemen, Sudan and Turkey but not in North and South America (Roistacher, 1991; Bové, 1995). In some regions of Turkey, 12-48.4 % of sweet orange trees budded on sour orange are affected (Gullu, 1989; Cinar *et al.,* 1993). Navel oranges infected with gummy bark disease of citrus in the Eastern Mediterranean region showed an incidence of 28% (Gullu, 1989). The disease is now popularly known as Gummy bark.

SYMTOMS

As the name suggests, citrus gummy bark pathogen causes problems in the bark of the sweet orange tree. Infected trees show stunting, wood pitting in the trunk and gum impregnation above the bud union in Turkey (Salibe, 1986). Trees affected with gummy bark are stunted. In affected trees, streaks of reddish–brown gum impregnated tissue are seen near bud union when bark of trunk is scrapped (Plate 50). In severe cases, the discoloration may extend above the bud union affecting the trunk and even the main branches. However, in milder cases gumming is restricted to bud union only. Stem pitting is also developed in affected trees which can be easily seen after removal of bark. Stem pitting caused by gummy bark resembles those of caused by cachexia on cachexia-sensitive hosts and those of Kassala disease of grapefruit. Two types of pits are observed in gummy bark disease 1) with grooved or channeled appearance to the cambial face of the wood and 2) round and conoid. Gum impregnates the inner tissue of the pegs. These symptoms are seen in trees older than 10 years. Early symptom of gummy bark is discolouration near bud union. Later this discolouration extends upwards up to one metre or more. Sweet orange trees grafted on rough lemon showing gummy bark symptoms in sweet orange scion also show bark pegging and gumming, and stem pitting on the rootstock. In addition to these symptoms bud-union crease can also be seen on such trees (Bové, 1995) but bud union crease in rough lemon may not be due to gummy bark agent. The amount of gum in the sweet orange bark above the bud union is quite high and may cause bark scaling (Bové, 1995).There are no leaf or fruit symptoms. In Yemen, gummy bark-affected sweet orange trees on rough lemon rootstock showed not only sweet orange bark gumming and bud-union crease, but also, sometimes show severe stem pitting and bark pegging on the rootstock. Graft transmission of gummy bark sources to sweet orange grafted on Orlando tangelo resulted in the development of symptoms on the sweet orange scion in the absence of cachexia symptoms on Orlando tangelo (Nour-Eldin, 1968). He, thus, concluded that the two diseases were not caused by the same agent. Conversely, transmission tests to sweet orange grafted on rough lemon induced a bud union constriction demonstrating that rough lemon is also sensitive to the putative gummy bark agent (Nour-Eldin, 1968).

TRANSMISSION

Gummy bark disorder is transmitted by grafting contaminated buds, shoots and bark to healthy test plants. This disease is not mechanically transmissible. Gummy bark disease agent is seed borne in Dortyol and Washington navel sweet orange seeds.

HOST-RANGE

Host-range of this disorder is restricted to citrus species and cultivars. Sensitive species are Sweet oranges and rough lemon (Nour-Eldin, 1968). Sweet oranges are comparatively more susceptible. Tolerant (symptomless carrier) are: Sour orange, Mexican lime, grapefruit, willowleaf mandarin, Orlando tangelo, trifoliate orange, citrange hybrids, *Citrus amblicarpa,* Rangpur lime and lemon.

INDICATOR HOST

Sweet orange scions on rough lemon or sour orange rootstocks are used as indicator host. Symptoms on inoculated indicator host develop as gumming of sweet orange bark and often of rough lemon bark and bud-union constriction. Development of typical symptoms of the disease takes several years.

THE PATHOGEN

Since the disease was graft transmissible, it was considered to be caused by a virus. Following the discovery that cachexia disease is caused by a viroid (Semancik *et al.,* 1988), it was assumed that gummy bark disease might also be caused by a viroid. Association of viroid with gummy bark disease was reported by some workers (Semansik and Duran Vila, 1991; Önelge *et al.,* 2004). The viroid associated with gummy bark disease was characterized as a specific variant of hop stunt viroid (HSVd) (Reanwarakorn and Semancik, 1998). HSVd variants inducing cachexia differed from non-cachexia variants in only 5–6 nucleotides located in the V domain (Reanwarakorn and Semancik, 1998). All symptomatic trees from Turkey, Egypt and the Sultanate of Oman showed several viroid species, *Citrus exocortis viroid* (CEVd), *Hop stunt viroid* (HSVd), *Citrus viroid III* (CVd-III) recently renamed *Citrus dwarfing viroid* (CDVd) and *Citrus viroid IV* (CVd-IV) recently renamed *Citrus bark cracking viroid* (CBCVd) (Önelge *et al.,* 1996; Bernad *et al.,* 2005). However, Studies conducted with sources from Turkey failed to identify a sequence variant of hop stunt viroid (HSVd) as the putative disease agent of gummy bark (Önelge *et al.,* 2004). In sPAGE, gummy bark affected trees showed a viroid profile consisting of CEVd, CVd-1, CVd-11 group, CVd-111 group and CVd-IV group. The presence of CVd-11 group component in all gummy bark samples and the similarities of gummy bark symptoms with those of cachexia disease caused by CVd-11b and CVd-11c suggested that a variant of CVd-11 may be either the causal agent or a factor involved in gummy bark (Onelge *et al.,* 1996, 2002). Samples from symptomatic sweet orange trees were grafted to etrog citron and analysed by sPAGE and slot-blot hybridization using viroid-specific probes. In addition to hop stunt viroid (HSVd), all samples also contained other viroid species, CEVd, CVd-III and CVd-IV. Sequence analysis of DNA amplicons generated by RT-PCR using specific primers for these viroids identified novel variants of CEVd (CEVd-gb) and CVd-III (CVd-III-gb) in all the gummy bark sources, whereas CVd-IV was identical to previously reported sequences (Bernard *et al.,* 2005). Electrophoresis, under denaturing conditions of total nucleic acid extracted from the grafted Etrog citron plants with symptoms, indicated the presence of viroid-like circular low molecular weight (LMW) RNA in all the samples. These circular LMW-RNAs were used in RT-PCR using a set of HSVd specific primers. Amplicons of ~300 bp were obtained in all the samples with primers specific to *Hop stunt viroid* (HSVd) (El-Dougdoug *et al.,* 2010). In order to study the degree of sequence variation among *Hop stunt viroid,* gel-purified RT-PCR product of the viroid was cloned and then sequenced (GenBank Accession No. FJ984562). CVd-II from diseased sweet orange trees is 299 nucleotides in length and shares 100% identity with CVd IIb (AF131249); (Reanwarakorn and Semancik, 1998). Finally, it has been established that citrus gummy bark viroid is a sub-species of the hop stunt viroid, and thus is a member of the genus Hostuviroid (Sofy *et al.,* 2010).

DIAGNOSIS AND MANAGEMENT

For diagnosis of the disease, sweet orange on sour orange rootstock indicator plants are grafted by bark patch. Inoculated plants will show a line of reddish-brown gum impregnated tissue just above the bud-union of sweet orange on sour orange. Sweet orange on rough lemon also develops the same brownish streak at the bud-union but with the formation of a continuous series of pegs at the bud-union forming a bud-union crease. These symptoms appear after 4–5 years of inoculation but Sweet orange on rough lemon develops the symptoms in about 3 years. The gummy bark disease can also be identified by sPAGE and slot-blot hybridization using viroid-specific probes and RT-PCR (Bernard *et al.*, 2005; El-Dougdoug *et al.*, 2010). The disease is managed by use of gummy bark-free budwood.

Gum Pocket and Gummy Pitting

A disorder of trifoliate orange rootstock was reported from Argentina (DuCharme and Knorr, 1954; Pujol, 1965; Foguet and Oste, 1968), South Africa (Schwarz and McClean, 1969;), Australia, Italy, Turkey and Chile (Fraser *et al.* 1976; Cartia *et al.*, 1984). The disorder in Argentina was called as 'wood pitting'. It was characterized by gum impregnated projections on the cambial face fitting into creases in the wood of trifoliate orange rootstock. In South Africa, it was called as 'Gum pocket' where it caused malformation of bud union in trifoliate orange rootstock with sweet orange scion. In Australia and other places, it was called as 'gummy pitting'. In gummy pitting, gum impregnated pits in the surface of wood associated with gumming in xylem and phloem of trifoliate orange rootstock is the main symptom (Plate 51). The terms 'Gum pocket', 'Wood pitting' and 'Gummy pitting' have been used to describe a characteristic disorder of trifoliate orange. These diseases commonly occur in old budwood lines. Fraser *et al.* (1976) showed that Gummy pitting of *Poncirus trifoliata* is associated with dwarfing of citrus in New south Wales (NSW), Australia. A method to make trees dwarfed on trifoliate orange by inoculating with a bud transmissible dwarfing factor was being tried in some countries as a possible means of producing high density planting of small trees on commercial basis (Long *et al.*, 1971). The dwarfing factor will be dealt with separately in following chapter. It has been observed that mixtures of strains of gummy pitting and exocortis may occur commonly (Fraser *et al.*, 1976). It is also known that more rapid wood-pitting symptom will develop in the presence of exocortis viroid (Foguet, 1973).

SYMPTOMS

Wood pitting disease produces gum impregnated pegs on the cambial face which fit in the pits of wood. The disease is common in trifoliate orange rootstock. It is usually associated with an indented ring at the bud-union, flattened and longitudinal fissures and laminate shelling of the bark (Du Charme and Knorr, 1954; Fernandez-Valiela *et al.*, 1965; Foguet and Oste, 1968). Gum pocket is considered to be similar to wood pitting. It forms gum pocket and malformation of bud union of trifoliate orange rootstock. Gum impregnated pits in the surface of wood associated with gumming in xylem and phloem of trifoliate orange rootstock are the main symptoms of gummy pitting. Symptoms are most severe in the shoulder of the stock below the bud union. Presence of severe pitting can be detected externally by uneven, flattened or depressed areas of bark (Fraser *et al.*, 1976). These pitting can be detected externally by uneven flattened or depressed areas in bark. Severe symptoms of trunk flattening normally appear towards the side of sun exposure. Affected trees are stunted. Symptoms on leaves or fruits have not been observed.

TRANSMISSION

Gummy pitting disease is graft transmissible from symptomatic trees to nucellar Eureka lemon on trifoliate orange rootstock and by direct propagation from contaminated trees (Fraser *et al.*, 1976).

HOST RANGE

Sweet orange, Marsh grapefruit, Clementine mandarin and Eureka lemon trees grafted on trifoliate orange rootstock develop better symptoms of the disease. Symptomless hosts of the disease are: Troyer citrange, Empress Mandarin and Rangpur lime. Rangpur lime is a tolerant host to this disease.

INDICATOR HOST

Sweet orange on *Poncirus trifoliata* is used as indicator host. Gum pockets are developed in the bark and wood of inoculated plants several years after graft inoculation.

THE PATHOGEN

Earlier observations on etiology of gum pocket disease suggested that the symptoms may be of physiological or stress origin (Du Charme and Knorr, 1954; Fernandez-Valiela *et al.*, 1965). The observations made on viroid infected trees maintained for at least ten years in field trials indicated that viroid infection may act simply by enhancing the development of non specific symptoms (Duran-Vila *et al.*, 2002). Later studies, however, revealed that gum pocket disease was associated with a viroid of the same apparent size as group I11 viroids as evident from the experiments of Polyacrylamide gel electrophoresis (PAGE) on non-denaturing 5% gels. Viroid-like RNAs with the electrophoretic mobility of CVd-III viroid different from exocortis was oberved (Marais *et al.*, 1996). CVd-III was also observed in Tahiti lime (*Citrus latifolia*) by RT-PCR and multiplex RT-PCR. In multiplex PCR in addion of CVd III, exocortis and hop stunt viroid were also detected (Almeyda-Leon *et al.*, 2002). Three viroid species, *Citrus exocortis viroid* (CEVd), *Hop stunt viroid* (HSVd) and *Citrus viroid-III* (CVd-III) isolated from a single naturally infected source from Africa were sequenced and their sequence variability was determined. This work confirms that naturally occurring viroid infections contain a mixture of sequence variants (Elleuch *et al.*, 2006).

DIAGNOSIS AND MANAGEMENT

The disease can be identified in field trees by observing symptoms of gum impregnated pits in the surface of the wood of trifoliate orange rootstock. Severe symptoms are observed in the shoulder of stock below bud union. Presence of pittings can be detected externally by uneven flattened or depressed areas in bark (Fraser *et al.*, 1976). The disorder can also be diagnosed by inoculating indicator host (sweet orange on trifoliate roots) under experimental condition. RT-PCR and multiplex PCR systems have also been suggested to detect the associated viroid (Almeyda-Leon *et al.*, 2002). Use of budwood derived from shoot tip grafting *in vitro* or nucellar embryony have been suggested to manage the disease.

Shell Bark

Shell bark disease was reported in old-line lemon trees in California (Calavan and Weathers, 1959). Symptoms of this disease appear in trees older than 10 years or more. Outer bark of affected trees cracks into vertical strips and dies. These symptoms initially start as small patches but later involve more branches and major portion of the trunk. After the old bark shell is out, new bark develops vertical strips but in severely affected trees even this new bark may dry out and crack allowing the decay organisms to enter the cambium and expose the wood resulting serious injury to infected tree. Lemon trees develop symptoms above the bud union. Shell bark symptoms are more severe in Eureka lemon than Lisbon lemon.

Shell bark symptoms develop when lemon trees are grafted on Seville orange, bitter sweet and bitter rough Seville oranges, Shaddock, Wheeny grapefruit and Poorman orange. Shell bark is a transmissible factor (Calavan and Weathers, 1959). The etiology of shell bark is yet to be determined but trees with shell bark symptoms usually carry a number of citrus viroids, including citrus exocortis viroid.

Shell bark is identified in field affected trees as cracking of outer bark into vertical strips (Calavan and Weathers, 1959). Shell bark disease can be eliminated by nucellar embryony or shoot tip grafting *in vitro* and thus healthy planting material is obtained. Budwood should be taken only from old parent trees free of shell bark. Temporary recovery from shell bark is obtained by pruning of affected trees.

Dwarfing Citrus Trees by Viroids

Dwarfing of citrus trees on trifoliate orange rootstock infected with a graft transmissible factor (GTF) without showing scaling symptoms on rootstock was reported from Australia (Fraser *et al.*, 1961). The yield was enhanced per ha through close planting of such dwarfed orange trees (Long *et al.*, 1972). Relationships between dwarfing and the effects of tree spacing on productivity have been investigated (Hutton and Cullis, 1981; Hutton, 1986). A factor contributing to the higher yield efficiency of inoculated trees was their denser canopy structure, which contributed to a greater number of potential fruit bearing sites relative to tree size. The dwarfing factor was graft transmissible and specific viroids were found associated with dwarfing trees in Australia and Israel (Schwinghamer and Broadbent, 1987; Gilling *et al.*, 1991; Bar-Joseph, 1993). Viroids causing dwarfing trees on *Poncirus trifoliata* and its hybrids are found in most of the major citrus growing countries. Israel and Australia are the major countries taking advantage of dwarfing technology and 115000 dwarfing buds were sold between 2003–2006 but it has now been realized that the use of graft transmissible dwarfing (GTD) is incompatible with the operations of a high health status budwood scheme. It has become obvious that GTD has benefits while conditions for tree growth and management are optimal, but any stress on trees (*e.g.*, drought, salinity) will have a greater adverse effect on GTD-inoculated trees than their viroid-free counterparts (Barkley and Hutton, 2007). This dwarfing factor is known as viroid dwarfing and graft transmissible dwarfing (GTD).

SYMPTOMS

The growth of GTD inoculated trees is reduced within two years after planting and canopy size is also affected within five years (Bevington and Bacon, 1977; Hutton, 2004). Total spring vegetative shoot growth and total vegetative growth during spring is significantly less. Morphology of autumn shoot growth is different, with more shoots of smaller internode length and fewer leaves developing on GTD trees than on non-inoculated trees whose autumn growth was sparse and appeared as rank water shoot growth in the top half of the tree canopy (Hutton *et al.*, 2000; Hutton, 2004). Stock is broader than the scion trunk and often predominantly shouldered or benched at the bud union, in contrast to non-inoculated trees in which the stock is fluted and evenly expanding from the bud union to the crown roots. The bark of the stock develops thin flaky bud scales or corky pustules which are visible after the bark is removed (Fraser and Broadbent,

1979). GTD inoculation has no effect on the timing, duration and intensity of flower production, patterns of fruit development and final fruit size at harvest, and internal fruit juice quality characteristics. As GTD trees produced a higher yield/ha, less water was used to produce every kg of fruit in GTD trees than in non-inoculated trees (Hutton, 2004).

TRANSMISSION

GTD is bud, graft and mechanically transmitted but not by vector or seed. Trees in field should be inoculated during the first year (Hutton *et al.*, 2000).

HOST RANGE

All *Citrus* spp. are susceptible to citrus dwarfing, but only *Poncirus trifoliata* and its hybrids (*e.g.*, citranges) and Rangpur lime provide a dwarfing response when inoculated with dwarfing factor. In Valencia sweet orange on *P. trifoliata,* deep pits with corresponding pegs develop in the bark with *Citrus bent leaf viroid* – CBLVd (variant CVd-Ia) infection, mild bark cracks with CVd-IIa and finger imprint and bark striations with CVd-IIIb infection (Roistacher *et al.*, 1993). Trifoliate orange seedlings and rootstocks infected with three different variants of HSVd (variants CVd-IIa, CVd-IIb and CVd-IIc) (Vernière *et al.*, 2002) and *Citrus viroid IV* - CVd-IV (Vernière *et al.*, 2002) showed characteristic bark cracks. CLBVd and CVd-III did not produce bark symptoms (Vernière *et al.*, 2002).

THE PATHOGEN

Natural isolates of [*Citrus viroid III*- CVd-III (variant CVd-IIIb) +/- *Hop stunt viroid*- HSVd (variant CVd-IIa)] are associated with dwarfing. All GTD isolates in Australia carry variant CVd-IIIa of CVd-III (Gilling *et al.*, 1991), sometimes in the presence of variant CVd-IIa of HSVd. CVd-IIIa variants present 99.3% sequence identity with the type member of CVd-III *i.e.* CVd-IIIb variant (Rakowski *et al.*, 1994). However, the reduction in tree size and fruit yield occurred mainly in trees infected with the two apscaviroids CVd-III and to a lesser extent with CBLVd (Vernière *et al.*, 2002; Duran-Vila *et al.*, 1988).

Citrus Bent Leaf Viroid

Several citrus viroids were obtained when Etrog citron was inoculated with graft transmissible citrus dwarfing agent (GTCDA) in Israel. GTCDA isolate 225# T showed dwarfing in some grapefruit trees in Israel. Citrus bent leaf viroid (CBLVd) previously designated as CV 1b was purified and circular RNA molecules were separated (Ashulin *et al.*, 1991). The PCR generated cDNA was cloned. Plasmid containing inserts which hybridized positively with CBLVd were sequenced. CBLVd consists of 318 nts which can be arranged in the viroid specific rod like structure (Keese and Simons, 1985). CBLVd is the first member of the apple scar skin viroid subgroup infecting citrus plants.

Citrus bent leaf viroid (CBLVd) has also been recored in Japan, the Philippines and China. CBLVd isolate sequenced in China consists of 318 nt and share 98% identity of CBLVd-225A from Israel. CVd-Ia was clearly separated from CBLVd by two 5-nt insertions located in upper (5'-AGCUG-3') and the lower (5'-CUUCU-3') strand of the right terminal region (which is also designated T2 domain) in rod-like secondary structure. Since both of the additional 5-nt sequences are similar to the adjacent sequences (5'-AGUUG-3' and 5'-CUUCU-3') (Hataya *et al.*, 1998), they hypothesized that CVd-Ia is a derivative of CBLVd caused by partial sequence duplications and substitutions taking place in the right terminal region.

A new viroid species consisting of 330–331 nucleotides with a central conserved region, characteristic of the genus *Apscaviroid* was reported from Japan. This viroid caused only mild petiole necrosis and mild leaf bending in Arizona 861–S1 Etrog citron (*C. medica*). These characteristic on this new viroid differed from other citrus viroids (Ito *et al.*, 2001). They concluded that this viroid could be a new citrus viroid species belonging to the genus *Apscaviroid* and was tentatively named *Citrus* viroid OS (CVd-OS) after the original sample. CVd-OS has high sequence similarity with CVd-III and *Apple dimple fruit viroid* in the putative central and terminal left domains, including a duplicative sequence from the lower CCR of the genus *Pospiviroid* in the left terminus. Further, CVd-OS shares high sequence similarity with *Citrus exocortis viroid* (CEVd) in the lower strand of the putative variable and terminal right domains, including the sequence identical to that of the right termini of CEVd and of *Citrus IV viroid* (Ito *et al.*, 2001). Citrus bent leaf viroid has been classified in the genus Apscaviroid in family Pospiviroidae and strain 80.001.0.04.004 *Citrus bent leaf* viroid (Index of Viruses, 2006).

REFERENCES

Ahlawat, Y.S. and Chakraborty, N.K. 1993. Yellow vein banding disease of citrus in India. *Indian Phytopath.* 46: 274.

Almeyda-Leon, I.H., Iracheta-cardenas, M.M., Jasso-argumedo, J., Kurti-Diaz, S.A., Ruiz- Beltran, P. and Roch-Pena, M.A. 2002. Reexamination of citrus viroids in Tahiti lime in Mexico. *Revista Mexicana de Fytopatologia* 20: 152–160.

Ashkenazi, S. and Oren, Y. 1988. The use of citrus exocortis virus (CEV) for tree size control in Israel – a practical aspect. *Proc. Interl. Soc. Citriculture* 2: 917–919.

Ashulin, L., Lachman, O., Hadas, R. and Bar-Joseph, M. 1991. Nucleotide sequence of a new viroid species, citrus bent leaf viroid (CBLVd) isolated from grapefruit in Israel. *Nucleic Acid Res.* 19: 47–67.

Azad, P. 1993. Reduction in yield of acid lime infected with citrus yellow corky vein virus in Assam. p. 458 *In*: Proc. 12th Conf. IOCV, IOCV, Riverside, CA.

Bagherian, Seyed Ali Akbar and Izadpanah, K. 2009. Two navel variants of hop stunt viroid with yellow corky vein disease of sweet orange and split bark disorder of sweet lime in Iran. p. 34. *In*: 21st Internl. Conf. on Virus and other Graft Transmissible Diseases of Fruit Crops. Berichte aus dem JKI 148 (Abs).

Bagherian, S.A.A. and Izadpanah, K. 2010. Two novel variants of hop stunt viroid associated with yellow corky vein disease of sweet orange and split bark disorder of sweet lime. 21st International Conference on Virus and other Graft Transmissible Diseases of Fruit Crops. *Julius-Kühn-Archiv,* 427: 105–113.

Baksh, N., Lee, R.F. and Garnsey, S.M. 1984. Detection of citrus exocortis viroid by polyacrylamide gel electrophoresis, pp. 343–352. *In*: Proc. 9th Conf. IOCV. IOCV, Riverside, CA.

Bani Hashemian, S.M., Barbosa, C.J., Pina, J.A. and Duran-Vila, N. 2010. Is desert lime (*Eremocitrus glauca*) resistant to viroid infection or only a poor viroid host? Pro. 17th Conf. IOCV, IOCV, Riverside.

Bar-Joseph, M. 1993. Citrus viroids and citrus dwarfing in Israel. *Acta Hort.* 349: 271–276.

Barkley, Particia (Broadbent) and Hutton, R.J. 2007. Dwarfing citrus trees by viroids (IOCV slide collection at www.ivia.es/iocv).

Benton, R.J., Bowman, F.T., Fraser, L. and Kebby, R.G. 1950. Stunting and scaly butt of citrus associated with *Poncirus trifoliata* rootstock. *N.S.W. Dept. Agric. Sci. Bull.* 70: 20.

Bernard, L., Moreno, P., Bove, J.M. and Duran-Vila, N. 2005. Viroids in gummy bark sources from the Sultanate of Oman. pp. 272–279. *In*: Proc. 16th Conf. IOCV, IOCV, Riverside

Bernad, L. and Duran-Vila, N. 2006. A novel RT-PCR approach for detection and characterization of citrus viroids. *Mol. Cell. Probes.* 20: 105–113.

Bevington, K.B. and Bacon, P.E. 1977. Effects of rootstocks on the response of navel orange trees to dwarfing inoculations. *Proc. Int. Soc. Citriculture,* 2: 567–570.

Bitters, W.P. 1952. Exocortis on trifoliate. *Citrus Leaves* 32(9): 14–16.

Bové, J.M. 1995. Virus and virus-like diseases of citrus in the Near East region. FAO, Rome. p. 518.

Childs, J.F.L. 1950. The cachexia disease of *Orlando tangelo*. *Plant Dis. Reptr*. 34: 295–298.

Childs, J.F.L. 1951. Cachexia, a bud-transmitted disease and the manifestation of phloem symptoms in certain varieties of citrus, citrus relatives, and hybrids. *Proc. Florida State Hort. Soc*. 64: 47–51.

Childs, J.F.L. 1952. Cachexia disease: Its bud transmission and relation to xyloporasis and tristeza. *Phytopathology* 42: 265–268.

Childs, J.F.L. 1958. Transmission experiments and xyloporosis-cachexia relations in Florida. *Plant Dis. Rep*. 40: 143–145.

Childs, J.F.L., Grimm, G.R., Grant, T.J., Knom, L.C., and Norman, G. 1955. The incidence of xyloporosis (cachexia) in certain Florida citrus varieties. *Proc. Fla. State Hort. Soc*. 68: 77–82.

Calavan, E.C., Carpenter, J.B. and Weathers, L.G. 1958. Observations on distribution of cachexia of citrus in California and Arizona. *Plant Dis. Rep*. 42: 1054–1056.

Calavan, E.C. and Weathers, L.G. 1959. Transmission of growth retarding factor in Eureka lemon trees. pp. 167–177. *In*: Citrus virus diseases. Wallace, J.M. *Ed*. Univ. California, *Div. Agric. Sci.* Berkeley.

Calavan, E.C. and Christiansen, D. 1985. Variability of cachexia reaction among varieties of rootstocks within clonal propagations of citrus, pp. 76–85. *In*: Proc. 3rd Conf. IOCV. IOCV, Riverside, CA.

Cartia, G., Rosa La Rosa and Catara, A. 1984. A gummy pitting of trifoliate orange in Italy. pp. 184–187. *In*: Proc. 9th Conf. IOCV, IOCV, Riverside, CA.

Cinar, A., Kersting, U., Onelge, N., Korkmaz, S. and Sas, G. 1993. Citrus virus and virus-like diseases in the eastern Mediterranean region, Turkey. pp. 397–401 *In*: Proc. 12th Conf. IOCV, IOCV, Riverside, CA.

Cohen, M. 1968. Exocortis virus as a possible factor in producing dwarf citrus trees. *Proc. Interl. Soc. Citriculture* 81: 115–119.

Du Charme, E.P. and Knorr, L.C. 1954. Vascular pits and pegs associated with diseases in citrus. *Plant Disease Reporter* 38: 127–142.

Duran-Vila, N., Flores, R. and Semancik, J.S. 1986. Characterization of viroid-like RNAs associated with the citrus exocortis syndrome. *Virology* 150: 75–84.

Duran-Vila, N., Pina, J.A., Balester, J.F., Juarez, J., Roistacher, C.N., Rivera-Bustamante, R. and Semancik, J.S. 1988. The citrus exocortis disease: a complex of viroid RNAs. pp. 152–164. *In*: Proc. 10th Conf. IOCV, IOCV, Riverside, CA.

Duran-Vila, N. Roistacher, C.N., Rivera-Bustamante, R. and Semancik, J.S. 1988. A definition of citrus viroid groups and their relationship to the exocortis disease. *J. Gen. Virol*. 69: 3069–3080.

Dura-Vila, N., Pina, J.A., Molins, M.I. and Navarro, L. 1991. A new indexing method for cachexia. pp. 224–229. *In*: Proc. 11th Conf., IOCV, IOCV, Riverside, CA.

Duran Vila, N., Pina, J.A. and Navarro, L. 1993. Improved indexing of citrus viroids. pp. 202–211. *In*: Proc. 12th Conf. IOCV, IOCV, Riverside, CA.

Duran-Vila, N., Botella, N.L. and Vernière, C. 2002. Wood pitting – gum pocket – gummy pitting of trifoliate orange: considerations about their etiology. p. 384-386. *In*: Proc. 15th Conf. IOCV, IOCV, Riverside, CA

Elleuch, A., Khouaja, F.D., Hamdi, I., Bsais, N. Perreault,J.P. Marrakchi, M. and Fakhfakh, H. 2006. Sequence analysis of three citrus viroids infecting a single Tunisian citrus tree (Citrus, reticulata, clementine). *Genet. Mol. Biol*. 29, No. 4, Sao Paulo.

El-Dougdoug, K.A., Osman, M.E., Hayam, S.A., Rehab, A.D. and Reham, M.E. 2010. Biological and molecular detection of HSVd-infecting peach and pear trees in Egypt. *Australian Journal of Basic and Applied Sciences* 4: 19–26.

Fernandez Valiela, M.V., Fortugno, C. and Corizzi, F. 1965. Incidence of bud-union crease in citrus trees grafted on trifoliate rootstock in the Delta del Parana and San Pedro areas of Argentina. pp. 182–186. *In*: Proc. 3rd Conf. IOCV, IOCV, Riverside, CA.

Flores, R. and Rodriguez, J.L. 1981. Altered pattern of root-formation on cuttings of *Gynura aurantiaca* infected by citrus exocortis viroid. *Phytopathology* 71: 964–966.

Flores, R., Duran-Vila, N., Pallas, V. and Semancik, J.S. 1985. Detection of viroid and viroid-like RNAs from grapevine. *J. Gen. Virol.* 66: 2095–2102.

Flores, R. 1988. Detection of Citrus Exocortis Viroid in Natural and Experimental Citrus Hosts by Biochemical Methods. pp. 192–196. *In*: Proc. 10th Conf. IOCV, IOCV, Riverside, CA.

Foguet, J.L. 1973. Transmission trials with trifoliate orange wood pitting and its relation to xyloporosis. *Revista Industrial y Agricola de Tucuman* 50: 1–25.

Foguet, J.L. and Oste, C.A. 1968. Disorder of trifoliate orange rootstock in Tacuman, Argentina. pp. 183–189. *In*: 4th Conf. IOCV, IOCV, Riverside, CA.

Fraser, L.R., Levitt, E.C. and Cox, J.E. 1961. Relationship between exocortis and stunting of citrus varieties on *Poncirus trifoliata* rootstock. pp. 34–39. *In*: Proc. 2nd Conf. IOCV, IOCV, Riverside, CA.

Fraser, L.R., Broadbent, P. and Cox, J.E. 1976. Gummy pitting of *Poncirus trifoliata*: its association with dwarfing of citrus in New South Wales. pp. 147–151. *In*: Proc. 7th Conf. IOCV, IOCV, Riverside, CA.

Fraser, L.R. and Broadbent, P. 1979. Virus and related diseases of citrus in New South Wales. NSW Department of Agriculture. Surrey Beatty & Sons Pty. Ltd., Chipping Norton.

Garnsey, S.M. and Jones, J.W. 1967. Mechanical transmission of exocortis virus with contaminated budding tools. *Plant Dis. Reptr.* 51: 410–413.

Garnsey, S.M. 1968. Exocortis virus in citrus can be spread by contaminated tools. *Citrus Indus.* 49: 13–16.

Garnsey, S.M., Lee, R.F. and Semancik, J.S. 1993. Lack of cross-protection between citrus exocortis viroid and citrus viroids associated with mild symptoms in Etrog citron. P. 187–195. *In*: Proc. 12th Conf. IOCV, IOCV, Riverside, CA.

Gillings, M.R., Broadbent, P. and Gollnow, B.I. 1988. Biochemical Indexing for *Citrus exocortis* Viroid. pp. 178–187. *In*: Proc. 10th Conf. IOCV, IOCV, Riverside, CA.

Gillings, M.R., Broadbent, P., Gollnow, B.I. and Lakeland, C. 1991. Viroids in Australian citrus. *Australian J. Plant Physiology* 18: 559–570.

Ghosh, D.K., Mathur, S., Gupta, K., Ahlawat, Y.S. and Ramachandran, P. 2002. Viroid infection of citrus in central India. *Indian Phytopath.* 55: 290–293.

Grant, T. J., Grimm, G.R. and Norman, Paul 1959. Symptoms of cachexia in *Orlando tangelo*, none in sweet lime and false symptoms associated with purple scale infestations. *Plant Dis. Rep.* 43: 1277–1278.

Gross, H.J., Krupp, G., Domday, H., Raba, M., Alberty, H., Lossow, C.H., Ramm, K. and Sanger, H.L. 1982. Nucleotide sequence and secondary structure of citrus exocortis and chrysanthemum stunt viroid. *Eup. J. Biochem.* 121: 249–257.

Gullu, M. 1989. Dogu Akdeniz Bolgesi Navel Grubu Portakavl e Satsuma Mandarin Agaclarinda Yaygin Virus ve Virus Benzeri Hastaliklarin Surveyi ve

Indekslenmesi Yzerinde Calismalar. Doktora Tezi, Arastirma Yayinlari Serisi Yayin, 70, Ankara.

Hataya, T., Nakahara, K., Ohara, T., Ieki, H. and Kano, T. 1998. Citrus viroid Ia is a derivative of citrus bent leaf viroid (CVd-Ib) by partial sequence duplications in the right terminal region. *Arch. Virol.* 143: 971–980.

Hutton, R.J. 1986. The influence of tree size control and plant density on citrus productivity. (*In*: "Physiology of Tree Fruits"). *Acta Horticulturae* 175: 249–254.

Hutton, R.J. 2004. Effects of cultural management and different irrigation regimes on tree growth, production, fruit quality and water relations of sweet orange (*Citrus sinensis* (L.) Osbeck). Ph. D Thesis, University of Sydney.

Hutton, R.J. and Cullis, B.R. 1981. Tree spacing effects on productivity of high density Dwarf orange trees. *Proc. Int. Soc. Citriculture* 1: 186–190.

Hutton, R., Broadbent, P. and Bevington, K. 2000. Viroid dwarfing for high density plantings. *Horticultural Reviews* 24: 277–317.

Index of Viruses - Pospiviroidae 2006. *In*: ICTVdB - The Universal Virus Database, version 4. Büchen-Osmond, C. (*Ed.*), Columbia University, New York, USA. http://www.ncbi.nlm.nih.gov/ICTVdb/Ictv/fs_index.htm

Ito, T., Ieki, H., Ozaki, K. and Tsurae, Ito 2001. Characterization of a new citrus viroid species tentatively termed *Citrus viroid* OS. *Archives of Virology* 146: 975–982.

Kapur, S.P., Dhillon, R.S., Cheema, H.S. and Deol, I.S. 1974. *Citrus exocortis* in the Punjab. *Punjab Agric. Univ. J. Res.* 11: 378–379.

Keese, P. and Symons, R.H. 1985. *Proc. Natl. Acad. Sci. USA* 82: 4582–4586.

Levy, L. and Hadidi, A. 1993. Direct Nucleotide Sequencing of PCR-Amplified DNAs of the Closely Related Citrus viroids IIa and IIb (Cachexia). pp. 180–186 *In*: Proc. 12th Conf. IOCV, IOCV, Riverside.

Long, J.K., Fraser, L.R. and Cox, J.E. 1971. Possible value of close-planted, virus-dwarfed orange trees. pp. 262–267. *In*: Proc. 5th Conf. IOCV, IOCV, Riverside, CA.

Long, J.K., Fraser, L.R. and Cox, J.E. 1972. Possible value of close planted, virus dwarfed orange trees. pp. 262–267. *In*: Proc. 5th Conf. IOCV, IOCV, Riverside, CA.

Marias, L.J., Lee, R.F., Breytenbach, J.H.J., Manicom, B.Q. and Van Vuuren, S.P. 1996. Association of a viroid with gum pocket disease of trifoliate orange. pp. 236–244. *In*: Proc. 13th Conf. IOCV, IOCV, Riverside, CA.

Mishra, M.D., Hammond, R.W., Owens, R.A., Smith, D.R. and Diener,, T.O. 1991. Indian bunchy top disease of tomato plant is caused by a distinct strain of citrus excocortis viroid. *Journal of General Virology*, 72: 1–5.

Moreira, S. 1955. *A. molestia* "exocortis" e o cavalo de limoeiro cravo. Rev. Agric. (Piracicaba) 30: 99–112.

Murcia, N., Bernad, L., Serra, P., Bani, S.M., Hashemian and Duran-Vila, N. 2009. Molecular and biological characterization of natural variants of Citrus dwarfing viroid. *Arch. Virol.* 152: 1283–1294.

Murcia, N., Serra, P., Olmos, A. and Duran-Vila, N. 2008. A novel hybridization approach for detection of citrus viroids. *Molecular and Cellular Probes* 23: 95–102.

Nariani, T.K., Raychaudhuri, S.P. and Sharma, B.C. 1968. Exocortis in citrus in India. *Pl. Dis. Reptr.* 52: 834.

Nour-Eldin, F. 1956. Phloem discoloration of sweet orange. *Phytopathology* 46: 238–239.

Nour-Eldin, F. 1959. Citrus virus research in Egypt. *In*: Citrus virus diseases, pp. 219–227. Berkeley, *Div. Agric. Sci. Univ. Calif.* Berkeley, CA.

Nour-Eldin, F. 1968. Gummy bark of sweet orange. *In*: Indexing procedures for 15 virus diseases of citrus trees, pp. 50–53. Agricultural handbook No. 33. Washington, D.C., Agr. Res. Service. USDA.

Onelge, N., Cinar, A., Kersting, U. and Semancik, J.S. 1996. Viroids associated with citrus gummy bark disease of sweet orange in Turkey. pp. 245–248. *In*: Proc. 13th Conf. IOCV, IOCV, Riverside, CA.

Onelge, N., Cinar, A., Szychowski, J.A. and Semancick, J.S. 2002. Cachexia-related viroids associated with citrus gummy bark. pp. 441. *In*: Proc. 15th Conf. IOCV, IOCV, Riverside, CA.

Önelge, N., Çinar, A., Szychowski, J.S., Vidalakis, G. and Semancik, J.S. 2004. Citrus viroid II variants associated with 'Gummy Bark' disease. *Eur. J. Plant Pathol.* 110: 1047–1052.

Palacio-Bielsa, A., Foissac, X., and Duran-Vila, N. 2000. Indexing of citrus viroids by imprint hybridization. *Eur. J. Plant Pathol.* 105: 897–903.

Palacio, A. and Duran-Vila, N. 2000. Citrus cachexia disease: Molecular characterization of its viroid agents. pp. 273–281. *In*: Proc. 14th Conf. IOCV, IOCV, Riverside, CA.

Patil, B.P. and Warke, D.C. 1968. A note on existence of exocortis virus in India. *Curr. Sci.* 37: 469–470.

Pujol, R.A. 1965. Exocortis and other problems with trifoliate orange rootstock, pp. 128–133. *In*: Proc. 3d Conf. IOCV, IOCV, Riverside, CA

Rakowski, A.G., Szychowski, J.A., Avena, Z.S. and Semancik, J.S. 1994. Nucleotide sequence and structural features of the Group III citrus viroids. *J. Gen. Virol.* 75: 3581–3584.

Ramachandran, P., Pandey, P.K., Ahlawat, Y.S., Varma, A. and Kapur, S.P. 1993.Viroid diseases of citrus in India P. 438-440. *In*: Proc. 12th Conf. IOCV, IOCV, Riverside.

Ramachandran, P. Mathur, S., Agarwal, J. and Ahlawat, Y.S. 2003. Diagnostics for *Citrus exocortis* and Hop stunt viroids associated with corky vein disease in citrus, *Indian Phytopath.* 56: 428–433.

Ramachandran, P. Agarwal, J., Roy, A., Ghosh, D.K. , Das, D.R. and Ahlawat, Y.S. 2005. First record of a hop stunt viroid variant of Nagpur mandarin and Mosambi sweet orange trees or rough lemon and Rangpur lime rootstocks New disease reports, British Society for Plant Pathology. Vol 10, August 04-January 05.

Raychaudhuri, S.P., Nariani, T.K. and Ahlawat, Y.S. 1977. Dieback of citrus in India. *Proc. Interl. Soc. Citriculture* 3: 914–918.

Reanwarakorn, K. and Semancik, J.S. 1998. Regulation of pathogenicity in hop stunt viroid- related group II. *J. Gen. Virol.* 79: 3163–3171.

Reichert, I. and Perlberger, P. 1934. Xyloporosis, the new citrus disease. *Jewish Agency for Palestine Agr. Expt. Sta.* (Rehovot, Bull). 12: 1–50.

Reddy, G.S., Dakshinamurti, V. and Reddy, V.R.K. 1974. Yellow corky vein: first report on a graft transmissible disorder of "Sathgudi" in Andhra Pradesh. *Indian Phytopath.* 27: 82–84.

Reddy, G.S. and Naidu, P.H. 1989. Transmission of citrus yellow vein. *Indian Phytopath.* 42: 455–456.

Roistacher, C.N. 1988. The cachexia and xyloporosis diseases of citrus - a review. pp. 116–124. *In*: Proc. 10th Conf. IOCV, IOCV, Riverside, CA.

Roistacher, C.N. 1991. Graft transmissible diseases of citrus. Handbook of detection and diagnosis. FAO Rome, p. 286.

Roistacher, C.N., Blue, R.L. and Calavan, E.C. 1973. A new test for cachexia. *Citrograph* 58: 261–262.

Roistacher, C.N., Nauer, E.M. and Wagner, R.C. 1980. Transmissibility of cachexia, Dweet mottle, psorosis, tatterleaf and infectious variegation viruses on knife blades and its prevention. pp. 225–229. *In*: Proc. 8th Conf. IOCV, I O C V Riverside, CA.

Roistacher, C.N., Gumpf, D.J., Nauer, E.M. and Gonzalea, R. 1983. Cachexia disease: virus or viroid. *Citrograph* 68: 111–113.

Roistacher, C.N., Bash, J.A. and Semancik, J.S. 1993. Distinct disease symptoms in *Poncirus trifoliata* indusced by three citrus viroids from three specific groups. pp. 173–179. *In*: Proc. 12th Conf. IOCV, IOCV, Riverside, CA.

Roy, A. and Ramchandran, P. 2003. Occurrence of a Hop stunt viroid (HSVd) variant in yellow corky vein disease of citrus in India. *Current Science* 85: 1608–1612.

Roy, A. and Ramachandran, P. 2006. Characterization of citrus exocortis viroid variant in yellow corky vein disease of citrus in India. *Curr. Sci.* 91: 798–803.

Rustem Ali, F.A. 1998. Investigation on Citrus yellow Corky Vein disease to establish its etiology. Ph.D. Thesis, I.A.R.I, New Delhi.

Rustem, F.A. and Ahlawat, Y.S. 1999. Yellow corky vein – a viroid disease of citrus in India. *Indian J. Agric. Sci.* 69: 789–794.

Rustem, F.A., Ramachandran, P. and Ahlawat, Y.S. 2000. pp. 778–779. *In*: Proceedings of the Golden Jubilee International Conference, New Delhi, 2000.

Salibe, A.N. 1986. A Programme for Citrus Improvement and Protection in Turkey. Report to the Government of Turkey. FAO Rome eds.

Schwarz, R.E. and McClean, A.P.D. 1969. Gum pocket, a new virus-like disease of *Poncirus trifoliata. Plant Dis. Reptr.* 53: 336–339.

Schwinghamer, M.W. and Broadbent, P. 1986. Detection of viroids in dwarfed orange trees by transmission to Chrysanthemum. *Phytopathology* 77: 210–215

Schwinghamer, M.W. and Broadbent, P. 1987. Association of viroids with graft transmissible dwarfing symptoms in Australian orange trees. *Phytopathology* 77: 205–207.

Semancik, J.S. and Weathers, L.G. 1972. Exocortis disease: Evidence for a new speices of infectious low molecular weight RNA in plants. *Nature New Bio.* 237: 242–244.

Semancik, J.S., Roistacher, C.N. and Duran-Vila, N. 1988. A new viroid is the causal agent of citrus cachexia disease. pp. 125–135. *In*: Proc. 10th Conf. IOCV, IOCV, Riverside, CA.

Semancik, J.S., Roistacher, C.N., Rivera-Bustamante, R. and Duran-Vila, N. 1988. Citrus cachexia viroid, a new viroid of citrus: Relationship to viroids of the exocortis disease complex. *J. Gen. Virol.* 69: 3059–3068.

Semancik, J.S. 1988. Citrus exocortis disease – 1976–1986. pp. 136–151. *In*: Proc. 10th Conf. IOCV, IOCV, Riverside, CA.

Semancik, J.S. and Duran-Vila, N. 1991. The grouping of citrus viroid: additional physical and biological determinants and relationship with diseases of citrus. pp. 178–189. *In*: Proc. 11th Conf. IOCV, IOCV, Riversise, CA.

Singh, R.P. and Clark, M.C. 1973. Similarity of host response to both potato spindle tuber and citrus exocortis viroids. *FAO Plant Protection Bulletin*. 21: 121–125.

Singh, R.P., Dilworth, A.D., Ao, X., Singh, M. and Baranwal, V.K. 2009. Citrus exocortis viroid transmission through commercially-distributed seeds of Impatiens and Verbena plants. *European J. Plant Pathol.* 124: 691–694.

Sharma, D.C. and Pandey, P.K. 1983. Leaf yellow mid vein a previously undescribed virus disease of citrus. *Interl. J. Trop. Plant Dis.* 1: 149–52.

Sofy, A.R., Soliman, A.M., Mousa, A.A., Ghazal, S.A. and El-Dougdoug, K.A. 2010. "First record of *Citrus* viroid II (CVd-II) associated with gummy bark disease in sweet orange (*Citrus sinensis*) in Egypt". *New Disease Reports* 21: 24.

Tanaka, H. and Yamada, S. 1969. Indexing for exocortis and its damage on citrus trees in *Japan. Bull. Hort. Res. Stn, Japan Series B,* 9: 181–195.

Terranova, G., Caruso, A. and Reforgiato Recupero, G. 1991. Susceptibility and symptomatology of mapo tangelo to xyloporosis. pp. 209–213. *In*: Proc. 11th Conf. IOCV, IOCV, Riverside, CA.

Vernière, C., Botella, L., Dubois, A., Chabrier, C. and Duran-Vila, N. 2002. Properties of citrus viroids: symptom expression and dwarfing. pp. 240–248. *In*: Proc. 15th Conf. IOCV, IOCV, Riverside, CA.

Vernière, C., Perrier, X., Dubois, C., Dubois, A., Botella, L., Chabrier, C., Bovè, J.M. and Duran-Vila, N. 2002. Interactions between citrus viroids affect symptom expression and field performance of Clementine trees grafted on trifoliate orange. *Phytopathology* 96: 356–368.

Vernière, C., Perrier, X., Dubois, C., Dubois, A., Botella, L., Chabrier, C., Bové, J.M. and Duran-Vila, N. 2004. Citrus viroids: Symptom expression and effect on vegetative growth and yield of clementine trees grafted on trifoliate orange. *Plant Disease* 88: 1189–1197.

Visvader, J.E., Gould, A.R., Bruening, G.E. and Simons, R.H. 1982. Citrus exocortis viroid: Nucleotide sequence and secondary structure of an Australian isolate. *FEBS letters* 137: 288–292.

Vogel, R. and Bove, J.M. 1976. La nouvelle technique d'indexation de la cachexia-xyloporosis: son utilization en Corse. *Fruits* 31: 93–96.

Weathers, L.G. 1965a. Transmission of exocortis virus of citrus by *Cuscuta subinclusa. Plant Dis. Reptr.* 49: 189–190.

Weathers, L.G. 1965b. Petunia, an herbaceous host of exocortis virus of citrus. *Phytopath.* 55: 1081 (Abs).

Weathers, L.G., Greer Jr., F.C. and Harjung, M.K. 1967. Transmission of exocortis virus of citrus to herbaceous plants. *Plant Dis. Reptr.* 51: 868–871.

8

MANAGEMENT OF CITRUS VIRUS AND VIRUS-LIKE DISEASES

Citrus is one of the major cultivated fruit crops in the world, grown in more than 100 countries throughout the tropical, sub-tropical and Mediterranean climate regions. Citrus belongs to family Rutaceae which is represented by approximately 150 genera and 1600 species of plants found in tropical and warm regions of the world. The genus *Citrus* is the most popular genus in subfamily *Aurantioideae*. For most of citrus growing countries, citrus is a crop which is significant for its export potential. South Africa, for example, exported over one million metric tonnes of oranges in 2008/2009 marketing year, whereas the world total export was over three million tonnes. In citrus, over 40 virus and virus-like diseases have been described from most citrus growing countries. In majority of cases, aetiology of these diseases have been established and the associated pathogens have been characterized and their taxonomy established. Management of these diseases was difficult earlier because of the non-availability of suitable diagnostic methods which is the major component in producing quality disease-free planting material. But during the last three decades, advanced methods of virus diagnosis have been developed worldwide including India (Ahlawat *et al.,* 1992, 2008).

Among citrus virus disease, 17 are well studied and their associated pathogens have been fully characterized and taxonomy established. About 14 diseases considered to be virus-like diseases based on their transmission and absence of other disease inciting agents. Among the important viruses associated with citrus diseases are (i) Ilarviruses associated with citrus crinkly leaf, citrus variegation and citrus leaf rugose diseases (ii) Rhabdoviruses associated with citus leprosis and zonate chlorosis and yellow vein mosaic of Pummelo (iii) Psorosis diseases (iv) Satsuma dwarf virus group (v) Tatter leaf-citrange stunt (vi) Closterovirus associated with tristeza disease (vii) Badnavirus associated with citrus yellow mosaic disease and (viii) Mandrivius associated with Indian citrus ring spot disease. Mollicutes are also associated with some serious diseases in citrus like citrus stubborn and lime witches broom and rubberywood. Diseases like exocortis, cachexia, and others associated with dwarfing of citrus trees are caused by viroids. Huanglongbing and variegated clorosis diseases are caused by bacteria. *Candidus* liberibacter species cause huanglongbing and *Xylella fastidiosa* causes citrus variegated chlorosis. Each and every disease is important and needs management. However, some diseases are disastrous such as tristeza and huanglonbing and need priority for management. Each pathogenic agent is spread and infects citrus in a different manner. All systemic citrus pathogens (*i.e.,* viruses, viroids, phytoplasmas, spiroplasmas and bacteria) are contained in propagation materials (buds, stocks and some of them in seeds) of the plant. Most of these systemic pathogens are also transmitted by specific insect vectors that feed on citrus plants.

There is lot of similarities in genome and morphology of viruses infecting animal, humans and plants. But animal and human viruses can form protective or defensive proteins against the virus infections, known as antibodies. These antibodies neutralize the infection in blooded animals and human. Normally, if an animal is infected by a virus once, it remains protected for long times against that virus, sometimes throughout his life because the antibodies against that virus remain in the blood stream. But contrary to this, plant viruses normally can't form antibodies in plants, hence a plant infected by a virus will remain infected with that virus throughout its life and if the virus is seed-borne, then the next progeny will also carry that virus. Plant viruses do make antibodies when injected in an animal like rabbit.

Viruses are difficult to manage as no chemical is effective against plant viruses. Before a strategy is evolved to manage a virus disease, one should know how it survives or spreads. Use of virus-free seed and planting material (for rootstock and scion species respectively) and control or minimize the vector population are two most important strategies for proper management of citrus virus diseases. A single management practice cannot be used to manage all citrus virus diseases because different diseases have distinct ecological and epidemiological characteristics. Vector populations, extent of the inoculum reservoir, and age of the grove at first infection, host-virus interaction and virus distribution are some of the important studies which should be carried out for proper management of a disease. Many viruses remain latent in their host or show symptoms only at particular environmental conditions like citrus tristeza (CTV) and Indian citrus ring spot virus (ICRSV) which show symptoms at low temperatures only. Management of citrus virus diseases depends on three most important factors:

1. Producion of virus-free planting material (both scion variety and rootstock).
2. To avoid/remove sources of infection/inoculums, and
3. Vector control

PRODUCTION OF VIRUS-FREE PLANTING MATERIAL

Pre-requisite to develop virus-free planting material is the development of suitable diagnostics. Earlier, indexing of planting material was done only by inoculating the indicator hosts as no serological and molecular detection techniques were available. Based only on indexing on indicator plants, budwood certification programmes were developed earlier in several countries. Psorosis virus was the first virus disease in citrus against which budwood certification programme was developed in 1938 in California and was followed by several countries where citrus is an important industry. At the time of inception of this programme no serological or molecular techniques were available for virus testing and biological indexing was the only method of testing. In spite of this, the programme was very successful to eradicate citrus psorosis in California. Budwood certification is the best programme through which virus-free budwood can be supplied indefinitely but diagnosis is a very important component to develop effective and viable budwood certification programmes. In general, citrus virus diseases fall in to two categories 1) those transmitted by vector(s), and 2) diseases for which no vector is known. The second category of infections can be easily avoided to develop virus free planting material for propagation by testing through effective diagnostics. Such virus-free planting material can be used for propagation for a long time as there will be no risk for its contamination. Diseases transmitted by vectors, however, need special strategies for their management such as vector control. These strategies will be discussed later in this chapter.

BUDWOOD CERTIFICATION

Many countries have implemented budwood certification programmes with the objective to produce and distribute high quality and pest-free planting materials. The citrus clonal protection programme being used in california is given in Fig. 35. A citrus certification programme producing disease-tested propagative material is one of the most efficient ways to eliminate or contain graft-transmissible and other citrus diseases (Fig. 36).

Budwood = seed

Budwood certification = seed certification

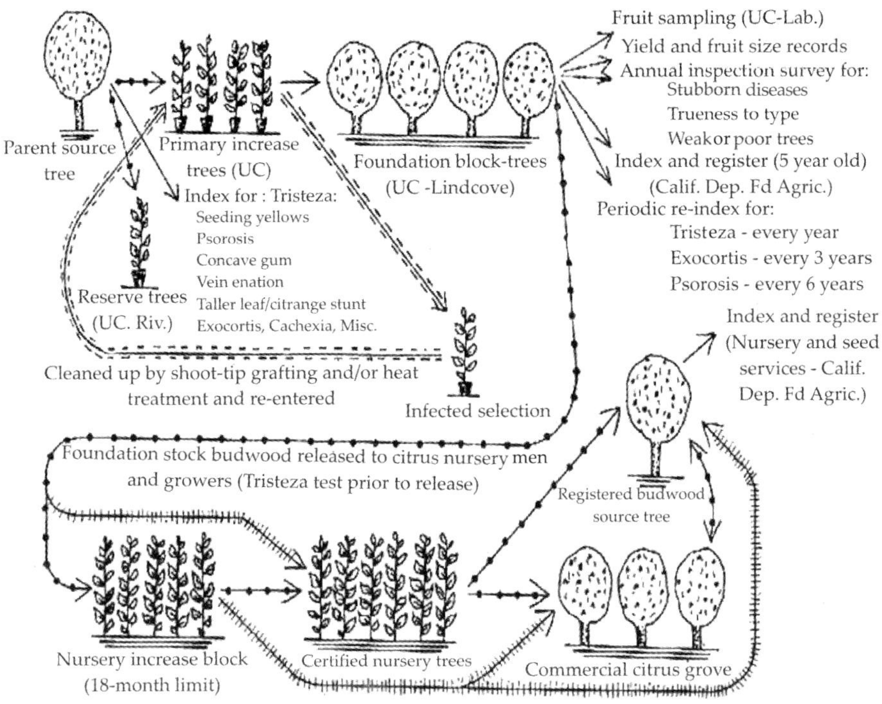

Fig. 35: Outline of Citrus clonal protection programme in California

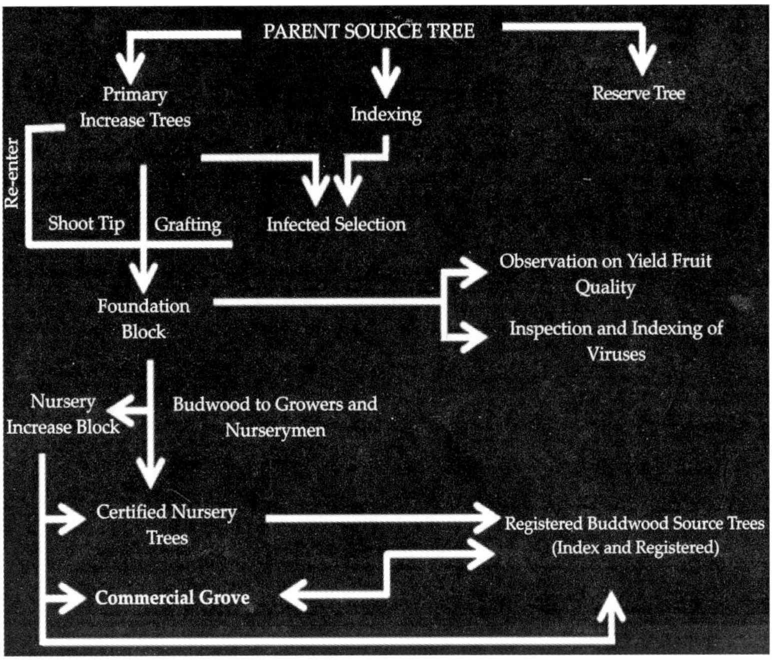

Fig. 36: Diagramatic presentation of budwood certification

In budwood cerification following steps should be taken:

1. Selection of apparently healthy tree (mother tree) by nurseryman/ grower. The tree should be true to type and free from pest and diseases.
2. Inspection of apparently healthy tree should be done by inspectors through visual inspections for two years, four trees around the mother tree are also inspected to avoid the possibility of transmission by root grafting.
3. Entry of the selected tree in registration programme if no obvious disease is found.
4. Propagation from selected tree on indicator host or use other methods of testing. If the virus is suspected to be seed transmitted, rootstock seedlings are also indexed.
5. Registration of candidate tree if found free.
6. Multiply from registered tree to establish scion groves.
7. Certification of nursery stock should be done under supervision of inspector and budwood taken from registered tree.

Budwood certification is an integrated programme for:

- Use of certified rootstock
- Tolerant selection of scion-rootstock
- Appropriate quarantine
- Detection techniques
- Preventing damage from vector borne viruses

General provisions required for certification programme:

- Responsibility of applicant
- Selection of tree for testing
- Selection of location (free from vectors, flooding, drought etc.)
- Registry of plants
- Identity of all plants entered in the programme
- Maintenance of plants
- Proper use of farming and nursery practices and routine inspection by budwood inspectors
- Removal of off types and infected plants
- Labeling of each selected registered tree and nursery stock
- Rootstock propagation from seed
- *Foundation block (FB)*: a tree raised from registered tree can only be planted in FB

- *Protected block*: Maintain in glasshouse, remove infected plant, if any
- *Registered outdoor tree*: Inspection and indexing should be done routinely.
- *Nursery increase block*: In this block all plants should come from registered tree
- Trees in nursery may be certified
- Scions used to propagate the nursery block and certification shall be; from nursery increase block or from Registered tree which should be given a number.

INSPECTION AND TESTING PROCEDURES

Following trees need inspection and testing for viruses, phytoplasmas, bacteria, and viroids:

- Trees in foundation block
- Trees in protected block
- Non-protected, selected trees in the programme
- Nursery increase block and certified blocks

FINAL REGISTRATION

Registration period is 3 years after testing and inspection. Registration may be cancelled or refused if the registration requirements are not properly met.

No resistance is known for several citrus virus and virus-like pathogens. Healthy planting material can be obtained by natural selection of healthy plants in highly infected crop or can be selected from horticulturally desirable clonal material (clonal selection). Use of nucellar clones is advantageous if no healthy source of a citrus variety is known. Nucellar seedlings of that variety can be obtained which will preserve the varietal characteristics and at the same time will be free from viruses. Heat treatment of infected plants is another option to get plants free from viruses. For example, psorosis virus was eliminated through heat treatment of infected citrus plants by keeping them for 2–3 months in hot air at alternating temperature of 40°C for 16h during day and 30°C for 8 h during night (Calavan *et al*., 1972). Normally plants can be made free from viral infections by heat treatment (normally 34–54°C for a few minutes to days). Heat treatment followed by meristem culture is a common method to get virus free material. The works of Murashige *et al.*

(1972) and Navarro, Roistacher and Murashige (1975) have led to the technique of shoot-tip grafting and its use in recovering citrus clones free of virus and virus-like agents. Over the last ten years, shoot-tip grafting has become the major, if not the only, technique used for producing citrus material free of graft-transmissible pathogens. Before the era of shoot-tip grafting, the production of citrus material free of disease agents followed three approaches: selection by visual inspection and indexing, the production of nucellar clones and recovery of virus-free material from infected plants by thermotherapy.

PRODUCTION OF NUCELLAR CLONES

Nucellar plants have been developed from unpollinated and unfertilized ovule. Nucellar embryony as a means of freeing citrus clones from viruses have been demonstrated by several workers (Weathers and Calavan, 1959; Cameron and Frost, 1968; Reuther *et al.*, 1968). Nucellar clone technique is based on the fact that most virus and virus-like pathogens of citrus are not seed borne and that most commercial citrus species and varieties have not only a zygotic embryo but also have several somatic, asexual embryos known as nucellar embryos. These are true to type and virus free as well. In general, ten to 15 years are necessary to evaluate nucellar clones. Commercial nucellar clones have been produced for the major commercial citrus species. Some of the better known nucellar clones were produced by Frost (1938). Some of citrus cultivars like shaddock, Citron, Clementine, King Mandarin, Meyer lemon and Tahiti lime are monoembryonic and have no nucellar embryos. However, techniques have been developed to induce nucellar embryos artificially in monoembryonic species (Rangan *et al.*, 1968; Esan, 1973; Button and Bornman, 1971; Kochba *et al.*, 1972). Navarro and his co-workers obtained nucellar plants of several navel cultivars by culturing whole ovules excised from ovaries at the flower bud stage and from developing fruits. However, with the development of better techniques for virus elimination like shoot tip grafting, use of nucellar clones is almost abandoned.

SHOOT-TIP GRAFTING

It was shown for the first time at the Rockefeller Institute in New York, that root tips of tomato plants infected with Tobacco Mosaic Virus (TMV) were free of the virus (White, 1934). Later, it was found that shoot tips could also be free of viruses (Morel and Martin, 1952; 1955). They suggested that shoot tips or meristems could be cultured in *vitro* into plants on special media, and that these plants would be free of viruses. Today, many cultivated plants have been "regenerated" by shoot-meristem culture. Unfortunately, meristems of woody plants such as citrus do not grow, or grow only poorly, on the media devised for herbaceous plants. This is why Murashige *et al.* (1972) replaced the medium and grow a root stock in a test-tube. The rootstock in test tube was decapitated and the meristem was placed on the decapitated rootstock where it bound to the rootstock by a graft union and, being adequately fed by the rootstock, grew into a shoot (Plate 52). The technique has been carefully standardized by Navarro and his co-workers in Spain (Navarro *et al.*, 1975, 1976, 1988; Navarro and Juarez, 1977) and in India (Vijaylaxmi *et al.*, 2006a, b; Hoa *et al.*, 2004). Shoot-tip grafted plants have all the characteristics of the mother tree from which the shoot tips have been taken. However, shoot-tip grafed plants must be tested for viruses as some shoot-tip grafted plants are still found to be infected by virus and virus-like pathogens. This is because, for the shoot tip to take and grow on the rootstock, it has to be of a minimum size. Normally, the grafted shoot tip comprises the meristem of two to three leaf primordia (0.10 to 0.15 mm). If the shoot tip is too small, it will not grow. If it is too large, it will not be free of pathogens. Shoot tip grafting is now used in many countries to obtain virus-free healthy plants. Shoot tip grafting have been used to eliminate several viruses, viroids and mollicutes (Roistacher and Kitto, 1977; Roistacher *et al.*, 1976; Su and Chu, 1984; Vogel *et al.*, 1988). Detection of viruses at various levels and in planting material is essential. Following method are used for detection of virus and virus-like pathogens in planting material.

METHODS OF DIAGNOSIS

Several methods of detection of virus and virus- like pathogens are known such as Biological, Serological, Electron microscopy, Immunological electron microscopy and Nucleic acid based diagnosis (Plate 53). These methods are briefly described below:

1. Biological Diagnosis

a. *Visual inspection*

Citrus species are susceptible to certain pathogens and tolerant to others. In some diseases such as citrus yellow mosaic, symptoms are quite conspicuous and are easily identified in field trees. There are other diseases like Indian citrus ringspot where symptoms are quite conspicuous but appear for a few months during the year mostly during cooler months. In such cases, frequent inspections would be required. Certain pathogens remain symptomless or latent in field trees like citrus tristeza and others. In such cases, field inspections would be of little value and laboratory tests will be needed. Citrus trees are most frequently grafted on a rootstock. Certain diseases will show symptoms on rootstock while others will show on scion variety and some will show on both. Some virus diseases like

Citrus tristeza (CTV) and several viroids will show stem pitting symptoms which will be difficult to identify in field unless an inspector is well acquainted with all such diseases otherwise laboratory tests will be necessary for proper diagnosis. During these situations, visual inspection is most often ineffective for the detection of all the pathogens present. Temperature conditions during field inspection are also important; for example, in a cool climate, symptom expression of exocortis takes longer time than it does under hot conditions. Inversely, psorosis young leaf flecking and Indian citrus ringspot show symptoms only in cool weather. It is always important to supplement visual inspection by laboratory indexing on indicator plants (biological indexing) and laboratory indexing such as electron microscopy, serological, molecular or culture of the agent (*e.g. Spiroplasma citri*) when possible.

b. *Biological indexing on Indicator plants*

Indexing of planting material is essential which will positively and consistently confirm the presence (or absence) of a transmissible pathogen, or that will identify the disease. For biological testing, it is important to establish indicator host(s) of a virus. Indicator plant(s) are established by way of host range studies.Some of the citrus viruses are transmitted by mechanical inoculation from citrus to herbaceous hosts like Indian citrus ring spot virus which can infect cultivars of French bean and *Chenopodium* species. In such cases, a herbaceous host as indicator will be advantageous. Citrus pathogens such as huanglongbing bacteria and phytoplasmas are transmitted by dodder (*Cuscuta* spp.) from citrus to periwinkle and several other hosts. Periwinkle is a good indicator host in such cases. Once a suitable indicator is identified, it is easy to reliably index the planting material. For example, tristeza virus is indexed on Kagzi lime *(Citrus aurantifolia)* seedlings (Bar-Joseph and Loebenstein, 1970). If the inoculum contains tristeza virus, the indicator lime seedling will become systemically infected and will then show the specific symptoms of tristeza leaf vein clearing (Plate 5) and stem pitting. Exocortis viroid will be indexed on Etrog citron (Calavan *et al.*, 1968; Flores, 1988; Roistacher *et al.*, 1977) and cachexia viroid by new indexing method (Duran-Vila *et al.*, 1991; Roistacher *et al.*, 1973). Different indicator plants are used for different viruses. Methods of indexing have been reviewed from time to time (Childs and Knorr, 1965; Roistacher, 1991). Indicator plants of some of the viruses and virus-like pathogens are given in Table 11.

C. *Laboratory indexing*

Several methods are now known for detection of virus and virus-like pathogens (Plate 53). Following methods of detection are frequently used for laboratory indexing of virus and virus-like pathogens:

Table 11. Indicator plants for detection of major virus and virus-like agents of citrus

Pathogen	*Indicator plant*	*Symptoms observed*	*Time required for symptom expression (temperature condition)*
Blight	Adult Valencia sweet orange on rough lemon, graft inoculated with root pieces from candidate tree	Blight syndrome	2 years at least
Cachexia (viroid)	Parson's Special mandarin on rough lemon[1]	Gumming at graft union	1-2 years (warm)
Concave gum-Blind pocket (uncharacterized)	Sweet orange seedlings	Vein flecking, oak-leaf patterns on young leaves	A few weeks (cool)
Cristacortis (uncharacterized)	Sweet orange seedling tree	Cortical symptoms	A few years
	Sweet orange seedlings	Vein flecking, oak-leaf patterns on young leaves	A few weeks (cool)
Exocortis complex (viroid complex)	Orlando or Wekiwa tangelo seedling	Stem pitting	1-2 years
	Etrog citron 861 -S1 on rough lemon[1]	Leaf epinasty and/or stunting midvein necrosis, petiole necrosis, leaf tip browning.	Weeks to months (warm)
Huanglongbing	Sweet orange seedlings	Leaf yellowing and mottling with zinc deficiency-like patterns	Several months (Asian form: warm; African form: cool)
Gummy bark	Sweet orange on rough lemon	Gumming of sweet orange and sometimes rough lemon bark, bud-union constriction	Several years
Gum pocket-Gummy pitting	Sweet orange on *Poncirus trifoliata*	Gum pockets in the bark and wood of *P. trifoliata*	Several years
Impietratura	Sweet orange seedling	Vein flecking, oak-leaf patterns on young leaves	A few weeks (cool)
	Fruiting grapefruit	Gum in fruit albedo	A few months

Table 11. (*Contd...*)

Table 11. (*Contd...*)

Pathogen	*Indicator plant*	*Symptoms observed*	*Time required for symptom expression (temperature condition)*
Indian citrus ring spot mandrivirus	Kinnow mandarin, French bean	Vein flecking on young leaves, ringspot on olders leaves. Local lesion and systemic in French bean	4-6 weeks at cool temperature 6-8 days in Frenchbean
Infectious variegation-Crinkly leaf (ilarvirus)	Lemon seedling, Etrog citron	Leaf variegation and/or crinkle	A few weeks
	Cowpea (cultivar Black eye)	Chlorotic lesion on mechanically infected leaves	A few days
Leaf rugose (ilarvirus)	Mexican time seedlings	Leaf puckering	A few weeks
	Eureka lemon seedlings	Pinpoint chlorotic spotting	A few weeks
	Red kidney, Bush bean	Local lesions	A few days
Psorosis A (scaly bark) (virus)	Sweet orange seedlings	Vein flecking, oak-leaf patterns on young leaves	A few weeks (cool)
	Sweet orange tree	Bark scaling and flaking	A few years
Ringspot (virus)	Sweet orange seedlings	Necrotic shoot reaction ring patterns, mottled chlorotic areas on young leaves. Chlorotic patterns on mature leaves. Stem lesions	A few weeks
	Chenopodium quinoa	Necrotic local lesions on mechanically infected, fully expended leaves. No systemic infection	A few days
Rubbery wood (MLO)	Lemon seedlings	Flexible, downward bent shoots	A few months
Satsuma dwarf (virus)	Satsuma mandarin	Stunting, boat- or spoon-shaped leaves	Several months (cool)
	Sesame, *C. quinoa*	Local lesions on mechanically infected leaves	A few days

Table 11. (*Contd...*)

Table 11. (*Contd...*)

Pathogen	*Indicator plant*	*Symptoms observed*	*Time required for symptom expression (temperature condition)*
Stubborn (*Spiroplasma citri*)	Madame Vinous or Hamlin sweet orange seedling	Leaf tip chlorosis, leaf mottle, leaf cupping, stunting	A few months (warm)
Tatterleaf-Citrange stunt(virus)	*C. excelsa* seedlings	Deformed leaves with chlorotic zones	A few months
	Rusk citrange seedlings	Deformed leaves with chlorotic leaf blotch. Stem with zigzag growth pattern	A few months
	C. quinoa	Chlorotic local lesions and systemic chlorotic mottle	A few days
Tristeza (closterovirus)	Small-fruited acid lime seedlings	Leaf vein clearing, leaf cupping, stem pitting, stunting	Weeks to months (cool)
Tristeza with seedling yellows reaction	Grapefruit, lemon, sour orange seedlings	Stunting, leaf yellows at seedling top	A few months
Vein clearing	lemon	Yellow Vein clearing	4-6 months
Vein enation-Woody gall (luteovirus)	Small-fruited acid lime seedlings	Vein enations	A few months
	Small-fruited acid lime or rough lemon seedlings	Galls on stems near thorns in association with wounds	A few months
Witches' broom (MLO)	Small-fruited acid lime seedlings	Small to very small leaves, witches' brooms	Several months (warm)
Yellow mosaic badnavirus	Sweet orange, *Citrus decumana*	Severe yellow mosaic on leaves	4-6 weeks at warm temperature
Yellow corky vein viroid(s)	Key lime	Yellowing of veins and corking of veins on ventral side and on twigs	4-6 weeks by graft transmission and 6-8 weeks by sap inoculation

Source: Professor J.M.Bove (1995). FAO Corporate Document Repository (partly by author)

- Detection or visualization of the viruses and agents like phloem-restricted prokaryotes (*Spiroplasma citri,* witches' broom MLO, rubbery wood MLO, Huaglongbing bacteria) directly by electron microscopy;
- Culture of *S. citri* and observation of the cultured *Spiroplasma* by dark field microscopy;
- sPAGE and R-PAGE and staining of the viroid RNAs of cachexia and exocortis.
- Production of antibodies (polyclonal or monoclonal) against proteins of the pathogens for use in enzyme-linked immuno-sorbent assays (ELISAs), immunofluorescence or western immunoblotting. Antibodies have been produced against the viral coat proteins of several plant viruses, and the outer membrane protein(s) of S. *citri,* the witches' broom MLO or the HLB bacteria.
- Use of DNA probes prepared against the nucleic acids of the virus and virus-like pathogens. These DNA probes are used most often in dot-blot hybridizations on nitrocellulose or nylon membranes, and they can also be used in southern hybridizations. The amplification of certain DNA sequences carried out by PCR, with the amplified DNA detected by ethidium bromide staining on PAGE.

Some of these methods are not yet routine indexing techniques. For instance, the momoclonal antibodies (MAbs) against huanglongbing bacteria (HLB) turn out to be highly specific and they will only detect the homologous serotypes of the bacterium. Hence, electron microscopy detection of the bacteria is still the most reliable technique. Major laboratory indexing techniques are briefly described below

C. 1. *Electron microscopy*

Viruses can not be seen in light microscope but are observed in electron microscope. A negative result in electron microscopy (EM) should not be taken as negative because a very small sample is viewed in which virus may or may not be present. Therefore, electron microscopy should be supported by other tests. Diagnosis of viruses in EM is normally based on virus morphology (Fig. 37) such as filamentous or flexuous particles (CTV), spherical (ilarviruses) or bullet shaped (citrus yellow mosaic badnavirus). Virus and virus-like pathogens are seen in EM by leaf dip preparation, thin sections and purified preparations.

C. 2. *Serological*

Following serological techniques are used for detection of virus and virus-like pathogens in citrus:

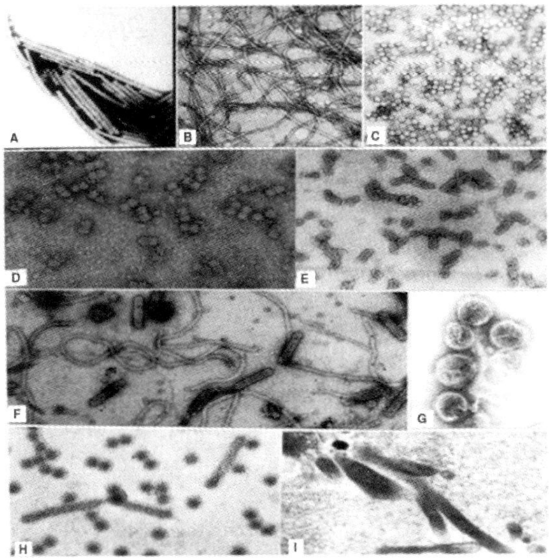

Fig. 37: Morphology of viruses. Last photograph is *Candidus* liberibacter asiaticus

- Microprecipitin test
- Tube precipitation test
- Agar double diffusion test*
- Immunofluorescence test*

* Enzyme-linked immunosorbent assay*

- Dot immunobinding assay*
- Immuno-tissue printing*
- Western blotting
- Immuno-gold labeling

* Commonly used for testing of pathogens infecting citrus

To conduct serological tests, it is essential to develop antibodies, polyclonal or monoclonal (MAbs). Polyclonal antibodies can be easily developed against viruses which can be purified. But monoclonal antibodies are prepared against viruses which are difficult to purify or have several strains (to develop MAbs against each strain) such as tristeza virus (Vela *et al.,* 1988). MAbs are also commonly prepared against pathogens which are pleomorphic and difficult to purify such as huanglongbing pathogens and mollicutes (Bove, 1980; Garnier *et al.,* 1987, 1991; Martin-Gros, 1987). Serological indexing was found useful to develop control strategy for important viruses like tristeza (Garnsey *et al.,* 1981). Polyclonal and monoclonal antibodies are used for detection of virus or virus-like pathogens in various serological techniques which will be briefly described below.

C. 2.1. Microprecipitin test

In this test, a small drop of extract having antigen (virus) is mixed with specific antibody on a glass slide and the precipitate is observed under a dark field microscope.

C. 2.2. Tube precipitation test

This test is carried out in thin wall tubes. 0.5 ml of each antigen and antibody are mixed in the tube. Antiserum should be carefully put on antigen in the tube. The precipitation will develop as a ring at the interface of antigen and antibody.

C. 2.3. Agar double diffusion test (Ouchterlony test)

In this test, antibody and antigen diffuse towards each other in a gel and form a white band or line due to precipitation at a place where the two reactants meet. The gel is made from 0.7–1% agar (noble agar, ion agar, bactoagar or agarose) in a petri-plate. The wells are cut in gel, one in centre and 4–6 near periphery. The antiserum is placed in central well and antigen (crude plant extract from virus infected leaves) on peripheral wells. The antigen and antibody will diffuse in gel and shall form a precipitate as white line at a place where they meet. This test can easily identify spherical viruses than long filamentous viruses.

C. 2.4. Immunofluorescence test (IF)

IF is normally used for phloem inhabited prokaryotes like HLB pathogens and can also be used for phloem restricted viruses such as CTV. In this test, antibodies are labeled with fluorescent dye, fluorescin isothiocynate. Thin longitudinal sections of infected midrib of leaves are cut in a cryostat and processed. The antigen and antibody complex is seen under fluorescent microscope. The infected sample will show apple green fluorescence in phloem tissue but no fluorescence will be visible in healthy samples (Plate 39a, b, c).

C. 2.5. Enzyme-linked immunosorbent assay (ELISA)

ELISA is more popular test for detection of virus and virus-like pathogens in citrus. ELISA procedure is done through several steps. The antibody is conjugated with an enzyme (Alkaline phosphatase, penicillinase or horseradish peroxidase) known as antibody-conjugate. The conjugate and its substrate {para-nitrophenyl phosphate (PNPP) for alkaline phosphatase enzyme} react in an ELISA plate and their reaction is augmented as the enzyme hydrolyses of the substrate to a chromogenic product which is seen by naked eye or can be measured by an ELISA reader. Several forms of ELISA are used but double antibody sandwich ELISA (DAS-ELISA) and direct antigen coating ELISA (DAC-ELISA) are most commonly used. ELISA techniques are very useful in detection of several citrus viruses like tristeza (Bar-Joseph *et al.*, 1979) and also prokaryotes like *Spiroplasma citri* (Saillard *et al.*, 1980).

C. 2.6. Dot immunobinding assay (DIBA)

This is also known as dot-ELISA. In this test, the antigen (crude plant extract) is dotted on nylon or nitrocellulose membrane and processed like ELISA. The substrate forms an insoluble coloured material on the membrane in infected samples which can be seen by naked eye. No colour is developed in healthy samples.

C 2.7. Immuno tissue printing (ITP)

This is similar to DIBA but instead of using antigen extract, cut surface of plant organs (mostly young shoots) are pressed on the nylon membrane. It is processed as DIBA and colour will develop in imprints of infected samples. This test is normally done to test field samples of citrus.

C 2.8. Western blotting

Western blotting involves electrophoresis, electroblotting and solid phase immunological detection as is done in DIBA. The proteins of the sample are separated in SDS-PAGE and shifted to nylon membrane by electroblotting. The protein free area is blocked on the membrane by serum albumin. Then it is processed like DIBA.

C 2.9. Immunological electron microscopy (IEM)

In this test serology is coupled with electron microscopy. Two type of test are carried out in IEM known as trapping and decoration.Trapping is done by first coating the EM grid with antibodies which is then put on an extract containing virus particles. The antibodies on the grid will trap large number of particles from sap as compared by normal leaf dip preparations when viewed in EM (Fig. 14) For decoration test, grids are first coated with antigen and then placed on antibodies. Large number of antibodies will attach to the virus particles which can be seen in EM as black coating around virus particle (Fig. 15). This test is highly sensitive and very useful for virus identification even of similar morphology and to test the specificity of the antibodies.

C 2.10. Immuno-gold labeling

This test is used to detect the virus particles in thin sections of infected tissue. Protein A is labeled with gold particles of about 10–15 nm diameter.

Protein-A- gold complex is then used to locate antibody bound in virus particle. The test is highly sensitive and is not used routinely.

Finally, it should be noted that more and more strains of virus and virus-like pathogens infecting citrus are being discovered and that diseases thought to be caused by a single pathogen now appear to involve several such agents. For instance, many strains of tristeza virus have been detected, but detection of these strains, mild or severe, requires the joint use of indicator plants and ELISA, showing that indexing on indicator plants and serological techniques are complementary.

D. *Nucleic acid based techniques*

Several detection techniques based on virus nucleic acid have been evolved and are used to detect virus and virus-like pathogens in citrus. The polymerase chain reaction (PCR) and nucleic acid hybridization are more popular among these techniques.

D. 1. *Nucleic acid hybridization*

In This test, nucleic acid from plant material is irreversibly bound to a nitrocellulose membrane (NCM) or nylon membrane, and is allowed to hybridize with labeled probe (both radioactive and non-radioactive probes are used) to a virus genome suspected to be present in the sample. Positive hybridization indicates the presence of suspected virus in test samples which can be assessed by autoradiography. The test is done with both viruses having RNA or DNA genomes and popularly known as DNA-DNA hybridization and DNA-RNA hybridization. It is also used to detect *Candidus* Liberibacter species and other prokaryotic pathogens infecting citrus (Calvert *et al.,* 1988; La Rosa *et al.,* 1988; Villechanoux *et al.,* 1992; Bove *et al.,* 1993; Varma *et al.,* 1993). It could be seen from Table 7 that 107 samples were collected from different orchards in India. Out of 107 samples, 19 showed typical symptoms of citrus yellow mosaic (badnavirus) but when tested by DNA-DNA hybridization 67 samples were positive to citrus yellow mosaic badnavirus suggesting that nucleic acid hybridization is an effective tool to detect viruses in citrus (Table 7). These results also suggested that visual inspection may be misleading.

D. 11. *Polymerase chain reaction (PCR)*

Among the detection technologies developed so far, PCR for detection of pathogens having DNA or RNA genomes has emerged as an effective and reliable method for detection of virus and virus-like pathogens. PCR is a primer mediated enzymatic amplification of a DNA segment lying between two regions of known sequences. Two oligonucleotides are used as primers, which on binding the template are extended with DNA polymerase in the presence of dNTPs and Mg ions. Multiplex PCR, nested PCR, Real time PCR and quantative real time PCR (qPCR) are becoming popular tools for quick and reliable detection of virus and virus-like pathogens infecting citrus.

D. 3. *Polyacrylamide gel electrophoresis (PAGE)*

This technique is used for detection of viroid pathogens. Electrophoretic analysis can detect even more than one viroid species in a sample. For example, viroid RNAs present in the various exocortis isolates have revealed that, besides citrus exocortis viroid RNA, other viroid RNAs are present and also produce a reaction on Etrog citron. Thus, the combined use of Etrog citron as an indicator plant and of the electrophoretic analysis of viroid RNAs results in a new understanding of exocortis as a disease caused by a complex of various viroid RNAs rather than only one. Reverse-PAGE (Fig. 34) and sPAGE are commonly used for viroid detection in citrus.

Although serological techniques such as ELISA is more popular technique for detection of citrus viruses but eventually, serological techniques will be replaced by the more sensitive molecular hybridization techniques and ultimately by PCR. While the MAbs against HLB are too specific to be used in routine indexing, they have been most useful as research tools in the purification of the bacterium by immunoaffinity chromatography. DNA from purified bacteria was extracted and cloned in *Escherichia coli* and sequenced, thus opening up the way for the detection of HLB bacterium by molecular hybridization and PCR (Bove, 1995).

Each plant virus forms inclusions within its host cells which are specific to that virus. These inclusions are used as diagnostic to that virus and can be seen even by light microscope (Christie and Edwardson, 1986). Other method of virus indexing is by observing dsRNA from virus infected plants (Dodds and Bar-Joseph, 1983; Dodds *et al.*, 1987). Sometimes, biochemical tests are also used for indexing of virus and virus-like pathogen (Gillings *et al.*, 1988). SDS immunodiffusion procedure has also been used for detection of CTV (Garnsey *et al.*, 1979). A comparative sensitivity of each detection technique is given in Table 12.

CULTURAL PRACTICES

Eradication of infected trees and replaced them with healthy ones is important to avoid virus spread to other trees in the orchard. Under eradication programme, sources of virus infection in weeds, ornamental plants, unrelated crops, plant remains and infected plants within the crop should be removed before and after planting of healthy trees in the field.

Table 12: Approximate sensitivity of virus detection

Techniques	*Minimum detectable virus concentration*
Host infectivity	100 ng
Electron microscopy	100 ng
Tube precipitation test	500 ng
Gel diffusion test	1000 ng
Latex test	5–20 ng
ELISA	1–10 ng
IEM	1–10 ng
Western Blot	1–10 ng
Nucleic acid hybridization (NAH)	1 pg
PCR	10 fg to attogram RNA

Cultural practices like change in planting dates, tree to tree spacing and isolation of planting in disease or vector free areas are some of the important methods especially for vector borne diseases of citrus. Another important measure is to remove or keep the insect vector population as low as possible by insecticidal treatment. Regular inspections of field plantations are necessary to see the build up of vector(s) population. At least four inspections per year would be required for identification and removal of symptomatic trees in grown up orchards. Interestingly, in the large groves of São Paulo state of Brazil and Florida, inspections of adult trees are conducted with "platforms" on which two or four inspectors can stand. The platform moves in between rows at a speed of 3 to 4 km/hour. Inspectors can inspect ~5000 trees/day. Five citrus viruses have been inactivated when plants held at warm glasshouse temperatures (Roistacher and Calavan, 1974). In case of serious diseases like huanglonbing and others following operations are recommended.

- ***Upto 5 yrs*:** Remove the infected tree
- ***6-10 yrs*:** Remove if 70% tree is infected otherwise remove infected branches
- ***>10 yrs*:** Remove 40% affected branches
- Wind breaks should be used
- Do not replant healthy stock in old infected orchards
- ***Antibiotic treatment*:** Treatment with Penicillin and other antibiotics including tetracyclines are effective against bacterial diseases such as huanglongbing, citrus variegated chlorosis and mollicute disease like stubborn but it is not a viable method of disease management.

PROTECTION OF PLANTS THROUGH CROSS PROTECTION

Citrus tristeza virus is known to have great genetic and biological diversity among strains (Yokomi *et al.,* 2010). Some of these strains cause very severe diseases like stem pitting and seedling yellows resulting heavy damage to the crop while others are of mild nature and cause no symptoms in sensitive citrus hosts and usually result in no economic damage or insignificant damage. It has been well established that if a plant is infected with a mild strain of a virus, it will provide protection against severe strains of that virus. This phenomenon is known as cross protection. Cross protection has been used in some countries to manage vector borne citrus tristeza virus which is otherwise difficult to manage and cause heavy damage to citrus crop throughout the world.

MANAGEMENT OF VIRUSES BY TRANSGENIC PROTECTION

Transgenic plants can be developed by manipulating genes which will show resistant against viruses. Following viral genes are normally used to develop virus resistant transgenic plants.

1. Natural resistance genes
2. Pathogen derived resistance
3. *Protein based*: viral coat protein, movement protein, Replicase proteins.
4. *Nucleic acid based*: RNA-mediated protection.
5. Protein-RNA mediated protection
6. Antisense RNAs expression by transgenics
7. Ribozyme derived resistance
8. Satellite and defective interfering nucleic acid-mediated protection.

In citrus, transgenics have been developed against some viruses but being a tree crop testing of these transgenics will take longer time before these are released for field planting (Pena *et al.*, 1995; Gutierrez *et al.*, 1997; Perez *et al.*, 1998).

CONTROL OF INSECT VECTORS

To minimize insect vector population, it is important to remove host plants of vector(s) in the vicinity of citrus plantations. Spraying contact and systemic insecticides (Monocrotophos, organophosphates and pyrethroids) and mineral oils are used to manage vector population. To identify vector-free areas for growing virus-free nurseries and mother trees are also important to avoid insect vectors of citrus viruses. Aphid vectors can be blasted off from plants with a strong spray of water. Bright yellow pan one quarter full with water will attract the aphid vectors. Spray of oil or milk can also avoid virus transmissibility by aphid vectors. Use of citrus species or varieties resistant to vector should be used. Sometimes natural barriers help to enter the vector(s) in a particular area. For example, in Assam state of India, there is a place known as Digaru which is surrounded by hills and is free form psyllid, *Diaphorina citri* - vector of huanglongbing (HLB) disease. Recently, it has been shown that volatiles of white seeded guava repel *Diphorina citri* - vector of Asian-type huanglongbing (Noronha *et al.,* 2008). Experiments to see the effect of guava on vector population are being conducted in Florida. Biological control of insect vector wherever is possible is an ecofriendly and inexpensive approach. In case of citrus, biological control of insect vectors has been achieved in some cases and is briefly described below.

BIOLOGICAL CONTROL OF INSECT VECTORS

Some aphid species like *Cladosporium spp* and *Capnodium citri* can parasitize the vectors of citrus viruses. Parasites, parasitoid, predators and microbes have been used for control of vectors of some viruses. Genera of *Aphelinus, Mesida* and *Mesidiopsis* are parasitoid of aphids. *Verticillium lecanii* and *Paecilomyces farinosus* are entomopathogenic fungi infecting vectors. *Coconellides* are predators of aphids. Parasites of psyllids have effectively controlled psyllid vectors of huanglongbing disease in Reunion Island. In Reunion Island, *Diaphorina citri* was present below 400–500 m, and *Trioza erytreae,* above 400–500 m. Two parasites for these two psyllids were imported in Reunion Island from South Africa and India for control of two psyllid vectors.

- *Tamarixia* (ex-*Tetrastichus*) *dryi,* an ectoparasite of *T. erytreae* was imported from Transvaal, South Africa (December, 1974).
- *Tamarixia radiata,* an ectoparasite of *D. citri,* was imported from Punjab in India (April 1978).

These parasites have completely eliminated the psyllid populations from Reunion Island. It could be possible because *Tamarixia* parasites (or Psyllid hyperparasites) were not present in Reunion Island. These ectoparasites (*Tamarixia* (ex-*Tetrastichus*) *dryi* and *Tamarixia radiata)* are not effective in countries where psyllid hyperparasites are available

In conclusion, citrus virus and virus-like diseases are managed by multiple approaches as described in this chapter. In brief, use of healthy planting material, control of insect vectors and protection of citrus plantation before or after planting will provide satisfactory management of citrus virus diseases.

REFERENCES

Ahlawat, Y.S., Chakraborty, N.K., Varma, A. and Raychaudhuri, S.P. 1992. Diagnosis of diseases caused by viruses and fastidious prokaryotes in citrus in India. *Proc. Intl. Soc. Citriculture* 2: 749–751

Ahlawat, Y.S., Das, A.K., Ghosh, D.K. and Baranwal, V.K. 2008. Recent trends in diagnosis and management of citrus diseases. pp. 119–130 *In*: Proc. National Symp. Citriculture; Emerging trends, held 24–26 July, 2008 at NRCC, Nagpur.

Bar-Joseph, M. and Loebenstein, G. 1970. Leaf flecking on indicator seedlings with citrus in Israel. *Plant Dis. Rep.*, 54: 643–646.

Bar-Joseph, M., Garnsey, S.M., Gonsalves, D., Moscovits, D.E., Clark, M.F. and Loebenstein, G. 1979. The use of enzyme-linked immunosorbent assay for detection of citrus tristeza virus. *Phytopathol.*, 69: 190–195.

Bové, J.M. 1980. Spiroplasma citri identification. *In*: Bové, J.M. and Vogel, R. (*Eds.*). Description and illustration of virus and virus-like diseases of citrus, p. 22. A collection of colour slides. Paris, IRFA SETCO-FRUITS.

Bove 1995. Virus and virus like diseases of citrus in the near east region. FAO Corporate Document Repository (Regional office for the near East).

Bove, J.M., Garnier, M., Ahlawat, Y.S., Chakraborty, N.K. and Varma, A. 1993. Detection of Asian strains of the greening BLO by DNA-DNA hybridization in Indian orchard trees and Malaysian *Diaphorina citri* psyllids. pp. 258–263 *In*: Proc. 12th Conf. IOCV, IOCV, Riverside, CA.

Button, J. and Bornman, C.H. 1971. Development of nucellar plants from unpollinated and unfertilized ovules of Washington navel orange *in vitro*. *J. S. Afr. Bot.*, 37: 127–134.

Calavan, E.C., Frolich, E.F., Carpenter, J.B., Roistacher, C.N. and Christiansen, D.W. Childs, J.F.L. *et al.* (*Eds.*) 1968. Indexing procedures for 15 virus diseases of citrus trees. Washington, DC, Agricultural Research Service, United States Department of Agriculture.

Calavan, E.C., Roistacher, C.N. and Nauer, E.M. 1972. Thermotherapy of citrus for inactivation of certain viruses. *Plant Dis. Reptr*. 56: 976–80.

Calvert, L.A., Lee, R.F. and Hiebert, E. 1988. Characterization of the RNA species of citrus variegation virus with complementary DNA clones. pp. 327–333. *In*: Proc. 10th Conf: IOCV, IOCV, Riverside, CA.

Cameron, J.W. and Frost, H.B. 1968. Genetics, breeding and nucellar embryony. *In*: Reuther, W., Batchelor, L.D. and Webber, H.J. (*Eds.*). The citrus industry, Vol. II. Anatomy, physiology, genetics and reproduction, pp. 325–370. Berkeley, *Div. Agric. Sci., Univ.* Calif.

Childs, J.F. L. and Knorr, L.C. 1965. Control of virus diseases of citrus trees: An evaluation of methods. *Phytopathology* 55: 675–80.

Christie, R.G. and Edwardson, J.R. 1986. Light microscopic techniques for detection of plant virus inclusions. *Plant Dis.*, 70: 273–279.

Dodds, J.A. and Bar-Joseph, M. 1983. Double-stranded RNA from plants infected with closteroviruses. *Phytopathol.*, 73(3): 419–423.

Dodds, J.A., Jarupat, T., Lee, J.G. and Roistacher, C.N. 1987. Effects of strain, host, time of harvest and virus concentration on double-stranded RNA analysis of citrus tristeza virus. *Phytopathol.*, 77(3): 442–447.

Duran-Vila, N., Pina, J.A., Molina, M.L. and Navarro, L. 1991. A new indexing method for cachexia. pp. 224–229. *In*: Proc. 11th Conf. IOCV, IOCV, Riverside, CA.

Esan, E.B. 1973. A detailed study of adventive embryogenesis in the Rutaceae. Riverside, Univ. Calif. (Ph.D. thesis)

Flores, R. 1988. Detection of citrus exocortis viroid in natural and experimental citrus hosts by biochemical methods. pp. 192–196. *In*: Proc. 10th Conf IOCV, IOCV, Riverside, CA.

Frost, H.B. 1938. Nucellar embryony and juvenile characters in clonal varieties of citrus. *J. Hered.*, 29: 423–432.

Garnier, M., Martin-Gros, G. and Bové, J.M. 1987. Monoclonal antibodies against the bacterium-like organism associated with citrus greening disease. *Ann. Inst. Pasteur/Microbiol.*, 138: 639–650.

Garnier, M., Zreik, L. and Bové, J.M. 1991. Witches' broom disease of lime trees: transmission of the mycoplasma-like organism (MLO) to periwinkle and citrus, production of monoclonal antibodies for the detection of the MLO. pp. 448–453. *In*: Proc. 11th Conf. IOCV, IOCV, Riverside, CA.

Garnsey, S.M., Gonsalves, D. and Purcifull, D.E. 1979. Rapid diagnosis of citrus tristeza virus infections by sodium dodecyl sulfate [SDS]-immunodiffusion procedures. *Phytopathol.*, 691: 88–95.

Garnsey, S.M., Bar-Joseph, M. and Lee, R.F. 1981. Application of serological indexing to develop control strategies for citrus tristeza virus. *Proc. Int. Soc. Citricult.*, 1: 448–452.

Gillings, M.R., Broadbent, P. and Gollnow, B.I. 1988. Biochemical indexing for citrus exocortis viroid. pp. 178–187. *In*: Proc. 10th Conf. IOCV, IOCV, Riverside, CA.

Gutierrez-E., M.A. D.L. Th., and Moore, G.A. 1997. Factor affecting Agrobacterium-mediated tranformation in citrus and production of orange (*Citrus aurantium* L.) plants expressing the coat protein gene of citrus tristeza virus. *Plant Cell Rept.* 743–753.

Hoa, N.V., Ahlawat, Y.S. and Pant, R.P. 2004. Production of virus-free Kinnow mandarin and Mosambi sweet orange nucleus planting material through shoot tip grafting. *Indian Phytopath.* 57: 482–487

Kochba, J., Spiegel-Roy, P. and Safran, H. 1972. Adventive plants from ovules and nucelli in citrus. *Planta*, 106: 237–245.

La Rosa, R., Albanese, G., Azzaro, A., Sesto, F. and Domina, F. 1988. Suitability of nucleic acid analysis to diagnose viroid infections in citrus. pp. 188–191. *In*: Proc. 10th Conf. IOCV, IOCV, Riverside, CA.

Martin-Gros, G., Garnier, M., Iskra, M.L., Gandar, J. and Bové, J.M. 1987. Production of monoclonal antibodies against phloem-limited prokaryotes of plants: a general procedure using extracts from infected periwinkles as immunogen. *Ann. Inst. Pasteur/Microbiol.*, 138: 625–637.

Morel, G. and Martin, C. 1952. Guérison de dahlias atteints d'une maladie ◊ virus. *C.R. Acad. Sci., Paris*, 235: 1324–1325.

Morel, G. and Martin, C. 1955. Guérison de pommes de terre atteintes de maladies à virus. *CR Acad. Agric., France*, 41: 472–475.

Murashige, T., Bitters, W.P., Rangan, E.M., Nauer, E.M., Roistacher, C.N. and Holliday, P.B. 1972. A technique of shoot apex grafting and its utilization towards recovering virus-free citrus clones. *Hort. Sci.*, 7: 118–119.

Navarro, L., Roistacher, C.N. and Murashige, T. 1975. Improvement of shoot-tip grafting *in vitro* for virus-free citrus. *J. Am. Soc. Hort. Sci.*, 100: 471–479.

Navarro, L., Roistacher, C.N. and Murashige, T. 1976. Influence of size and source of shoot-tips on psorosis A and exocortis content of navel orange plants obtained by shoot-tip grafting *in vitro*. pp. 194–197. *In*: Proc 7th Conf: IOCV, IOCV, Riverside, CA.

Navarro, L. and Juarez, J. 1977. Elimination of citrus pathogens in propagative budwood. 11. *In vitro* propagation. *Proc. Int. Soc. Citricult.*, 3: 973–981.

Navarro, L., Juarez, J., Pina, A., Ballester, J.F. and Arregui, J.M. 1988. The citrus variety improvement program in Spain after eleven years. pp. 400–406. *In*: Proc. 10th Conf. IOCV, IOCV. Riverside, CA.

Noronha Jr, N.C., Bento, J.M.S. and Parra, J.R.P. 2008. Guava and citrus plant volatiles. Fundecitrus workshop on HLB-resistant citrus. Araraquara, February 2008.

Pena, L., Cervera, M., Antonio, J.J., Pina, J.A., Duran-Vila, N. and Navaro, L. 1995. Agrobacterium-mediated transformation of sweet orange and regeneration of transgenic plants. *Plant Cell Rept.* 14: 616–619.

Perez-Molphe-Balch, E. and Ochoa, A.N. 1998. Regeneration of transgenic plants of Mexican lime from Agrobacterium rhizogenes transformed tissues. *Plant Cell Rept.* 17: 591–596.

Rangan, T.S., Murashige, T. and Bitters, W.P. 1968. *In vitro* initiation of nucellar embryos in monoembryonic citrus. *Hort/ Sci.*, 3: 226–227.

Reuther, W., Batchelor, L.D. and Webber, H.J. (*Eds.*) 1968. The citrus industry, vol. 11. Anatomy, physiology, genetics and reproduction. Berkeley, *Div. Agric. Sci., Univ.* Calif.

Roistacher, C.N., Blue, R.L. and Calavan, E.C. 1973. A new test for cachexia. *Citrograph*, 58: 261–262.

Roistacher, C.N. and Calavan, E.C. 1974. Inactivation of five citrus viruses in plants held at warm glasshouse temperatures. *Plant Dis. Rep.*, 58: 850–853.

Roistacher, C.N., Navarro, L. and Murashige, T. 1976. Recovery of citrus selections free of several viruses, exocortis viroid and *Spiroplasma citri* by shoot-tip grafting *in vitro*. pp. 186–193. *In*: Proc. 7th Conf. IOCV, IOCV, Riverside, CA.

Roistacher, C.N. and Kitto, S.L. 1977. Elimination of additional citrus viruses by shoot-tip grafting *in vitro*. *Plant Dis. Rep.*, 61: 594–596.

Roistacher, C.N., Calavan, E.C., Blue, R.L., Navarro, L. and Gonzales, R. 1977. A new more sensitive citron indicator for detection of mild isolates of citrus exocortis viroid (CEV). *Plant Dis. Rep.*, 61: 135–139.

Roistacher, C.N. 1991. Graft-transmissible diseases of citrus. handbook for detection and diagnosis. FAO. Rome. p. 286.

Saillard, C., Garcia-Jurado, O., Bové, J.M., Vignault, J.C., Moutous, G., Fos, A., Bonfils, J., Nhami, A., Vogel, R. and Viennot-Bourgin, G. 1980. Application of ELISA to the detection of *Spiroplasma citri* in plants and insects. pp. 145–152. *In*: Proc. 8th Conf IOCV, Riverside, CA.

Su, H.J. and Chu, J.Y. 1984. Modified technique of citrus shoot-tip grafting and rapid propagation method to obtain citrus budwoods free of citrus viruses and likubin organism. *Proc. Int. Organ. Citricult.*

Varma, A., Ahlawat, Y.S., Chakarborty, N.K., Garnier, M. and Bove, J.M. 1993. Detection of greening BLO by electron microscopy, DNA hybridization in citrus leaves with and without mottle from various regions in India. pp. 280–285, *In*: Proc. 12th Conf. IOCV, IOCV, Riverside, CA.

Vela, C., Cambra, M., Sanz, A. and Moreno, P. 1988. Use of specific monoclonal antibodies for diagnosis of citrus tristeza. pp. 55–61. *In*: Proc. 10th Conf: IOCV, IOCV, Riverside, CA.

Vijayakumari, N., Ghosh, D.K., Das, A.K. and Singh, S. 2006a. Elimination of CTV and greening pathogens through *in vitro* shoot grafting from citrus exotic germplasm. *Indian J. Agric. Sci.* 76: 209–210.

Vijayakumari, N., Ghosh, D.K., Das, A.K. and Singh, S. 2006b. Production of disease-free planting material of Nagpur mandarin (*Citrus reticulata* Blanco) by shoot tip grafting. *Indian Phytopath.* 59: 48–51.

Villechanoux, S., Garnier, M., Renaudin, J. and Bové, J.M. 1992. Detection of several strains of the bacterium-like organism of citrus greening disease by DNA probes. *Curr. Microbiol.*, 24: 89–95.

Vogel, R., Nicoli, M. and Bové, J.M. 1988. Le microgreffage de méristèmes *in vitro,* son utilisation en Corse pour la génération des agrumes. *Fruits*, 43(3): 167–173.

Weathers, L.G. and Calavan, E.C. 1959. Nucellar embryony: a means of freeing citrus clones of viruses. *In*: Wallace, J.M., (*Ed.*). Citrus virus diseases, pp. 197–202. Berkeley, Div. Agric. Sci., Univ. Calif.

White, P.R. 1934. Multiplication of the viruses of tobacco and acuba mosaics in growing excised tomato root tips. *Phytopathol.*, 24: 1003–1011.

Yokomi, R.K., Polek, M. and Gumpf, D.J. 2010. Transmission and spread of Citrus tristeza virus in Central California. pp. 151–165. *In*: Karasev, A.V. and Hilf, M.E. (*Eds.*). Citrus Tristeza Virus Complex and Tristeza Diseases. American Phytopathological Society, St. Paul, MN

Annexure-1

CITUS SPECIES, CULTIVARS AND HYBRIDS OF CITRUS

- *Citrus aurantifolia* – Key Lime, Omani Lime, from India
- *Citrus maxima* – Pomelo (pummelo, shaddock), from the Malay Archipelago
- *Citrus medica* – Citron, from India
- *Citrus reticulata* – Mandarin oranges, from China
- *Citrus trifoliata* – Trifoliate Orange, from Korea and adjacent China (often separated as *Poncirus*)
- Australian limes
    - o *Citrus australasica* – Finger Lime (sometimes separated in *Microcitrus*)
    - o *Citrus australis* – Australian Round Lime (sometimes separated in *Microcitrus*)
    - o *Citrus glauca* – Desert Lime (sometimes separated in *Eremocitrus*) and 3 other *Microcitrus*
- Kumquats, 4-5 species from East Asia ranging into Southeast Asia (often separated as *Fortunella*)
- Papedas, including
    - o *Citrus halimii* – *limau kadangsa, limau kedut kera,* from Thailand and Malaya
    - o *Citrus indica* – Indian Wild Orange, from India

CITRUS MAXIMA-BASED

- Amanatsu, natsumikan – *Citrus* × *natsudaidai* (*C. maxima* × unknown)
- Cam sành (*C. reticulata* × *C.* × *sinensis*)

- Grapefruit – *Citrus* × *paradisi* (*C. maxima* × *C.* × *sinensis*)
- Imperial Lemon (*C.* × *limon* × *C.* × *paradisi*)
- Kinnow, Wilking (*C.* × *nobilis* × *C.* × *deliciosa*)
- Kiyomi (*C. sinensis* × *C.* ×*unshiu*)
- Minneola Tangelo (*C. reticulata* × *C.* × *paradisi*)
- Orangelo, Chironja (*C.* × *paradisi* × *C.* × *sinensis*)
- Oroblanco, Sweetie (*C. maxima* × *C.* × *paradisi*)
- Sweet Orange – *Citrus* × *sinensis* (probably *C. maxima* × *C.* × *reticulata*)
- Tangelo – *Citrus* × *tangelo* (*C. reticulata* × *C. maxima* or *C.* × *paradisi*)
- Tangor – *Citrus* × *nobilis* (*C. reticulata* × *C.* × *sinensis*)
- Ugli (*C. reticulata* × *C. maxima* or *C.* × *paradisi*)

CITRUS MEDICA-BASED

- Buddha's Hand – *Citrus medica* var. *sarcodactylus*
- Citron varieties with sour pulp: Diamante citron, Florentine citron, Greek citron and Balady citron
- Citron varieties with sweet pulp: Corsican citron and Moroccan citron.
- Yemenite citron – a pulpless citron.
- Fernandina – *Citrus* × *limonimedica* (probably *C. medica* × *C.* × *limon*)
- Ponderosa Lemon (probably *C. medica* × *C.* × *limon*)

CITRUS RETICULATA-BASED

- Bergamot Orange – *Citrus* × *aurantium* ssp. *bergamia* or *Citrus* × *bergamia*
- Bitter Orange, Seville Orange – *Citrus* × *aurantium*
- Blood Orange – *Citrus* × *sinensis* cultivars
- Cam sành (*C. reticulata* × *C.* × *sinensis*)
- Chinotto – *Citrus* × *aurantium* var. *myrtifolia* or *Citrus* × *myrtifolia*
- ChungGyun – *Citrus reticulata* cultivar[verification needed]
- Clementine – *Citrus* × *clementina*
- Cleopatra Mandarin – *Citrus* × *reshni*
- Dekopon – *Citrus reticulata* cv. 'Siranui' (ChungGyun × Ponkan)

- Daidai – *Citrus* × *aurantium* var. *daidai* or *Citrus* × *daidai*
- Grapefruit – *Citrus* × *paradisi* (*C. maxima* × *C.* × *sinensis*)
- Hermandina – *Citrus reticulata* cv. 'Hermandina'
- Imperial Lemon (*C.* × *limon* × *C.* × *paradisi*)
- Kinnow, Wilking (*C.* × *nobilis* × *C.* × *deliciosa*)
- Kiyomi (*C. sinensis* × *C.* × *unshiu*)
- Mediterranean Mandarin, Willow Leaf – *Citrus* × *deliciosa*
- Meyer Lemon, Valley Lemon – *Citrus* × *meyeri* (*C.* × *limon* × *C.* × *paradisi* or *C.* × *sinensis*)
- Michal Mandarin – *Citrus reticulata* cv. 'Michal'
- Mikan, Satsuma – *Citrus* × *unshiu*
- Minneola Tangelo (*C. reticulata* × C. × paradisi)
- Nova Mandarin, Clemenvilla
- Orangelo, Chironja (*C.* × *paradisi* × *C.* × *sinensis*)
- Oroblanco, Sweetie (*C. maxima* × *C.* × *paradisi*)
- Ponkan – *Citrus reticulata* cv. 'Ponkan'
- Rangpur, Lemanderin, Mandarin Lime – *Citrus* × *limonia* (*C. reticulata* × *C.* × *limon*)
- Sweet Orange – *Citrus* × *sinensis* (probably *C. maxima* × *C.* × *reticulata*)
- Tangelo – *Citrus* × *tangelo* (*C. reticulata* × *C. maxima* or *C.* × *paradisi*)
- Tangerine – *Citrus* × *tangerina*
- Tangor – *Citrus* × *nobilis* (*C. reticulata* × *C.* × *sinensis*)
- Ugli (*C. reticulata* × *C. maxima* or *C.* × *paradisi*)
- Yuzu – *Citrus* × *junos* (*C. reticulata* × *C.* × *ichangensis*)

UNRESOLVED

- Alemow, Colo – *Citrus* × *macrophylla*
- Djeruk limau – *Citrus* × *amblycarpa*
- Gajanimma, Carabao Lime – *Citrus* × *pennivesiculata*
- Ichang Lemon, Ichang Papeda – *Citrus* × *ichangensis*
- Imperial Lemon (*C.* × *limon* × *C.* × *paradisi*)
- Iyokan, *anadomikan* – *Citrus* × *iyo*
- Kabosu – *Citrus* × *sphaerocarpa*
- Kaffir Lime, *makrut* – *Citrus* × *hystrix*
- Khasi Papeda – *Citrus* × *latipes*

- Lemon – *Citrus* × *limon*
- Limetta, Sweet Lemon, Sweet Lime, mosambi – *Citrus* × *limetta*
- Palestine Sweet Lime – *Citrus* × *limettioides* Tanaka
- Persian lime, Tahiti Lime – *Citrus* × *latifolia*
- Pompia ("Citrus mostruosa", a *nomen nudum*)
- Rough Lemon – *Citrus* × *jambhiri* Lush.
- Shekwasha, Hirami Lemon, Taiwan Tangerine – *Citrus* × *depressa*
- Sudachi – *Citrus* × *sudachi*
- Sunki, Suenkat – *Citrus* × *sunki*
- Tachibana Orange – *Citrus* × *tachibana* (Mak.) Tanaka
- Volkamer Lemon – *Citrus* × *volkameriana*

For hybrids with kumquats, see × *Citrofortunella*. For hybrids with the Trifoliate Orange, see citrange.

- *Citrus macroptera* – grown in the Sylhet region of Bangladesh, where it is called *shatkora*

Annexure-2

CITRUS GERMPLASM AND ITS DISTRIBUTION IN INDIA

Species	*Common name*	*Distribution*
Citrus assamensis (dutta and Bhatt)	Ada-jamir	North-east India
C. aurantifolia (Christm) Swing.	Abhayapuri lime, Lime, Kagzi lime	Widely distributed
C. aurantium L.	Karun-jamir, Godh Hauntra, Herale, mole, Kaipuli	All over the Country
C. grandis (L.) Osback	Dowali, Nowgong Burni, Gagar, Zenaboawk, Aijal	All over the Country
C. ichangenis Swing	Kesta Shepfu	North-east India
C. indica Tan	Indian wild orange	North-east India
C. jambhiri Lush	Soh-myndong, Sohijhahlia, Kata jamir mithi, Mitha Tulia, Jambher,Kotangiri, Renuka lemon, Jatti khatti, Jambheri Kodur, Jullandhuri Khatti, Khatta Patiala, Jambhiri local, Moogu Nimbe	All over India
C. karna Raf.	Soh-sarkar, Nibu, Karna Khatta, Karna	North-eastern and Western India
C. latipes (Swing.) Tan.	Soh-kymphorshrich, Soh-shyrkhoit	North-east India
C. limetta Risso	Mitha-Kagzi	North-east India
C. limettoides Tan.	Sweet lime, Mitha Nimbu, sakka nimbu, sarbati chikna	North east India
C. limon (L.) Burm F.	Assam lemon, Pati lebu, Joratenga, Naga changney, Soh-systong, Pani-Jamir	North-east India
C. limon var. decumana	Soh-long, Hill lemon, Gal-gal or kil-kil	North-eastern region and foot hills of Western India
C. limon var. limon	Gandhraj lemon, Godhapti lebu, Katajamuri, Elachi-lebu	All over India
C. limon Osbeck	Rangpur lime, Marmalade orange, Kusai lime, Sylhet lime	North-east and central India
C. macroptera Montr.	Tith-kara, Sat-kara	North-east India
C. maderspatana Hort ex y.Tan.	Kichli, Vadalapudi, Belladikithuli	South India

Table (*Contd...*)

Table *(Contd...)*

Species	*Common name*	*Distribution*
C. madurensis Lour.	Calamondin	North-east India
C. medica L.	Bira-jora, Soh-mango, Mitha-jora, Yemekaipuli, Citron Bira-jora, Soh-mango, Mitha -jora, Yemekaipuli, Citron	All over India
C. grandis (L.) Osbeck	Bor-tenga, Hukma-tenga, Holong-tenga, Jamir-tenga, Ambilbed	North-east and some other parts of India
C. nobilis Lour.	Jeneru-tenga	North-east India
C. paratangerina Hort. Ex. Tan	Laddu, Sylhet Billikichili, Karpura tenga	Northwest India
C. reshmi Hort. Ex. Tan. (small fruited mandarin)	Nagasantra, Kodaikithuli, Kokni, Soh-seim	All over India
C. pennivasculata Tan.	Gajanimma, Baduvapulli	South India
C. reticulata Blanco.	Nagpur mandarin, Khasi mandarin, Coorg mandarin, Soh-niamtara, Lahore local, Deshi, Darjeeling mandarin	All over India
C. reticulata var. Crenatifolia	Kaula mandarin	Northwest India
C. rugulosa Hort. Ex. Tan.	Attani small, Indian grapefruit	Northwest India
C. semiperfloreens Tan.	Sadaphal	Several parts of India
C. sinensis (L.) Osbeck	Soh-niariang	Northeastern India

Source: S.P. Ghosh

CITRUS GERMPLASM COLLECTION IN NCGR, CHINA

Species	*Number*	*Indigenus*	*Bud mutation*	*Nucellars*	*Introductions*
Pummelo	97	81	–	6	10
Mandarin	272	155	10	4	103
Sweet orange	253	131	17	2	102
Trifoliate	111	83	–	9	19
Others	309	157	–	–	152
Total	1041	607	27	21	386
%	100	58.3	2.6	2.0	37.1

Source: S.P. Ghosh

Common and Botanical names of important Citrus species, hybrids and relatives

- *Aeglopsis chevalieri* Swing.
- Alemow (*Citrus macrophylla* Webster)
- Bergamot (*C. bergamia* Risso and Poit.)
- Calamondin (*C. madurensis* Lour = *C. mitis* Blanco)
- Chironja (*C. paradisi* hybrid)
- Citrange (*C. sinensis X Poncirus trifoliata*): Carrizo, Rusk, Troyer, Yuma
- Citron (*C. medica* L.): Etrog Arizona 861-S-1
- Citrumelo (*C. paradisi X P. trifoliata*) Swingle
- *C. bigaradia* = *C. aurantium* L.
- *C. deliciosa* Ten.
- *C. erythrosa* Hort. Ex. Ten.: San-hu-hong-ju, Zhuju
- *C. excelsa* Webster
- *C. halimii* Stone
- *C. succosa* Hort. Ex. Ten.
- *C. sunki* Hort. Ex. Ten.: Suanyu
- Clementine (*C. clementina* Hort. Ex. Ten. Or *C. reticulata* Blanco): Algerian tangerine, Arrufatina, Commune, Fina, Monreal, Nules, Oroval
- Cleopatra mandarin (*C. renshi* Hort. Ex. Ten.)
- *Eremocitrus glauca* (Lindl.) Swing.
- *Esenbeckia* sp.
- Grapefruit (*C. paradisi* Macf.): Cecily, Duncan, Henderson, Marsh, Rio Red, Ruby Red, Star Rubi
- Hassaku (*C. hassaku* Hort. Ex. Ten.)
- Hongju (*C. tangerina*)
- Hongyukan (*C. succosa X C. sinensis*)
- Hyungatsu (*C. tamurana* Hort. Ex. Ten.)
- Ichang orange (Ichang papeda) (*C. ichangensis* Swing)
- Karna (*C. karna* Raf.)
- Kumquat (Jindan) (*Fortunella crassifolia*)
- Kumquat (Marumi, Luowen) {*F. japonica* (Thumb) Swing.}

- Kumquat (Nagami, Luofu) {*F. margarita* (Lour.) Swing}
- Lemon (*C. limon* (L.) Burn.f.): Eureka, Eureka cascade, Femminello Siracusano, Feminello, Frost, Lisbon, Hong-li-meng, Lapithos, Ross Eureka, Tu-li-meng
- Lime (*C. aurantifolia* (Christm.) Swing.): Key, Mexican, West Indian, Kagzi
- Lime (*C. latifolia* Tan.): Bearss, Persian, Tahiti
- Limetta (*C. limetta* Risso)
- Mandarins and hybids (*C. reticulata* Blanco (see also Clementine, Cleopatra, Satsuma): Angeru, Baillinzi, Balye 2, Bendizhao, Canh, Cha-Zhi-gan, Coorg, Cravo, Dancy, Duang, Ellendale, Fiju, Harry, Homg Nhieu, Hongju, Kara, Khasi, King, Kinnow, Ladu, Loose Jacket, Mangju, Mexerica do Rio, MexericaTardia A, Miyakawa, Murcott, Nagpur, Nan-feng-mi-ju, New Ben No.1, Nova, Oneco, Parson's special, Ponkan, Ruju, Shaohe Ponkan, Shehue Kan, Sikkim, Som-keo-wan, Som-pan, Szinkon, Tankan, Tianma Ponkan, Thieu, Thorney Zaoju, Wenmi, Willow leaf, Yanhou, Ponkan, Zhanna, Zhuhong.
- Meyer lemon (*C. limon* hybrid?)
- *Microcitrus australis* (Planch) Swing.
- Orange jessamine (*Murraya exotica* L. = *M. paniculata, M. paniculata* L. (Jack.), *M. koenigii L.*
- Pummelo (Shaddock) (*C. grandis* (L.) Osb. = *C. decumana* L.): Duong da Lang, Fenghuan, Guan-xi-mi-you, Nam Roi, Nippon yu, Peiyu, Sha-tian-you, Siamese (Siamelo), Taibei you, Thong Dee, Wanbai- you, Wentan, Wu-bu-you, Yu-yuan-you
- Rangpur lime (*C. limonia* Osb,)
- Rough lemon (*C. jambhiri* Lush.)
- *Ruta graveolens* L.
- Satsuma (*C. unshiu* (Macf.) Tenore: Ishikawa, Kamei, Kusumoto, Long-hui, Matsuyama, Okitsu, Owari, Shebon, Wakiyama, Yonezawa
- *Severinia buxifolia* (Poir.) Tenore
- Shaddock (see pummelo)
- *Skimmia reevesiana* (Fort.) N.P. Tayl. & Airy Shaw
- Sour orange and hybrids (*C. aurantium* L.): Bittersweet Seville, Gou Tou, Zhuluan
- Sweet lime (*C. limettoides* Tan.): Palestine
- Sweet orange (*C. sinensis* L.): An-liu-cheng, Bing-tang-chang, Blood, Cadenera, Caipira, Chini, Cutter, Dahongtiancheng,

Dortyol, Fuling, Gai-liang-cheng, Hamlin, Jaffa, Jincheng, Junar, Kamala, Madam Vinous, Marrs, Mat, Meishan, Moro, Mosambi, Navel, Bahianinha, Benyenda, Dream, Gilette, Gillemberg, Leng, Meishia, N 33, Navelina, Newhall, Robertson, Seike, Shirayanogi, Skagg's Bonanza, Summerfield, Washington, Washington sanguine, Paak ling, Mung, Pera, Pera-Ipigua, Pera-Rio, Pineapple, Ruby red, Satgudi, Shekeriko, Somtra, Sucrena, Tarocco, Trovita, Valencia (Campbell, Carter, Delta, Delta late, Rhode red, Trepadeira), Vandu, Xa Doai, Xian-feng-cheng, Xin-hui-cheng, Xuegan.

- Tangelo (*C. reticulata* × *C. paradisi*): Minneola, Orlando
- Tangor (*C. reticulata* × *C. sinensis*): Dweet, Homely, Murcott, Ortanique
- *Toddalia lanceolata* Lam.
- *Triphasia trifolia* (Burm.f) P. Wils.
- Trifoliate orange (*Poncirus trifoliata* (L.) Raf.: Flying Dragon
- Volkamer lemon (*C. volkameriana* Ten. and Pasq)
- Wampee (Clausena lansium (Laur) Skeels
- Wood apple (*Limonia acidissima* L. =*Feronia limonia*
- Yuzu (*C. yuzu* Sieb ex. Tan)

Source: IOCV

SUBJECT INDEX

A

B

C

D

E

I

J

K

L

M

Q

R

S

W

X

Y

Z